王伟德 著

Elasticsearch
数据分析与实战应用

中国铁道出版社有限公司
CHINA RAILWAY PUBLISHING HOUSE CO., LTD.

内 容 简 介

本书使用一套金融数据以及Elasticsearch搜索技术和聚合框架,用来进行数据分析,是一本技术全面、案例丰富、注重实操的入门书。

书中不仅细致地讲解Elasticsearch基础知识和核心接口,还进一步讲述数据建模和实际应用,例如金融舆情分析及机器学习等技术应用。最后逐步详细讲解了如何以编程方式构建RESTful服务接口,实践所学技术。由于书中有很多金融数据分析例子,读者也可了解相应应用场景以及学会其解决方案。

书中示例采用一套开源的金融数据和文本分析插件进行编写。本书非常适合搜索工程师、数据分析师、数据库工程师阅读使用,还适合Elasticsearch初学者以及想要进阶提升为中高级技术的读者使用。

图书在版编目(CIP)数据

Elasticsearch数据分析与实战应用/王伟德著.—北京:中国铁道出版社有限公司,2021.9
ISBN 978-7-113-27886-1

Ⅰ.①E… Ⅱ.①王… Ⅲ.①搜索引擎-程序设计 Ⅳ.①TP391.3

中国版本图书馆CIP数据核字(2021)第066625号

书　名	Elasticsearch 数据分析与实战应用 Elasticsearch SHUJU FENXI YU SHIZHAN YINGYONG
作　者	王伟德
责任编辑	张 丹　编辑部电话:(010)51873028　邮箱:232262382@qq.com
封面设计	MX DESIGN STUDIO
责任校对	焦桂荣
责任印制	赵星辰
出版发行	中国铁道出版社有限公司(100054,北京市西城区右安门西街8号)
印　刷	北京铭成印刷有限公司
版　次	2021年9月第1版　2021年9月第1次印刷
开　本	787 mm×1 092 mm　1/16　印张:23.25　字数:550千
书　号	ISBN 978-7-113-27886-1
定　价	99.80元

版权所有　侵权必究

凡购买铁道版图书,如有印制质量问题,请与本社读者服务部联系调换。电话:(010)51873174
打击盗版举报电话:(010)63549461

前 言

搜索引擎在人们的日常生活中发挥着越来越重要的作用。近年来由于开源软件的不断普及和发展，涌现出许多优秀的搜索软件。其中，Elasticsearch 就是以大规模分布式搜索见长。

本书选择 Elasticsearch 作为实现搜索引擎的工具。Elasticsearch 具有强大的分布式搜索能力和易于与 Kibana 集成而提供可视化功能，不仅丰富了实现搜索引擎的方法，而且还使复杂抽象的数据结构与算法变得直观而鲜活。在国外被迅速地引入到人工智能的相关应用行业领域中。

书中全面、系统地介绍了分布式搜索引擎的相关内容及实现 Elasticsearch 在 Java 和 Python 的应用代码。本书内容既注重基础知识，又非常注重实践，每章都提供了大量的实例程序。读者可以通过这些实例快速上手，并迅速提高搜索引擎开发技术。通过对本书内容的学习，读者不仅可以掌握搜索引擎开发的基本知识，而且还可以灵活地将 Elasticsearch 运用到解决实际问题当中，从而提升工作效率。

■ 这本书是写给谁的

本书主要针对从没有使用 Elasticsearch 经验的初学者，书中内容逐步介绍进阶主题，以帮助读者构建和掌握这个强大的搜索引擎。本书尝试使用一套金融数据的分析技术，有系统性地详细说明如何采用 Elasticsearch 在金融数据的分析和应用。

书中所有示例使用一套国内开源的金融数据和文本分析插件。内容主线是如何使用 Elasticsearch 打造一个金融信息搜索与分析系统，由于使用这套金融数据，能帮助读者有效地理解各种核心技术的使用技巧。因此读者阅读并练习所有示例操作后，可以很好地掌握 Elasticsearch 新功能并具备专业知识。

■ 本书架构和内容

在这本书开始写作之前，在我脑海中已经构建一个蓝图，希望读者可以很容易地吸收本书所要传达的信息和知识。大多数章节都有许多示例，采用同一套国内开源的金融数据，务使理论与实践并举。

全书分为四篇：第一部分为基础知识和核心接口，包括第 1 章到第 7 章。大体内容从 Elasticsearch 概述开始，接着是实验数据的取得和索引与文档管理。进而深入文本内容分析与插件的介绍。最后，引进 Elasticsearch 重点之一，搜索和查询。通过第一部分的学习过程，让读者已经具备基础能力，可以有兴趣地继续阅读本书。

第二部分为数据建模、聚合框架、管道处理和探索性数据分析，包括第 8 章到第 11 章。首先介绍 Elasticsearch 数据建模的各类型方法，然后是 Elasticsearch 第二个重点，主要用于

I

数据分析的聚合框架，其广泛地应用于数据分析。随后是增强文档索引功能的管道处理。最后，引进探索性数据分析案例。

第三部分为 Java 和 Python 客户端编程介绍，包括第 12 章和第 13 章。介绍如何使用 Elasticsearch Java 客户端和 Python 客户端在应用程序中编程。这两章大多数的示例都是从前几章命令行指令改写而成的。因此，可以很容易地理解它是如何以编程方式进行工作的。如果读者可以理解并且熟练掌握这一部分内容，则可以更轻松地运用到实际应用中。

第四部分为进阶功能和数据分析实战，包括第 14 章到第 16 章。以数据分析实战，帮助读者融会贯通并掌握技能。我们将讨论两个热点话题，通过 Elasticsearch 支持金融舆情分析及如何进行机器学习等技术的应用，在最后一章构建金融数据分析服务 RESTful 接口。希望读者能够消化并掌握前面所学的知识，在构建 RESTful 服务接口中实践所学技术，从而提升自己达到更高的水平。

■ 本书的主要特色

1. 内容全面，注重实操

本书首先介绍 Elasticsearch 的基本配置和基本使用方法，然后介绍从搜索到数据分析等方面的知识。

2. 讲解详尽，实例丰富

本书对每个技术要点都做了非常细致的讲解，在介绍过程中还提供了丰富的实例，特别是本书最后一篇的实战案例，更是对相关技术的一个全面应用。另外，书中所有实例的实现代码，读者都可以加以修改，以解决自己的实际问题。

3. 语言通俗，图文并茂

本书用通俗易懂的语言进行讲解，尽量避免生疏的专业术语。在讲解一些重要知识点时，书中给出了大量的图示及实例运行结果，读者可以更加直观、高效地理解所学内容。

■ 本书的读者对象

本书读者对象为数据搜索工程师、软件工程师、数据分析师、数据库工程师和希望将 Elasticsearch 的基础知识提高到另一个新水平的读者。本书还适用于想将技术扩展到用于金融数据分析领域，并在日常核心任务中将其运用到最佳水平的读者阅读。

虽然我已对书中所述内容都尽量核实，并进行了多次校对，但由于写作时间仓促，加之作者水平所限，书中可能还存在疏漏和错误之处，恳请广大读者批评、指正。

■ 整体下载包

为了方便不同网络环境的读者学习，也为了提升图书的附加价值，本书重要案例数据和源代码，请读者在电脑端打开链接下载获取。

出版社网址：http://www.m.crphdm.com/2021/0630/14361.shtml

网盘网址：https://pan.baidu.com/s/1o-M3HtUc7o_9X5Q3mPA0dg

提取码：6gl8

扫一扫，复制网址到电脑端下载文件

王伟德

2021 年 6 月

目 录

第一篇 基础知识和核心接口

第 1 章 Elasticsearch 概述

1.1 准备环境 1
1.2 运行 Elasticsearch 7.5.1 2
    1.2.1 Elasticsearch 基本配置 2
    1.2.2 重要的系统配置 4
1.3 与 Elasticsearch 7.5.1 进行对话 4
1.4 了解 Elasticsearch 的架构 5
1.5 熟悉关键概念 7
    1.5.1 Elasticsearch 映射概念 7
    1.5.2 比较 SQL 与 Elasticsearch 的相关概念 7
    1.5.3 分析器（Analyzer）..8
    1.5.4 标准分析器（Standard Analyzer）...8
1.6 接口用法约定说明 8

第 2 章 获取本书实验数据

2.1 了解实验数据集 11
    2.1.1 公募基金简介 11
    2.1.2 公募基金类型介绍 11
    2.1.3 公募基金数据归类 12
2.2 通过 Python SDK 获取多维数据 15
    2.2.1 注册及安装 Tushare 15
    2.2.2 Tushare 数据接口的调用方法 16

第 3 章　索引管理

3.1 基础索引管理 19
3.1.1 基本索引操作 19
3.1.2 设置索引 22
3.1.3 索引模板 25
3.1.4 索引别名 26
3.1.5 索引别名应用场景 27

3.2 进阶索引管理 29
3.2.1 索引存储原理 29
3.2.2 索引的底层信息接口 30
3.2.3 优化索引 33

第 4 章　文档管理

4.1 文档管理 37
4.1.1 了解 Elasticsearch 文档管理原理 37
4.1.2 认识单文档接口功能 38
4.1.3 认识批量多文档接口功能 43

4.2 文档结构 47
4.2.1 认识映射类型 47
4.2.2 映射的元字段简介 52
4.2.3 映射数据类型简介 52
4.2.4 映射数据类型的参数简介 54
4.2.5 更新显式映射内容并刷新文档索引 56

第 5 章　分析文本内容

5.1 分析器的结构 59
5.1.1 分析接口 59
5.1.2 字符过滤器 60
5.1.3 分词器 61
5.1.4 词汇单元过滤器 64

5.2 利用内置分析器进行分析 68
5.3 利用自定义分析器进行分析 69
5.4 归一化处理器 70

第 6 章　文本分析插件

6.1　Elasticsearch 插件是什么 71
　　　安装插件 72
6.2　使用 ICU 分析插件 72
　　　使用 icu_analyzer 分析器示例 73
6.3　使用 Smart Chinese 分析插件 74
　　6.3.1　安装 Smart Chinese 分析插件 74
　　6.3.2　使用 smartcn 分析器示例 74
6.4　使用 IK 分析插件 75
　　6.4.1　安装 IK 分析插件 75
　　6.4.2　使用 ik_smart 分析器示例 76
6.5　使用 HanLP 分析插件 76
　　6.5.1　安装 elasticsearch-analysis-hanlp 分析插件 77
　　6.5.2　使用 hanlp 分析器示例 77
　　6.5.3　使用 hanlp 自定义词典热更新 78
　　6.5.4　使用 hanlp 分词器自定义分析器 79
　　6.5.5　简评 hanlp 分词器的对称性 81
6.6　使用 Aliws 分析插件 81

第 7 章　搜索数据和查询表达式

7.1　索引样本文件 83
7.2　基础搜索接口 85
　　7.2.1　通过 URI 进行搜索 85
　　7.2.2　通过请求主体（request body）进行搜索 89
7.3　进阶搜索 98
　　　认识查询表达式 98
7.4　其他相关功能 114
　　7.4.1　搜索多重目标接口 114
　　7.4.2　搜索结果试算接口 115
　　7.4.3　评分说明接口 117
　　7.4.4　字段功能接口 117
　　7.4.5　搜索查询评估接口 118
　　7.4.6　性能分析设置参数 119
　　7.4.7　查询建议器 120

第二篇 数据建模、聚合框架、管道处理和探索性数据分析

第 8 章 数据建模

8.1 数据建模及方法 125
    8.1.1 使用非规范化方法 125
    8.1.2 使用对象数据类型方法 128
    8.1.3 使用嵌套数据类型方法 132
    8.1.4 使用父子类关联数据类型方法 134
    8.1.5 父子类关联数据类型查询方法 137
8.2 实际应用场景操作 140

第 9 章 聚合框架

9.1 基金净值和基金持仓样本文档 142
9.2 聚合查询语法 144
9.3 矩阵统计聚合 145
9.4 度量指标聚合 146
    9.4.1 最大值聚合 146
    9.4.2 最小值聚合 147
    9.4.3 总和聚合 147
    9.4.4 值计数聚合 148
    9.4.5 平均值聚合 148
    9.4.6 加权平均值聚合 149
    9.4.7 基数聚合 149
    9.4.8 统计聚合 149
    9.4.9 扩展统计聚合 150
    9.4.10 中位数绝对偏差聚合 151
    9.4.11 百分位聚合 151
    9.4.12 百分位等级聚合 152
    9.4.13 地理重心聚合 152
    9.4.14 地理边界聚合 153
    9.4.15 最热点聚合 153
    9.4.16 脚本式度量指标聚合 154
9.5 存储桶聚合 155
    9.5.1 范围聚合 156

	9.5.2	IP 范围聚合	157
	9.5.3	日期范围聚合	158
	9.5.4	直方图聚合	159
	9.5.5	日期直方图聚合	160
	9.5.6	自动间隔日期直方图聚合	161
	9.5.7	词条聚合	161
	9.5.8	稀有词条聚合	162
	9.5.9	显著词条聚合	163
	9.5.10	显著文本聚合	164
	9.5.11	采样器聚合	164
	9.5.12	多元化采样器聚合	165
	9.5.13	过滤器聚合	166
	9.5.14	多过滤器聚合	166
	9.5.15	地理距离聚合	167
	9.5.16	地理哈希网格聚合	167
	9.5.17	地理瓦片网格聚合	169
	9.5.18	缺失字段聚合	169
	9.5.19	全局聚合	170
	9.5.20	邻接矩阵聚合	170
	9.5.21	复合聚合	171
	9.5.22	子文档聚合	173
	9.5.23	嵌套聚合	174
	9.5.24	父文档聚合	175
	9.5.25	反向嵌套聚合	176
9.6	管道聚合		178
	9.6.1	桶平均值聚合	179
	9.6.2	桶最小值聚合	180
	9.6.3	桶最大值聚合	181
	9.6.4	桶百分位聚合	182
	9.6.5	桶统计聚合	182
	9.6.6	桶扩展统计聚合	183
	9.6.7	桶总和聚合	184
	9.6.8	桶脚本聚合	185
	9.6.9	桶选择器聚合	186

9.6.10 桶排序聚合 187
9.6.11 累计基数聚合 188
9.6.12 累计总和聚合 189
9.6.13 导数聚合 189
9.6.14 移动函数聚合 191
9.6.15 串行差分聚合 194
9.7 后置过滤器 195

第 10 章 摄取节点管道处理接口

10.1 摄取节点接口 197
  10.1.1 创建或更新接口 197
  10.1.2 读取接口 198
  10.1.3 模拟接口 198
  10.1.4 删除接口 199
10.2 摄取管道处理器 199
  10.2.1 附加处理器 200
  10.2.2 删除处理器 200
  10.2.3 重命名处理器 201
  10.2.4 小写处理器 202
  10.2.5 大写处理器 202
  10.2.6 拆分处理器 203
  10.2.7 连接处理器 204
  10.2.8 修剪处理器 205
  10.2.9 设置处理器 205
  10.2.10 日期处理器 206
  10.2.11 脚本处理器 207
  10.2.12 丢弃处理器 207
  10.2.13 管道委托处理器 207
  10.2.14 故障处理器 208
  10.2.15 字节处理器 209
  10.2.16 转换处理器 209
  10.2.17 循环处理器 210
  10.2.18 geoip 处理器 211
  10.2.19 Grok 处理器 211

   10.2.20 分解处理器 212

   10.2.21 Gsub 处理器 213

   10.2.22 HTML Strip 处理器 214

   10.2.23 URL 解码处理器 215

   10.2.24 JSON 处理器 215

   10.2.25 键值对处理器 216

   10.2.26 用户代理处理器 218

   10.2.27 排序处理器 219

   10.2.28 点扩展器处理器 219

   10.2.29 丰富处理器 220

   10.2.30 日期索引名称处理器 221

 10.3 处理管道中的故障 222

第 11 章 使用 Elasticsearch 进行探索性数据分析

 11.1 数据处理 224

   11.1.1 日线行情显式映射 224

   11.1.2 创建 ohlc_avg_price_pipeline 摄取节点管道 225

   11.1.3 批量处理索引文档 225

   11.1.4 公募基金交易行情文档索引操作 225

 11.2 指标数据分析 226

   11.2.1 执行扩展统计聚合 227

   11.2.2 执行矩阵统计聚合 227

   11.2.3 执行百分位聚合和百分位等级聚合 228

   11.2.4 执行导数聚合 229

   11.2.5 执行移动函数聚合 229

 11.3 投资组合 230

   投资组合季度清单 230

第三篇 Java 和 Python 客户端编程介绍

第 12 章 Java 客户端编程

 12.1 Elasticsearch Java REST 客户端概览 234

 12.2 Java 低级别 REST 客户端 236

   12.2.1 Java 低级别 REST 客户端操作流程 236

   12.2.2 REST 客户端初始化 236

12.2.3　执行 REST 客户端请求 237
　　12.2.4　处理 REST 客户端响应 238
　　12.2.5　关闭 REST 客户端 238
　　12.2.6　封装低级别 REST 客户端请求与处理其响应 238
　　12.2.7　调用自定义的 performSyncRequest 和 performAsyncRequest 方法 240
　12.3　使用 Swagger UI 测试低级别 REST 客户端 240
　12.4　Java 高级别 REST 客户端 244
　　12.4.1　封装高级别 REST 客户端 244
　　12.4.2　提供间接调用方法的 RestHighLevelClient 成员 245
　　12.4.3　间接调用方法 245
　　12.4.4　直接调用方法 248
　　12.4.5　构造查询请求 251
　　12.4.6　自定义 searchSync 与处理其响应 253
　　12.4.7　自定义 searchASync 与处理其响应 254
　　12.4.8　自定义 getIndexSettingsSync 方法与处理其响应 255
　　12.4.9　自定义 getIndexSettingsAsync 方法与处理其响应 256
　12.5　使用 Swagger UI 测试高级别 REST 客户端 256
　12.6　Java 高级别 REST 客户端聚合操作简介 259
　12.7　使用 Swagger UI 测试 Java 高级别 REST 客户端聚合操作 262

第 13 章　Python 客户端编程

　13.1　Elasticsearch Python 客户端概览 264
　13.2　elasticsearch-py 软件包 265
　　13.2.1　提供间接调用方法的成员 265
　　13.2.2　间接调用方法 266
　　13.2.3　直接调用方法 267
　　13.2.4　操作流程 269
　13.3　使用 Swagger UI 测试调用方法 270
　13.4　elasticsearch-dsl 软件包 274
　　13.4.1　提供特定接口的类 275
　　13.4.2　elasticsearch-dsl 软件包中 Index 类提供的调用方法 276
　　13.4.3　elasticsearch-dsl 软件包中 Search 类提供的调用方法 276
　　13.4.4　构造查询请求 277
　　13.4.5　执行请求方法与处理其响应 279

13.5	使用 Swagger UI 测试 elasticsearch-dsl 软件包	280
13.6	elasticsearch-dsl 聚合操作简介	283
13.7	使用 Swagger UI 测试 elasticsearch-dsl 聚合操作	285

第四篇 进阶功能和数据分析实战

第 14 章 Elasticsearch 与金融舆情分析

14.1	文本情感分析简介	287
14.2	文本情感分析软件服务	287
14.3	文本情感分析开源项目	288
	14.3.1 TextBlob	288
	14.3.2 SnowNLP	288
	14.3.3 BosonNLP	289
	14.3.4 Stanford CoreNLP	289
	14.3.5 百度 Senta	289
14.4	文本情感分析插件开源项目	291
	14.4.1 ESAP 开源项目简介	291
	14.4.2 ESAP 开源项目安装与测试	294
14.5	中文金融领域文本情感分析	295
14.6	应用 Elasticsearch 进行股票分析和预测	296
	14.6.1 安装与运行相关软件	296
	14.6.2 Stocksight 开源项目的 sentiment_analysis 程序	299

第 15 章 使用 Elasticsearch 进行机器学习

15.1	Kibana 简介	301
	15.1.1 准备环境和运行	301
	15.1.2 测试 Kibana	301
15.2	Kibana、Elasticsearch 与机器学习	304
	15.2.1 安装试用许可证	304
	15.2.2 机器学习测试数据	305
	15.2.3 Elastic 机器学习工作流程	308
15.3	Elasticsearch 机器学习异常检测接口	322
	15.3.1 异常检测任务资源	322
	15.3.2 数据馈送	324

  15.3.3 特定时间段资源 326
  15.3.4 过滤器 329
 15.4 Elasticsearch 机器学习数据框分析接口 330

第 16 章 构建金融数据分析服务 RESTful 接口

 16.1 基金业绩指标 333
  16.1.1 基金业绩分类 333
  16.1.2 投资类型数据丰富处理器 334
 16.2 测试样本文件 335
  16.2.1 准备测试环境 336
  16.2.2 检验测试环境 337
 16.3 使用 Spring Boot 构建 RESTful 接口服务 342
  16.3.1 AnalyticsController 类简介 343
  16.3.2 AnalyticsServiceImpl 类简介 343
  16.3.3 使用 Swagger UI 测试 poof-analytics 项目 348
  16.3.4 聚合结果解析简介 353

第一篇 基础知识和核心接口

第 1 章

Elasticsearch 概述

本章首先介绍如何设置开发环境；然后介绍启动新版本的 Elasticsearch 服务器的主要操作步骤，以帮助初学者快速了解一些基本功能；接着介绍 Elasticsearch 的架构和核心概念，以帮助读者了解 Elasticsearch 内部的工作流程，并且将极大地简化以后的学习过程；最后，带领读者熟悉应用程序接口（以下简称接口）用法。

1.1 准备环境

新手的第一步是设置 Elasticsearch 服务器，启动 Elasticsearch 服务器的方法因安装方式而异。Elasticsearch 支持许多流行的操作系统，例如 CentOS/RHEL、Oracle Enterprise Linux、Ubuntu、SLES、OpenSUSE、Windows、Debian 和 Amazon Linux。有关 Elasticsearch 支持的操作系统和兼容性的信息，请参阅官方网站。

开始撰写本书时，Elasticsearch 官方支持的最新版本是 7.7.0。但是由于需要使用的中文文本分析器 HanLP 仅支持 7.5.1 版本。因此本书采用的 Elasticsearch 版本也是 7.5.1。

虽然有很多方法可以在不同的操作系统上安装 Elasticsearch，但是对于新手来说，从命令行运行 Elasticsearch 会相对简单一些。在本书中，我们将使用 Ubuntu 18.04 操作系统托管 Elasticsearch 服务器。

以下是从官网下载、安装并运行 7.5.1 版本的操作步骤：

（1）如图 1-1 所示，选择适合的操作系统如 Windows、MacOS、Linux、DEB、RPM 或 MSI 的软件包，然后下载 7.5.1 版本。对于 Linux 64-bit，文件名是 elasticsearch-7.5.1-linux-x86_64.tar.gz。

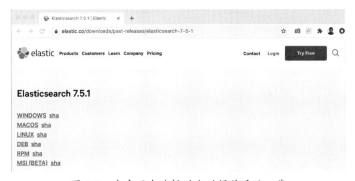

图 1-1　在官网中选择适合的操作系统下载

（2）如图1-2所示，将压缩文件解压缩，用以下命令生成一个名为elasticsearch-7.5.1的文件夹，并移至目标目录（假设位于用户的主目录下）。

```
tar -zxvf elasticsearch-7.5.1-linux-86_64.tar.gz; mv . /elasticsearch-7.5.1 ~/
```

```
[wai@wai:~/Downloads$ tar -zxvf elasticsearch-7.5.1-linux-x86_64.tar.gz; mv ./elasticsearch-7.5.1 ~/
elasticsearch-7.5.1/
elasticsearch-7.5.1/lib/
elasticsearch-7.5.1/lib/elasticsearch-7.5.1.jar
elasticsearch-7.5.1/lib/elasticsearch-x-content-7.5.1.jar
elasticsearch-7.5.1/lib/elasticsearch-cli-7.5.1.jar
elasticsearch-7.5.1/lib/elasticsearch-core-7.5.1.jar
```

图1-2 将压缩文件解压缩并移至目标目录

（3）Elasticsearch是使用Java构建的，7.5.1版本已经内含OpenJDK 13.0.1。默认情况下，Elasticsearch所使用的Java版本是用JAVA_HOME环境变量来定义的。在程序运行之前，需要检查是否已设置好JAVA_HOME环境。

```
export JAVA_HOME=~/elasticsearch-7.5.1/jdk
```

（4）如图1-3所示，在指定的路径（elasticsearch-7.5.1的文件夹），使用"-p"参数（代码如下）运行Elasticsearch以创建一个PID文件。执行指令后，Elasticsearch将在前台运行。

```
cd ~/elasticsearch-7.5.1; ./bin/elasticsearch -p pid
```

```
[wai@wai:~$ cd ~/elasticsearch-7.5.1; ./bin/elasticsearch -p pid
OpenJDK 64-Bit Server VM warning: Option UseConcMarkSweepGC was deprecated in version 9.0 and will likely b
e removed in a future release.
[2020-10-21T11:05:05,317][INFO ][o.e.e.NodeEnvironment    ] [wai] using [1] data paths, mounts [[/ (/dev/sd
a3)]], net usable_space [137.2gb], net total_space [435.1gb], types [ext4]
[2020-10-21T11:05:05,320][INFO ][o.e.e.NodeEnvironment    ] [wai] heap size [989.8mb], compressed ordinary
object pointers [true]
```

图1-3 执行指令运行Elasticsearch

（5）如果需要将Elasticsearch程序运行在后台模式下，请在命令行上指定"-d"：

```
./bin/elasticsearch -d -p pid
```

（6）如果要关闭Elasticsearch进程，可以按【Ctrl + C】组合键将它停止，也可以使用工作目录中"pid"文件的进程ID，终止该进程，代码如下。

```
kill -15 'cat pid'
```

1.2 运行Elasticsearch 7.5.1

Elasticsearch安装后不会自动启动。在Ubuntu上，最好使用Debian软件包，该软件包会为配置Elasticsearch服务安装所需的一切步骤。

1.2.1 Elasticsearch基本配置

在elasticsearch-7.5.1/config目录中，使用Linux的"ls"指令，可以看到一些配置文件如：elasticsearch.yml、log4j2.properties、roles.yml、users_roles、jvm.options、role_mapping.yml和users。

Elasticsearch的配置文件提供了很好的默认值，开发人员很少需要修改它。下面让我们看一下elasticsearch.yml、jvm.options和log4j2.properties文件的具体内容。

1.elasticsearch.yml

这是主要的配置文件，其中主要包含集群（cluster）、节点（node）、网络地址（network）和路径（path）的设置。如果要指定项目值，请取消该行注释。

以下代码是我们从文件内容中提取的部分配置参数的设置：

```
# ---------------------------------- Cluster -----------------------------------
# Use a descriptive name for your cluster:
#cluster.name: my-application
# ------------------------------------ Node ------------------------------------
# Use a descriptive name for the node:
#node.name: node-1
# ----------------------------------- Paths ------------------------------------
# Path to directory where to store the data (separate multiple locations by comma):
#path.data: /path/to/data
```

```
# Path to log files:
#path.logs: /path/to/logs
# ---------------------------------- Network -----------------------------------
# Set the bind address to a specific IP (IPv4 or IPv6):
#network.host: 192.168.0.1
# Set a custom port for HTTP:
#http.port: 9200
```

在上述注释掉的设置范例中，集群名称（cluster.name）为 my-application，节点名称（node.name）为 node-1，节点数据路径（path.data）为 /path/to/data，节点日志文件路径（path.logs）为 /path/to/logs，网络主机地址（network.host）为 192.168.0.1，超文本传输协议服务端口（http.port）为 9200。由于这些设置被注释掉而失效，待启用 Elasticsearch 服务后，我们可以看到这些设置的默认值。

2.jvm.options

Elasticsearch 是用 Java 开发的，jvm.options 文件是设置 Java 虚拟机（JVM）选项的首选位置。Elasticsearch 服务器除非用在生产环境，否则开发人员很少需要更改这些选项，这些设置可用于提高性能。在配置堆内存时请记住，Xmx 最大只能设置为 32 GB，而且不能设置超过可用内存（RAM）的 50%。

以下是我们从该文件内容中提取部分配置参数的设置。Xms 和 Xmx 分别代表堆内存（heap）空间的初始值和最大值，默认值为 1G byte。

```
# Xms represents the initial size of total heap space
# Xmx represents the maximum size of total heap space
-Xms1g
-Xmx1g
```

3.log4j2.properties

Elasticsearch 使用 log4j2.properties 文件来定义记录日志相关信息。日志文件的产生位置可以由 3 个属性来决定：${sys: es.logs.base_path}、$ {sys:es.logs.cluster_name} 和 $ {sys:es.logs.node_name}。Elasticsearch 预设生成的日志文件的位置如下：

```
appender.rolling_old.fileName=${sys:es.logs.base_path}${sys:file.separator}${sys:es.logs.cluster_name}.log
```

我们的安装目录为 ~/ elasticsearch-7.5.1，波浪号 ~ 字符指用户的主目录。由于未指定基本路径（base_path），因此 sys：es.logs.base_path 默认为 ~/elasticsearch-7.5.1/logs。Linux 文件分隔字符（file.separator）为 /。由于未指定集群名称（cluster_name），因此 sys:es.logs.cluster_name 默认为 elasticsearch。连接相关字串后，日志文件即为"~/elasticsearch-7.5.1/logs/elasticsearch.log"。

1.2.2 重要的系统配置

Elasticsearch 有两种工作模式：开发模式和生产模式。在本书中，我们仅使用开发模式演示所有示例。如果更改 network.host 设置值，则 Elasticsearch 将切换到生产模式。在生产模式下，必须充分了解这些设置才能使生产环境安全无误地运行。

在开发模式中，我们需要确认两个系统设置：文件描述符和虚拟内存设置。

1. 文件描述符

因为 Elasticsearch 使用了大量文件描述符（file descriptor），所以可能会用尽文件描述符而导致数据丢失。为了避免这种情况发生，我们可以用"ulimit"命令，设置当前会话（session）可打开的最大文件数量。用以下命令将文件描述符的数量增加到 65536。

```
ulimit -n 65536
```

如果要将 Ubuntu 永久设置为该值，请将以下指令添加到 /etc/security/limits.conf 文件中。

```
Elasticsearch - nofile 65536
```

对于由 init.d 启动的进程，因为 Ubuntu 会忽略 limits.conf 文件，可以在文件 /etc/pam.d/login 中注释掉以下指令以启用"ulimit"命令功能。

```
# Sets up user limits according to /etc/security/limits.conf
# (Replaces the use of /etc/limits in old login)
# session required pam_limits.so
```

2. 虚拟内存设置

默认情况下，Elasticsearch 允许使用内存映射文件系统（mmapfs）类型文件存储其索引，但是操作系统的 mmap 默认计数设置比较低。如果你的系统设置低于标准，请用以下指令将限制提高到 262 144 或更高。更改的设置仅在 Elasticsearch 服务器实例重新启动后才生效。

```
sudo sysctl -w vm.max_map_count=262144
sudo sysctl -p
cat /proc/sys/vm/max_map_count
262144
```

1.3 与 Elasticsearch 7.5.1 进行对话

尽管许多编程语言都支持 Elasticsearch 客户端，但实际上官方仅支持两种协议：HTTP（通过 RESTful API）和原生（native）协议。

可以通过以下方式之一与 Elasticsearch 进行对话：

- 传输客户端：连接到 Elasticsearch 的原生方法之一（不推荐使用，它将在 8.0 版本中被完全删除）。
- 节点客户端：类似于传输客户端。在大多数情况下，如果使用 Java，则应该选择 Java

高级 REST 客户端而不是节点客户端。
- HTTP 客户端：对于大多数编程语言，HTTP 是连接到 Elasticsearch 的最常见的一种方法。
- 其他协议：只需编写插件即可为 Elasticsearch 创建新的客户端接口。

在 Ubuntu 上，可以使用"curl"命令发送 HTTP（RESTful API）（GET）指令，并通过默认的 9200 端口与 Elasticsearch 通信。请确认 Elasticsearch 服务器正在运行，然后执行以下代码，返回的响应中将看到服务器的详细信息及群集信息，在运行代码命令之前。

```
curl -XGET localhost:9200?pretty=true
{
  "name" : "wai",
  "cluster_name" : "elasticsearch",
  "cluster_uuid" : "r6NHL_oNRb-6Pq2z2I8vPA",
  "version" : {
    "number" : "7.5.1",
    "build_flavor" : "default",
    "build_type" : "tar",
    "build_hash" : "3ae9ac9a93c95bd0cdc054951cf95d88e1e18d96",
    "build_date" : "2019-12-16T22:57:37.835892Z",
    "build_snapshot" : false,
    "lucene_version" : "8.3.0",
    "minimum_wire_compatibility_version" : "6.8.0",
    "minimum_index_compatibility_version" : "6.0.0-beta1"
  },
  "tagline" : "You Know, for Search"
}
```

在响应中，机器的主机名是 wai，集群名称（cluster_name）默认是 elasticsearch，正在运行的版本（version.number）是 7.5.1，下载的 Elasticsearch 软件格式（build_type）是 tar，Lucene 版本（lucence_version）是 8.3.0，而 tar 档案的建立日期（build_date）是 2019-12-16。

1.4 了解 Elasticsearch 的架构

Elasticsearch 具有高可用性实时分布式搜索引擎和分析引擎的功能，适合用于全文本搜索、结构化搜索和数据分析。

搜索引擎建立在 Apache Lucene 软件库之上，是无模式（schemaless）和面向文档（object-oriented）的数据库。而分析引擎建立于从搜索结果返回的大量数据中，继续进行快速分析。除非完全了解用例，否则一般建议不要用于存储主要数据。其优点之一是所提供的 RESTful API 使用了 HTTP 和 JSON，这样就可以通过多种方式集成、管理和查询索引数据。

Elasticsearch 集群是指一个节点或多个 Elasticsearch 节点连接在一起。尽管每个节点都有自己的目的和任务，但是每个节点都可以将客户端的请求转发到适当的节点。

以下是 Elasticsearch 集群中使用的节点种类。
- 符合主节点（master-eligible node）资格的节点：主节点的任务主要是负责轻量级集群范围内的操作，包括创建或删除索引，追踪集群节点，以及确定所分配的分片位置。默认情况下，主节点角色处于启用状态。任何符合主节点,资格的节点都可以通过选举过程成为主节点。可以通过 elasticsearch.yml 文件将 node.master 设置为 false，来禁用该节点成为主节点。

- 数据节点（data node）：包含索引文档的数据，处理相关的操作。例如创建、读取、更新、删除、搜索和聚合。默认情况下，数据节点角色处于启用状态。可以通过 elasticsearch.yml 文件将 node.data 设置为 false，来禁用该节点成为数据节点。
- 摄取节点（ingest node）：在摄取节点中，可以管道模式在文档索引之前，预先处理文档。默认情况下，摄取节点角色处于启用状态。可以通过在 elasticsearch.yml 文件中将 node.ingest 设置为 false，来禁用该节点成为摄取节点。
- 机器学习节点（machine learning node）：默认情况下，机器学习节点角色处于启用状态，各个节点都提供机器学习功能。可以通过 elasticsearch.yml 文件将 node.ml 设置为 false，来禁用该节点成为机器学习节点。
- 转换节点（transform node）：如果想使用转换功能，使现有的 Elasticsearch 索引转换为汇总（summary）索引，产生新的见解和分析。条件为必须至少在一个节点上将 node.transform 设置为 true，并在所有符合主节点资格和所有数据节点上，将 xpack.transform.enabled 设置为 true。默认情况下，转换点角色处于启用状态。
- 协调节点：如果上述角色均被禁用，则该节点将仅充当执行路由请求，处理缩减搜索结果，并通过批量索引操作来分发工作的协调节点。

启动 Elasticsearch 实例，实际上是启动 Elasticsearch 节点。在我们的安装中，是以单个节点 Elasticsearch 集群运行的。让我们使用 Elasticsearch 集群节点应用界面（nodes API）从已安装的服务器中获取该节点的信息。

执行以下代码可以看到响应中的节点信息。以下是我们从响应中提取部分内容。

```
curl -XGET localhost:9200/_nodes?pretty=true
{
  "_nodes" : {
    "total" : 1, "successful" : 1, "failed" : 0
  },
  "cluster_name" : "elasticsearch",
  "nodes" : {
    "DiHKEH33Q3egOY4r_KSomA" : {
      "name" : "wai",
      "transport_address" : "127.0.0.1:9300",
      "host" : "127.0.0.1",
      "ip" : "127.0.0.1",
      "version" : "7.5.1",
      "build_flavor" : "default",
      "build_type" : "tar",
      "build_hash" : "3ae9ac9a93c95bd0cdc054951cf95d88e1e18d96",
      "total_indexing_buffer" : 103795916,
      "roles" : [
        "ingest", "master", "transform", "data", "remote_cluster_client", "ml"
      ],
......
}
```

在响应中，集群名称（cluster_name）是 elasticsearch。节点总数（_nodes.total）为 1。节点识别码为 DiHKEH33Q3egOY4r_KSomA。节点名称为 wai。节点具有六个角色（roles），分别是摄取节点（ingest）、主节点（master）、转换节点（transform）、数据节点（data）、远程集群客户端（remote_cluster_client）和机器学习节点（ml）。在节点上运行的 Elasticsearch 版本（version）为 7.5.1。

1.5 熟悉关键概念

从前节中学习了一些核心概念，例如集群和节点。本节将介绍其他关键概念，然后在后续章节中深入研究细节。

1.5.1 Elasticsearch 映射概念

在传统的关系数据库中，必须显式指定字段和字段类型。Elasticsearch 被认定为无模式的数据库。无模式的简单含义是无须预先指定模式即可对文档建立索引。从第一个文档的索引操作开始，若未指定显式静态映射（static mapping）时，在 Elasticsearch 内部会决定如何对这文档进行索引，并动态（dynamic）产生一个文档索引映射来进行索引。Elasticsearch 称模式为映射，其含义是描述 Lucene 如何存储索引文档及其包含的字段。当向文档添加新字段时，映射也将自动更新。

从 Elasticsearch 6.0 开始，每个索引仅允许一种映射类型。映射类型由数据类型和元字段定义的字段组成。Elasticsearch 支持许多不同数据类型。每个文档都包含与之关联的元字段。创建映射类型时可以定义元字段的数据类型。下面我们将在第 4 章文档管理对此进行介绍。

1.5.2 比较 SQL 与 Elasticsearch 的相关概念

SQL 数据库广泛用于处理结构化数据。尽管 SQL 和 Elasticsearch 对于如何组织数据用了不同的术语，但本质上它们的目的是相同的。

为了帮助读者了解 Elasticsearch 中的概念，表 1-1 中显示了 SQL 和 Elasticsearch 之间的相似术语。

表 1-1　SQL 和 Elasticsearch 之间的相似术语

SQL	Elasticsearch
列（Column） 列是具有相同数据类型的一组数据值。每个数据库列的每一行都有一个值	字段（Field） 字段是 Elasticsearch 中最小的数据单位，它可以包含相同类型的多个值的数组
行（Row） 行表示结构化数据项，包含表的每一列中的一系列数据值	文档（Document） 文档就像是将字段分组的行。一个文档在 Elasticsearch 中以 JSON 对象表达
表（Table） 一个表由列和行组成	索引（Index） 索引是 Elasticsearch 中最大的数据单位。索引是文档索引的逻辑分区，是搜索查询的目标
模式（Schema） 在关系数据库管理系统（RDBMS）中，模式包含模式对象，例如表、列、数据类型和视图等。模式通常由数据库用户拥有	隐含（Implicit） Elasticsearch 没有为其提供相同的概念
目录或数据库（Catalog/database） 在 SQL 中，目录或数据库表示一组模式	集群（Cluster） 在 Elasticsearch 中，集群包含一组索引

1.5.3 分析器（Analyzer）

Elasticsearch 随附带有各种内置分析器，无须进一步配置即可在任何索引中使用。如果内置分析器不适合用例使用，则可以自定义一个分析器。

无论是内置分析器还是自定义的分析器，包括以下 3 种构建模块的组合。

- 字符过滤器（character filter）：以字符流的形式接收原始文本，并且可以进行添加、删除或更改其字符。
- 分词器（tokenizer）：将给定的字符流拆分为词元流。
- 词元过滤器（token filter）：以词元流的形式接收原始文本，并且可以进行添加、删除或更改其词元。

在索引过程和搜索过程中通常使用同一种分析器，但是可以在字段的映射中设置搜索分析器（search_analyzer）的属性，以便在搜索时执行与索引过程不同的分析方法。

1.5.4 标准分析器（Standard Analyzer）

标准分析器是默认的分析器，包括以下内容：

（1）字符过滤器：无。

（2）分词器：使用标准分词器生成基于语法的词元。

（3）词元过滤器：使用小写词元过滤器（lowercase）和停止词元过滤器（stop）。但停止词元过滤器默认情况下是禁用的。

- 小写词元过滤器，将词元文本规范化为小写。
- 停止词元过滤器，从词元流中删除停用词（stop words）。有关英语停用词的列表，请参阅官方网站。

1.6 接口用法约定说明

这里仅讨论一些主要约定。其他约定请参阅官方网站。以下清单项目可以在 REST API 参数中使用。

1. 跨多个索引访问约定

- _all：代表所有索引。
- 逗号（,）：用于两个索引之间的分隔符。
- 通配符（*）：星号字符（*）可匹配索引名称中的任何字符序列，用于索引的前缀、中缀和后缀。
- 破折号（-）：排除（不包括）随后的索引，用于索引的前缀。

2. 常用选项约定

- 漂亮的格式（?pretty=true）：若在请求网站后附加此参数设定，则响应中的 JSON 字符串将采用漂亮的打印格式。pretty 默认值为 false。
- 人类可读的输出格式（?human=true）：若在请求网站后附加此参数设定，则响应中的 JSON 字符串的数值将转换为人类可读的值，例如 1h 是 1 hour（1 小时），1kB 是

1 kilobytes（1024 bytes）。human 默认值为 false。
- REST 参数：遵循使用下画线的约定。
- 布尔值（boolean）：false 表示假值，true 表示真值，而其他的值将引发错误。
- 数字值（numeric）：基于 JSON 数字类型的字符串。
- 持续时间的单位：支持的时间单位有 d（天）、h（小时）、m（分钟）、s（秒）、ms（毫秒）、micros（微秒）及 nanos（纳秒）。
- 日期字符串：日期的表示方式，可以使用这种方式来设置。
  - 格式化日期的字符串，例如 yyyy-MM-dd HH:mm:ss，或者 yyyy-MM-dd。
  - 以毫秒为单位的整数，表示自 epoch 以来的毫秒数。
  - 以秒为单位的整数，表示自 epoch 以来的秒数。
- 格式化日期的数学表示方式：在日期范围查询或聚合时，可以使用这种方式来设置日期字段的格式。
  - 日期的数学表示方式可以锚点日期（now，或者是以双竖线 || 结尾的日期字符串）开头，然后由一个或多个子表示方式，例如 +1h、-1d 和 /d，分别表示加 1 小时、减去一天和四舍五入到最近一天。
  - 支持的时间单位与前述提到的持续时间的单位不同。其中，y 是数年，M 是数月，w 是数周，d 是数天，h 或 H 是数小时，m 是数分钟，s 是数秒，+ 表示加法，- 表示减法，以及 "/" 用于四舍五入到最接近的时间单位。例如，"/d" 表示四舍五入到最接近的一天。
- 字节大小单位：支持的数据单位有，B 表示字节，kB 表示 1 024 字节，MB 表示 1024^2 字节，GB 表示 1024^3 字节，TB 表示 1024^4 字节，PB 表示 1024^5 字节。
- 无单位的数量：如果表示的值足够大，则可以用数量作为乘数。支持的数量有 k（1000）、M 为（10002）、G 为（10003）、T 为（10004）、P 为（10005）。例如，10M 代表 1000 万。
- 距离单位：支持的距离单位有，mi 表示英里，yd 表示码，ft 表示英尺，in 表示英寸，km 表示公里，m 表示米，cm 代表厘米，mm 代表毫米，nmi 或 NM 代表航海英里。
- 堆栈追踪（error_trace = true）：若在请求网站后附加此参数设定，则在引发异常时将追踪调用函数的堆栈包含在响应中。error_trace 的默认值为 false。
- 查询字符串中的请求正文：如果客户端不接受 non-POST 请求的正文，则可以使用源查询字符串（source_content_type）参数传递请求正文，并使用受支持的媒体类型指定参数。
- 内容类型（content-type）：请求正文中的内容类型。必须在请求标头中指定使用内容类型名称，以检查所使用的内容类型是否被支持。在本书中所有的 POST/UPDATE/PATCH 请求示例中，都使用 application/json 内容类型。

3. 索引名称中的日期数学约定

如果要索引时间序列数据（例如日志），则可以用具有不同日期字段的模式作为索引名称，以管理每日日志记录的信息。Elasticsearch 提供了一种搜索一系列时间序列索引的方法。

索引名称的日期数学语法如下：

```
<static_name{date_math_expr{date_format|time_zone}}>
```

上述语法中，

- static_name：索引名称指定的固定文本部分。
- date_math_expr：索引名称根据日期数学表示方式而变化的文本部分。
- date_format：日期格式默认为 YYYY.MM.dd，其中 YYYY 代表年份，MM 代表月份，dd 代表日期。
- time_zone：时区偏移量，默认时区为 UTC+0。例如，PST 时区使用 UTC 时间偏移量来表示，即为 UTC-08:00。

4. 基于 URL 的访问控制约定

Elasticsearch 中有许多接口，以便在请求正文中指定索引，例如多次搜索（multi-search）、多次获取（multi-get）和批量请求（bulk）。

默认情况下，请求正文中指定的索引将覆盖 URL 中指定的 index 参数。如果使用具有基于 URL 的访问控制的代理人，来保护对 Elasticsearch 索引的访问，则可以将以下设置添加到 elasticsearch.yml 配置文件，以禁用默认操作：

```
rest.action.multi.allow_explicit_index: false
```

到目前为止，我们已经在 Elasticsearch 服务器上执行了一些简单的测试，还从架构的角度熟悉了一些基本概念。在下一章中我们将开始获取本书的实验数据。

第 2 章 获取本书实验数据

Elasticsearch 可以为任何有意义的数据提供搜索和聚合。由于应用场景范围比较大，使用者也很多，本书尝试使用一套金融数据，有系统性地详细说明，如何运用 Elasticsearch 进行金融数据的分析和应用。有别于外语的中文翻译本沿用国外的数据，本书所有示例采用的是一套国内开源的金融数据。主线是如何使用 Elasticsearch 打造一个信息搜索与分析系统，使读者易于理解其中各种核心技术及其应用。

在金融投资过程中，公募基金在信息披露、估值、流动性、管理费用、投资范围等方面具有非常严格的限制，在证监会的严格监管下可以对信息不对称所引起的风险进行最小化。本章先为读者介绍公募基金以及公募基金的类型和数据分类，再向读者介绍如何通过市场开放的数据平台获取相关数据，并作为本书的实验数据。

2.1 了解实验数据集

2.1.1 公募基金简介

公募基金是指以对市场上的投资者以公开的形式募集资金，并以证券为主要投资对象的投资基金。截至 2019 年末，公募基金的总规模从 2008 年的 4 万亿元增长到 14.8 万亿元，其中指数基金的规模从 2000 亿元增加到了近万亿元。越来越多的投资者开始关注公募基金，认同公募基金的投资方法，并将公募基金作为一种极为重要的投资工具。管理公募基金的基金经理及团队常年对宏观环境、市场动态、行业发展和上市公司的研究有着敏锐的把控能力，可以科学地为不同类型的资产生成最优的配置方案，做到收益风险最优化。这样的公募基金可以为普通投资者省去需要自己去配置一篮子"鸡蛋"，但又不知如何去配置的烦恼，同时避免了频繁换仓、追涨杀跌等不合理的操作，符合长期投资、价值投资的投资理念。

2.1.2 公募基金类型介绍

不同的公募基金对于投资的方向和标的是有区别的。从投资类型来区分公募基金的类型，一般分为股票型基金、债券型基金、混合型基金和货币市场基金四种类型。股票型基金是主要以仓位投资于股票市场上的基金；债券型基金是以主要仓位投资于利率债、信用债、企业债和地方政府债等债券市场的基金；混合型基金是以主动灵活的调整仓位，同时投资于股票市场和债券市场的基金；货币基金是以全部资产投资于极低风险的货币市场上的基金。

从投资者交易基金的方式来区分，公募基金可以分为场外基金和场内基金。场外基金主要是通过申购赎回进行交易的。当投资者需要申购某一基金时，需要将现金交给基金公司，基金公司会给投资者对应的基金份额。场内基金则需要在证券交易所中完成其购买或卖出。场外基金的优势在于交易渠道便捷，不需要在证券交易所开户，通过基金公司、银行、券商或者第三方交易平台便可直接申购。场内基金的优势在于交易手续费率低，几乎是场外基金的1/10，并且支持日内交易，可以以日内的目标价位进行买卖场内基金。

2.1.3 公募基金数据归类

公募基金的数据主要可以分为基金信息数据、基金公司信息数据、基金净值数据和场内/外基金行情数据等。Tushare是一个开放和免费的金融数据平台，包含沪深股票数据、指数数据、基金数据、期货数据、期权数据、债券数据、外汇数据、港股数据、行业经济数据、宏观经济数据以及新闻快讯等数据。Tushare提供不同接口，配合输入参数，可以获取各个公募基金数据列表。以下介绍本书采用的相关数据列表内容和获取方法的详细信息。

1. 基金信息数据

公募基金在发行前需要向证监会报备公开募集基金的基本信息，证监会审核通过后，才可在规定的日期，以报备的信息对市场开放。Tushare提供fund_basic接口和输入参数market，可以获取各个公募基金数据，包括场内基金（market="E"）和场外基金（market="O"）。返回的数据内容如名称、类型和描述等详细信息，见表2-1。

表2-1　公募基金信息数据列表

数据描述	数据名称	数据类型
基金代码	ts_code	
基金简称	name	
基金管理人	management	
基金托管人	custodian	
投资类型	fund_type	
成立日期	found_date	
到期日期	due_date	
上市时间	list_date	
发行日期	issue_date	
退市日期	delist_date	str（字符串）
业绩比较基准	benchmark	
存续状态 D 摘牌 I 发行 L 已上市	status	
投资风格	invest_type	
基金类型	type	
受托人	trustee	
日常申购起始日	purc_startdate	
日常赎回起始日	redm_startdate	
E 场内 O 场外	market	

续表

数据描述	数据名称	数据类型
发行份额	issue_amount	
管理费	m_fee	
托管费	c_free	
存续期	duration_year	float（浮点数）
面值	p_value	
起点金额	min_amount	
预期收益率	exp_return	

2. 基金公司信息数据

公募基金管理公司的相关信息也是公开的，Tushare 提供 fund_company 接口（无输入参数），可以获取各个公募基金公司和管理人信息。返回的数据内容如名称、类型和描述等详细信息，见表 2-2。

表 2-2 公募基金公司数据列表

数据描述	数据名称	数据类型
基金公司名称	name	
简称	shortname	
英文缩写	short_enname	
省份	province	
城市	city	
注册地址	address	
电话	phone	
办公地址	office	str（字符串）
公司网址	website	
法人代表	chairman	
总经理	manager	
成立日期	setup_date	
公司终止日期	end_date	
主要产品及业务	main_business	
组织机构代码	org_code	
社会信用代码	credit_code	
注册资本	reg_capital	float（浮点数）
员工总数	employees	

3. 基金净值信息数据

公募基金净值数据包含了每一个基金的每日的净值、分红等相关数据的披露，Tushare 提供 fund_nav 接口和输入参数（ts_code 或 end_date，和 market），可以获取各个公募基金每日与净值相关的信息，包括场内基金（market="E"）和场外基金（market="O"）。返回的数据内容如名称、类型和描述等详细信息，见表 2-3。

表 2-3 公募基金净值数据列表

数据描述	数据名称	数据类型
基金代码	ts_code	str（字符串）
公告日期	ann_date	
截止日期	end_date	
单位净值	unit_nav	float（浮点数）
累计净值	accum_nav	
累计分红	accum_div	
资产净值	net_asset	
合计资产净值	total_netasset	
复权单位净值	adj_nav	

4. 基金交易行情信息数据

公募基金场内行情数据，就与股票日内数据比较相似，是场内基金每天的开盘、收盘、成交量等相关数据的收集整理。Tushare 提供场内基金日线行情接口 fund_daily，再配合输入参数（ts_code、trade_date、start_date、和 end_date），可以获取各个公募基金每日与价格与成交量相关的信息。返回的数据内容如名称、类型和描述等详细信息，见表 2-4。

表 2-4 日线行情数据列表

数据描述	数据名称	数据类型
基金代码	ts_code	str（字符串）
交易日期	trade_date	
开盘价	open	float（浮点数）
最高价	high	
最低价	low	
收盘价	close	
昨收价	pre_close	
涨跌额	change	
涨跌幅	pct_chg	
成交量	vol	
成交额	amount	

5. 基金持仓数据信息数据

一般而言，公募基金将资金投资于某些金融产品，例如股票、信托、债券和固定收益产品。根据中国证监会规定，公募基金对其投资目标、投资组合和其他信息，要在每个季度结束之日起 15 个工作日内进行披露。从披露的基金持仓的资产配置情况，可以总结该基金在上一季度的市场风格策略。

Tushare 提供季度更新的持仓数据接口 fund_portfolio，再配合输入参数（ts_code、ann_date、start_date、和 end_date），可以获取各个公募基金持有股票市值和数量等相关信息。返回的数据内容如名称、类型和描述等详细信息，见表 2-5。

表 2-5 基金持仓数据列表

数据描述	数据名称	数据类型
基金代码	ts_code	str（字符串）
公告日期	ann_date	
截止日期	end_date	
股票代码	symbol	
持有股票市值（元）	mkv	float（浮点数）
持有股票数量（股）	amount	
占股票市值比	stk_mkv_ratio	
占流通股本比例	stk_float_ratio	

2.2 通过 Python SDK 获取多维数据

Tushare 用户目前可以通过 Http、Python、Matlab 和 R 语言方式来获取数据。在可拓展性方面，Tushare 接口基于标准的 RESTful API 设计，用户可以通过任何程序语言获取数据。Tushare 社区已经开放了主流编程语言的软件开发工具包（SDK）。本节介绍如何通过 Python SDK 获取基金的各种数据。

2.2.1 注册及安装 Tushare

这里向大家介绍一下 Tushare 的注册以及安装方法。为避免操作软件工具版本的差异，当前采用 Virtualenv 生成 Python 3 的操作环境。使用的 pip3 版本为 20.1.1，而 Python 3 的版本为 3.6.7。

1. 注册方法

用户需要访问 Tushare 官方网站，进行新用户注册。在成功注册后，可以在个人主页中接口 Token 选项获取 Token（令牌），如图 2-1 所示。令牌是一系列字符，是 Tushare 提供用户在接口调取数据的唯一凭证，本书假设令牌为 abc123，并用于以后所有有关示例。

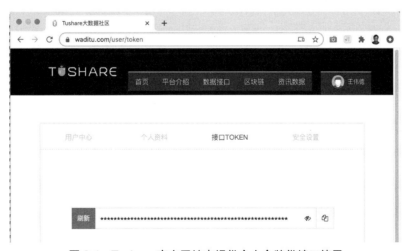

图 2-1 Tushare 官方网站中提供个人令牌供接口使用

2. 下载安装 Tushare Python SDK 方法

Tushare 支持 Python 2.x 和 Python 3.x。本书使用 Python 3.6.7 进行 Python 编程。下载安装有三种方式：

（1）通过 pip3 指令

在 terminal 中对 PyPI 源网站进行 Tushare 安装，截至撰写本书时，Tushare 在 PyPI（The Python Package Index）的最新版本是 1.2.54，而使用的 pip3 版本为 20.1.1。

```
pip3 install tushare
```

（2）通过镜像

访问 pypI 源网站找 tushare 的镜像下载，并执行下列代码进行安装。

```
python setup.py install
```

（3）通过访问 github

在 github 中搜索 waditu，将项目下载或者克隆（clone）到本地，进入项目的目录下执行指令。

```
python setup.py install
```

2.2.2　Tushare 数据接口的调用方法

本节会向大家介绍一下 Tushare，在 Python 中如何去调用接口，以及获取数据的方法。

（1）导入 tushare

```
import tushare as ts
```

（2）初始化接口

```
ts_pro = ts.pro_api('abc123')
```

（3）调取公募基金(场内)列表

使用接口 fund_basic 和输入参数 market="E"，可以获取如表 2-6 所示的数据（只列出第一个数据）。

```
df = ts_pro.fund_basic(market = 'E')
print(df)
```

表 2-6　场内基金数据

数据名称	数据内容
ts_code	159809.SZ
name	恒生湾区
management	博时基金
custodian	招商银行
fund_type	股票型
found_date	20200430
due_date	None
list_date	20200521
issue_date	20200330
delist_date	None
benchmark	恒生沪深港通大湾区综合指数收益率
status	L

续表

数据名称	数据内容
invest_type	被动指数型
type	契约型开放式
trustee	None
purc_startdate	20200521
redm_startdate	20200521
market	E
issue_amount	None
m_fee	3.9117
c_free	0.15
duration_year	0.05
p_value	NaN
min_amount	1.0
exp_return	0.100

（4）调取公募基金公司数据列表

使用接口 fund_company，可以获取如表 2-7 所示的数据（只列出第一个数据）。

```
df = ts_pro.fund_company()
print(df)
```

表 2-7　公募基金公司数据列表

数据名称	数据内容
name	北京广能投资基金管理有限公司
shortname	广能基金
short_enname	（默认不显示）
province	北京
city	北京市
address	北京市朝阳区北四环中路 27 号院 5 号楼 2712-2715A
phone	None
office	北京市朝阳区北四环中路 27 号院 5 号楼 2712-2715A
website	www.gnfund.cn
chairman	刘锡潜
manager	杨运成
setup_date	20111031
end_date	None
main_business	None
org_code	584419680
credit_code	None
reg_capital	1.000000e+04
employees	5347.0

（5）调取公募基金净值数据列表

使用接口 fund_nav 和输入参数 ts_code="159809.SZ"（恒生湾区）和 market="E"，可以获取如表 2-8 所示的数据（只列出第一个数据）。

```
df = ts_pro.fund_nav(ts_code= '159809.SZ', market= 'E')
print(df)
```

表2-8 公募基金净值数据列表

数据名称	数据内容
ts_code	159809.SZ
ann_date	20200530
end_date	20200529
unit_nav	0.9828
accum_nav	0.9828
accum_div	None
net_asset	NaN
total_netasset	NaN
adj_nav	0.9828

（6）调取公募基金交易行情数据列表

使用接口 fund_daily 配合输入参数 ts_code="159809.SZ"、start_date="20200525" 和 end_date="20200529"，可以获取如表2-9所示的数据(只列出第一个数据)。

```
df = ts_pro.fund_daily(ts_code='159809.SZ', start_date='20200525', end_date='20200529')
print(df)
```

表2-9 公募基金净值数据列表

数据名称	数据内容
ts_code	159809.SZ
trade_date	20200529
open	0.972
high	0.982
low	0.971
close	0.979
pre_close	0.980
change	-0.001
pct_chg	-0.1020
vol	106073.89
amount	10377.487

第 3 章 索引管理

在第一章曾经谈到索引是文档索引的逻辑分区，是搜索查询的目标范围。本章先学习如何使用索引接口来管理各个索引、索引设置、别名和模板，然后在下一章再介绍文档索引。本章除了说明索引管理操作包括刷新、清空和清除缓存，也对索引统计和状态监测进行了探讨，以了解发生在服务器上的操作记录，供进一步的分析。

3.1 基础索引管理

索引管理操作负责管理整个索引生命周期，包含创建、读取、打开、关闭、更新、删除、设置、别名、模板和映射等操作。

3.1.1 基本索引操作

基本索引操作包括以下 8 个：

1. 创建索引

创建索引时，索引名称必须遵循以下约定：

- 仅小写；
- 不能包含 \、/、*、?、"、<、>、|、空格、逗号或 # 字符；
- 不能使用冒号 : 字符；
- 不能使用 -、_ 或 + 字符开始；
- 不能使用 . 或 .. 字符；
- 索引名称的最大字节为 255。

下面通过"curl"命令向统一资源定位符（URL）http://localhost:9200/fund_basic 发出 PUT 请求，以创建一个名为"fund_basic"基金列表的索引。如果操作成功，则返回的响应状态值为 200（OK），如以下所示：

```
curl -XPUT localhost:9200/fund_basic
```

2. 检查索引是否存在

通过"curl"命令发出 HEAD 请求，以检查索引是否存在。若索引存在，则返回的响应状态值为 200（OK），否则响应状态值为 404（INVALID REQUEST）。此接口也适用于索引别名（alias）的操作。

```
curl --head localhost:9200/fund_basic
HTTP/1.1 200 OK
```

3. 获取索引内容

使用此接口可以检索有关一个或多个索引的信息。下面通过"curl"命令发出 GET 请求，以获取刚刚创建的 fund_basic 索引。如果操作成功，返回的响应状态值为 200（OK）。响应主体包含索引的 3 部分，即别名（aliases）、映射（mappings）和设置（settings）。设置和别名将分别在 3.1.2 及 3.1.4 节介绍，而映射部分将在第 4 章的文档结构中进行说明。

```
curl -XGET localhost:9200/fund_basic?pretty=true
{
  "fund_basic" : {
    "aliases" : { },
    "mappings" : { },
    "settings" : {
      "index" : {
        ......
      },
      "provided_name" : "fund_basic"
    }
  }
}
```

如果该索引不存在，则获取索引请求失败，返回的响应状态值为 404（Not Found），并且在响应主体中报告的错误原因（reason）的异常类型（type）为找不到索引（index_not_found_exception）。

例如，让我们通过"curl"命令发出 GET 请求，以请求一个不存在的索引 fund-basic，结果如下所示：

```
curl -XGET localhost:9200/fund-basic?pretty=true
{
  "error" : {
    "root_cause" : [{…}],
    "type" : "index_not_found_exception",
    "reason" : "no such index [fundbasic]",
    "resource.type" : "index_or_alias",
    "resource.id" : "fundbasic",
    "index_uuid" : "_na_",
    "index" : "fundbasic"
  },
  "status" : 404
}
```

4. 打开与关闭索引

Elasticsearch 支持索引在线或离线的状态。使用离线模式时，可以在群集上几乎没有消耗资源的情况下维护数据。关闭索引后，读 / 写操作将停止。

提示：关闭索引会占用大量的磁盘空间。可以通过将 cluster.indices.close.enable 的默认值从 true 更改为 false 来禁止使用关闭索引功能，以避免发生磁盘空间溢出意外。

下面是以 _close 接口的示例。此操作使用 POST 请求，如果操作成功，则返回的响应状态值为 200（OK）。

```
curl -XPOST localhost:9200/fund_basic/_close?pretty=true
{
  "acknowledged" : true,
  "shards_acknowledged" : true,
```

```
    "indices" : {
      "fund_basic" : {
        "closed" : true
      }
    }
}
```

当需要索引重新在线时，只需打开该索引即可。以下是 _open 接口的示例。此操作使用 POST 请求，如果操作成功，则返回的响应状态值为 200（OK）。

```
curl -XPOST localhost:9200/fund_basic/_open?pretty=true
{
  "acknowledged" : true,
  "shards_acknowledged" : true
}
```

5. 更新索引

Elasticsearch 不支持更新索引的操作，用户不可以直接更新索引的内容。但可以间接地更新索引的映射，例如 PUT /fund_basic/_mappings，或更新索引的设置，例如 PUT /fund_basic/_settings。

6. 删除索引

通过此接口，可以删除一个或多个现有索引。

如以下代码所示，若删除操作成功，则返回响应状态值为 200（OK）。

提示：如果已成功删除 fund_basic 索引，请再次创建该索引以供后续章节使用。

```
curl -XDELETE localhost:9200/fund_basic?pretty=true
{ "acknowledged" : true }
```

已经了解基本的索引操作后，可以根据需要自定义索引的设置来调整索引的属性。

7. 冻结索引

虽然关闭索引操作可使索引进入离线模式（该模式不需要内存空间），但若需要重新打开它，操作成本则很高。

Elasticsearch 支持冻结索引（_freeze）操作。当索引被冻结时，它变为可供搜寻的唯读模式，临时数据结构会从内存中删除，操作成本较关闭索引低。以下是 _freeze 接口的示例。此操作使用 POST 请求来冻结 fund_basic 索引，如果操作成功，则返回的响应状态值为 200（OK）。

```
curl -XPOST http://localhost:9200/fund_basic/_freeze?pretty=true
{
  "acknowledged" : true,
  "shards_acknowledged" : true
}
```

8. 解冻已冻结的索引

如果需要在已经冻结的索引上再次进行文档索引操作，则需要先解冻它。

以下是 _unfreeze 接口的示例。此操作使用 POST 请求来解冻已冻结的 fund_basic 索引，如果操作成功，则返回的响应状态值为 200（OK）。

```
curl -XPOST http://localhost:9200/fund_basic/_unfreeze?pretty=true
{
  "acknowledged" : true,
  "shards_acknowledged" : true
}
```

3.1.2 设置索引

索引的设置可以对个别索引或是对全局索引。基本上，全局级别的目的是避免逐个地配置它们的设置。

- 全局级别的索引设置包括断路器（Circuit breakers）、字段数据缓存（Fielddata cache）、节点查询缓存（Node query cache）、索引缓冲区（Indexing buffer）、分片请求缓存（Shard request cache）、恢复（Recovery）和搜索设置（Search settings）。它们由所谓的索引模块（Indices Module）控制。有关更多详细信息，请参阅官方网站。
- 个别索引的设置可以分为静态和动态。静态设置只能在创建索引或关闭索引时设置，而动态设置可以随时更改。

表 3-1 是个别索引静态设置的简要说明，而表 3-2 是动态设置的简要说明。

表 3-1 个别索引静态设置

名 称	描 述
分片数 index.number_of_shards	这是索引的主要分片的数量，可以设置为 1（默认）至 1024（上限）
打开分片之前检查 index.shard.check_on_startup	这是打开分片前检查损坏的选项，可以设置为 true、false（默认）或 checksum。true 会检查逻辑分片和物理分片，而 checksum 只会检查逻辑分片
压缩算法解码器 index.codec	使用的压缩方法，可以设置为 LZ4（默认）或是 best_compression
路由分区数 index.routing_partition_size	这是文档以扇形方式发出搜索请求的分片数量。可以设置为 1（默认）至 index.number_of_shards（上限）
提前加载过滤器 load_fixed_bitset_filters_eagerly	是否为嵌套查询预加载缓存的过滤器。可以设置为 true（默认）或是 false

表 3-2 个别索引动态设置

名 称	描 述
副本数 index.number_of_replicas	这是每个主分片的副本数。预设值为 1
自动扩展副本 index.auto_expand_replicas	设置自动增加副本数的范围，以适应读取繁重环境中的峰值；选项为 false（默认），x-y 和 x-all，其中 x-y 为范围，all 为最大上限
搜索分片允许闲置时间段 index.search.idle.after	这是分片被变更为闲置分片的许可停止搜索活动时间段。默认值为 30s
刷新操作频率 index.refresh_interval	这是执行刷新操作的频率。默认值为 1s。可以设置为 -1 以禁止使用刷新
最大返回结果窗口 index.max_result_window	这是从搜索请求返回的结果所允许的最大数目。默认值为 10000
最大内部返回结果窗口 index.max_inner_result_window	这是从内部嵌套对象或顶部匹配聚合返回的结果所允许的最大数目。默认值为 100
最大重评分返回结果窗口 index.max_rescore_window	这是从重新评分请求返回的结果所允许的最大数目；它与 index.max_result_window 相同
docvalue 字段最大允许搜索数目 index.max_docvalue_fields_search	这是匹配返回的 doc_values 字段所允许的最大数目。默认值为 100
脚本字段最大数目 index.max_script_fields	这是查询中 script_fields 个数所允许的最大数目。默认值为 32
n 元语法最大差异数 index.max_ngram_diff	当使用 NgramTokenizer 和 NGramTokenFilter 时，max_ngram 和 min_ngram 之间所允许的最大差异。默认值为 1
shingle 最大差异数 index.max_shingle_diff	当使用 ShingleTokenFilter 时，max_shingle_size 和 min_shingle_size 之间允许的最大差异。默认值为 3

续表

名　　称	描　　述
唯读模式 index.blocks.read_only	启用唯读操作。默认为 false
唯读并允许删除模式 index.blocks.read_only_allow_delete	类似于 index.block.read_only，当设定为 true 时，索引是唯读并允许删除操作
读取模式 index.blocks.read	禁止使用读取操作。默认为 false
元数据操作 index.blocks.metadata	禁止使用元数据。默认为 false
刷新侦听器的最大数目 index.max_refresh_listeners	这是索引的每个分片上可用的刷新侦听器的最大数目
最大分词个数 index.analyze.max_token_count	这是 _analyze 接口中允许产生的最大分词个数。默认值为 10000
高亮请求分析的最大字符数 index.highlight.max_analyzed_offset	这是高亮请求中允许分析的最大字符数。默认值为 1000000
最大词条数 index.max_terms_count	这是词条查询中允许的最大词条数。默认值为 65536
正则表达式字串最大长度 index.max_regex_length	这是正则表达式查询中允许的字串最大长度。默认值为 1000
分片启用分配选项 index.routing.allocation.enable	它指定在哪个节点上控制此索引的分片分配。选项为 all（默认为全部）、primaries（主分片）、new_primaries（新创建的主分片）和 none（无）。
分片启用重新平衡选项 index.routing.rebalance.enable	它指定在哪个节点上控制此索引的重新平衡。选项为 all（默认为全部）、primaries（主分片）、new_primaries（新创建的主分片）和 none（无）。
删除的文件回收期限 index.gc_deletes	这是垃圾回收删除期限。默认值为 60 秒
默认摄取节点管道 index.default_pipeline	这是此索引的默认摄取节点管道名称。设定 _none 表示没有管道

以下介绍索引设置（_settings）接口。

1. 获取设置

以下是使用 GET _settings 请求来获取索引的设置接口的示例。如果操作成功，则返回的响应状态值为 200（OK）。在返回的响应中，fund_basic 的所有设置均为默认值。默认设置包括：索引创建日期（creation_date）为 1582907164131、分片数（number_of_shards）为 1、分片副本数（number_of_replicas）为 1、通用唯一识别码（uuid）为 jsACaAWdRx6VtiFZvgD55A、Elasticsearch 版本（version）为 7060099，以及给定的索引名称（provided_name）为 fund_basic。

```
curl -XGET http://localhost:9200/fund_basic/_settings?pretty=true
{
  "fund_basic" : {
    "settings" : {
      "index" : {
        "creation_date" : "1582907164131",
        "number_of_shards" : "1",
        "number_of_replicas" : "1",
        "uuid" : "jsACaAWdRx6VtiFZvgD55A",
        "version" : {
          "created" : "7060099"
        },
        "provided_name" : "fund_basic"
```

```
      }
    }
  }
}
```

2. 更新个别动态索引设置

以下是使用 PUT _settings 请求来更新索引的动态设置。示例为更新分片副本数（number_of_replicas）。如果操作成功，则返回的响应状态值为 200（OK）。fund_basic 索引上的 number_of_replicas 从 1 更新为 2。

若要验证分片副本数已从 1 更新为 2，则需要可以发出 GET _settings 请求，并设置 Content-Type 标头为 application/json。

```
curl -XPUT -H "Content-Type: application/json" http://localhost:9200/fund_basic/_settings?pretty=true --data $'{"index": {"number_of_replicas": 2 }}'
  { "acknowledged" : true }
```

3. 更新个别静态索引设置

如果在打开（open）的索引上更新静态设置，操作会失败。返回的响应状态值为 400（Bad Request），错误原因（reason）的异常类型（type）为无效参数（invalid_argument_exception）。

```
curl -XPUT -H "Content-Type: application/json" http://localhost:9200/fund_basic/_settings?pretty=true --data $'{"index": {"codec":"best_compression" }}'
{
  "error" : {
    "root_cause" : [
      {
        "type" : "illegal_argument_exception",
        "reason" : "Can't update non dynamic settings [[index.codec]] for open indices [[fund_basic/Sy7_uq3wQX21oIt2y4hCjA]]"
      }
    ],
    "type" : "illegal_argument_exception",
    "reason" : "Can't update non dynamic settings [[index.codec]] for open indices [[fund_basic/Sy7_uq3wQX21oIt2y4hCjA]]"
  },
  "status" : 400
}
```

正确的方法是首先使用 3.1.1 节中介绍的关闭索引，然后再发出更新请求。以下是 PUT _settings 请求来更新静态设置的示例。如果操作成功，返回的响应状态值为 200（OK）。操作成功后，请参考 3.1.1 节重新打开 fund_basic 索引，用于后续操作，同时也可以使用 GET _settings 请求检查设置是否已更新。

```
curl -XPUT -H "Content-Type: application/json" http://localhost:9200/fund_basic/_settings?pretty=true --data $'{"index": {"codec":"best_compression" }}'
  { "acknowledged" : true }
```

4. 删除索引设置

Elasticsearch 不支持删除索引设置操作。

5. 重置索引设置

若是在更新索引设置操作中指定某设置为空值（null），则会将其重置为默认值。

3.1.3 索引模板

Elasticsearch 提供了一种简单的方法，可以在创建索引时，重复使用定义好的索引设置，即索引模板。模板中的设置会被创建索引接口中的设置覆盖。

以下介绍基本的索引模板操作，并用不同示例分别说明。

1. 创建索引模板

以下示例使用 PUT _template 发出请求，以创建索引模板。并且使用选项 number_of_replicas（设置为 2）和 codec（设置为 best_compression）来创建索引模板 fund_basic_template。在模板中，有一个名为 index_patterns（索引模式）的字段，在示例中设置为 fund_basic*。这种设置可以指定为全局样式，即任何与索引模式匹配的索引都将应用模板的设置，同时 index_patterns 字段可以设置多个名称。

以下是 PUT _template 接口的示例。如果操作成功，返回的响应状态值为 200（OK）。

```
curl -XPUT -H "Content-Type: application/json" http://localhost:9200/_template/fund_basic_template?pretty=true --data $'{"index_patterns": ["fund_basic*"], "settings": {"number_of_replicas":"2", "codec": "best_compression" }}'
{ "acknowledged" : true }
```

2. 获取索引模板内容

使用 GET _template 发出请求以获取索引模板 fund_basic_template 的信息。

以下是 GET _template 接口的示例。如果操作成功，则返回的响应状态值为 200（OK）。

```
curl -XGET http://localhost:9200/_template/fund_basic_template/?pretty=true
{
  "fund_basic_template" : {
    "order" : 0,
    "index_patterns" : [
      "fund_basic*"
    ],
    "settings" : {
      "index" : {
        "codec" : "best_compression",
        "number_of_replicas" : "2"
      }
    },
    "mappings" : { },
    "aliases" : { }
  }
}
```

3. 检查索引模板是否存在

使用 HEAD _template 发出请求，以获取索引模板是否存在的信息。可以通过 curl 命令发出 HEAD _template 请求，以检查索引模板 fund_basic_template 是否存在的信息。

以下是 GET _template 接口的示例。如果操作成功，则返回的响应状态值为 200（OK）。

```
curl --head http://localhost:9200/_template/fund_basic_template/?pretty=true
HTTP/1.1 200 OK
content-type: application/json; charset=UTF-8
content-length: 272
```

4. 使用索引模板创建索引

创建一个新索引时，若使用与索引模板定义的索引名称相匹配的索引，索引模板中定义的设置将会应用到新索引中。

以下示例使用索引模板 fund_basic_template 及其定义中相匹配的索引名称 fund_basic_large 来创建新索引。如果操作成功，则返回的响应状态值为 200（OK）。可以使用 GET _settings 来检查索引 fund_basic_large 是否已应用 fund_basic_template 的设置。

```
curl -XPUT -H "Content-Type: application/json" http://localhost:9200/fund_basic_large/?pretty=true --data $'{}'
{
  "acknowledged" : true,
  "shards_acknowledged" : true,
  "index" : "fund_basic_large"
}
```

5. 更新索引模板

以下示例通过 curl 命令发出 PUT _template 请求以更新索引模板 fund_basic_template 定义的 number_of_replicas 设置，将其数值从 2 更新为 3。如果操作成功，返回的响应状态值为 200（OK）。更改模板时不会影响任何现有索引的设置。

```
curl -XPUT -H "Content-Type: application/json" http://localhost:9200/_template/fund_basic_template/?pretty=true --data $'{"index_patterns": ["fund_basic*"], "settings": {"number_of_replicas":"3"}}'
{ "acknowledged" : true }
```

6. 删除索引模板

以下示例通过 curl 命令发出 DELETE _template 请求，以删除索引模板 fund_basic_template。如果操作成功，则返回的响应状态值为 200（OK）。

```
curl -XDELETE http://localhost:9200/_template/fund_basic_template
{ "acknowledged" : true }
```

3.1.4 索引别名

Elasticsearch 提供了索引别名（允许多个别名），在索引操作中将用其作为备用名称。对于简单的操作情况，可使用 _alias 接口。对于复杂的操作情况，可使用 _aliases 接口。

以下介绍基本的索引别名操作。

1. 创建索引别名

为一个或多个索引创建一个替代名称，索引使用的语法可以为 *、_all、全局与列举式如 name1、name2 等。

以下示例通过 curl 命令发出 PUT _alias 请求，为索引 fund_basic 创建索引别名 fund_basic_1 的过程。如果操作成功，则返回的响应状态值为 200（OK）。

```
curl -XPUT -H "Content-Type: application/json" http://localhost:9200/fund_basic/_alias/fund_basic_1?pretty=true --data $'{}'
{ "acknowledged" : true }
```

2. 获取索引别名

以下示例通过 curl 命令发出 Get _alias 请求，以获取索引 fund_basic 别名的过程。如果操作成功，则返回的响应状态值为 200（OK）。

```
curl -XGET http://localhost:9200/fund_basic/_alias?pretty=true
{
  "fund_basic" :
```

```
    "aliases" : {
      "fund_basic_1" : { }
    }
  }
}
```

3. 检查索引别名是否存在

可以用 HEAD _alias 发出请求，以获取索引别名是否存在的信息。以下示例通过 curl 命令发出 HEAD _alias 请求，以检查索引别名 fund_basic_1 是否存在于 fund_basic 索引的信息。以下是 GET _alias 接口的示例。如果操作成功，则返回的响应状态值为 200（OK）。

```
curl --head http://localhost:9200/fund_basic/_alias/fund_basic_1?pretty=true
HTTP/1.1 200 OK
content-type: application/json; charset=UTF-8
content-length: 78
```

4. 删除索引别名

可以用 DELETE _alias 发出请求，以删除索引别名。以下示例通过 curl 命令发出 DELETE _alias 请求，以删除索引别名 fund_basic_1。如果操作成功，则返回的响应状态值为 200（OK）。

```
curl -XDELETE http://localhost:9200/fund_basic/_alias/fund_basic_1
{ "acknowledged" : true }
```

5. 使用 _aliases 接口执行多个操作

可以用 POST _aliases 发出请求，以对一个或多个索引执行多个操作，所有操作为一个单元。支持的操作包括添加（add）索引别名、删除（delete）索引别名和删除索引（remove_index），以下通过两个步骤演示。

（1）在演示 _aliases 示例前，需要先创建一个备用索引 index_to_be_deleted 和它的备用别名 alias_to_be_deleted，如下所示：

```
    curl -XPUT -H "Content-Type: application/json" http://localhost:9200/index_to_
be_deleted
    {"acknowledged":true,"shards_acknowledged":true,"index":"index_to_be_deleted"}
    curl -XPUT -H "Content-Type: application/json" http://localhost:9200/index_to_
be_deleted/_alias/alias_to_be_deleted
    {"acknowledged":true}
```

（2）通过 curl 命令发出 POST _aliases 请求，为 fund_basic 添加一个备用别名 fund_basic_3，同时删除索引别名 alias_to_be_deleted 及索引 index_to_be_deleted。如果所有操作成功，则返回的响应状态值为 200（OK）。

```
    curl -XPOST -H "Content-Type: application/json" http://localhost:9200/_
aliases?pretty=true --data $'{"actions": [{"add": {"index": "fund_basic", "alias":
"fund_basic_3"}}, {"remove": {"index": "fund_basic", "alias": "alias_to_be_
deleted"}}, {"remove_index": {"index": "index_to_be_deleted"}}]}'
    { "acknowledged" : true }
```

3.1.5 索引别名应用场景

索引别名在不同的应用场景下能够提供适当的灵活性，例如使用零停机时间重新建立索引和聚合多个索引。

1. 用零停机时间重新建立索引

官方建议在生产环境中使用别名代替索引。由于开始时,索引的设计可能不完美,例如文档的某些字段日后需要更改,因此需要重新编制索引。解决方案是通过使用别名,无需停机而可将应用程序从使用旧索引切换到新索引。

方法描述如下;假设在应用程序中已经使用索引别名 fund_basic_3 来替代 fund_basic,可将数据从 fund_basic 重新索引到新索引 fund_basic_new。重新索引成功后,从旧索引中删除别名 fund_basic_3,并为新索引创建相同的别名 fund_basic_3。这样,应用程序中已自动切换成使用新索引了,以下通过两个步骤演示。

(1) 演示示例前需要先创建一个备用索引 fund_basic_new。

```
curl -XPUT -H "Content-Type: application/json" http://localhost:9200/fund_basic_new
    {"acknowledged":true,"shards_acknowledged":true,"index":"fund_basic_new"}
```

(2) 用 _aliases 接口来同时实现这两个命令,使应用程序切换成使用新索引。

```
curl -XPOST -H "Content-Type: application/json" http://localhost:9200/_aliases?pretty=true --data $'{"actions": [{"remove": {"index": "fund_basic", "alias": "fund_basic_3"}}, {"add": {"index": "fund_basic_new", "alias": "fund_basic_3"}}]}'
    { "acknowledged" : true }
```

2. 聚合多个索引

如果索引下的文档过多,就可能会降低搜索效率。如果大多数查询都基于索引的同一字段,则可以按该字段的类别对数据进行逻辑分组。

假设文档中有一个类别字段(fund_type)有 5 个类别,分别是债券型(bond)、商品型(commodity)、混合型(hybrid)、股票型(stock)和货币市场型(money_market)。

如果要将文档平均划分为每个类别,则需要先基于该类别创建 5 个索引以对每个文档向所属类别进行索引,然后创建一个名为 fund_type_alias 的别名以包含所有索引。

提示:如果要搜索所有类别的记录,只需要使用索引别名 fund_type_alias。

以下步骤通过 curl 命令发出 PUT 请求,以创建各个类型基金列表的索引示例。如果操作成功,则返回的响应状态值为 200(OK),以下通过 7 个步骤演示。

(1) 创建 fund_type_bond 索引。

```
curl -XPUT http://localhost:9200/fund_type_bond
{"acknowledged":true,"shards_acknowledged":true,"index":"fund_type_bond"}
```

(2) 创建 fund_type_commodity 索引。

```
curl -XPUT http://localhost:9200/fund_type_commodity
{"acknowledged":true,"shards_acknowledged":true,"index":"fund_type_commodity"}
```

(3) 创建 fund_type_hybrid 索引。

```
curl -XPUT http://localhost:9200/fund_type_hybrid
{"acknowledged":true,"shards_acknowledged":true,"index":"fund_type_hybrid"}
```

(4) 创建 fund_type_stock 索引。

```
curl -XPUT http://localhost:9200/fund_type_stock
{"acknowledged":true,"shards_acknowledged":true,"index":"fund_type_stock"}
```

(5) 创建 fund_type_money_market 索引。

```
curl -XPUT http://localhost:9200/fund_type_money_market
{"acknowledged":true,"shards_acknowledged":true,"index":"fund_type_money_market"}
```

(6) 创建索引别名 fund_type_alias 并包含所有 5 个索引。

```
curl -XPUT http://localhost:9200/fund_type*/_alias/fund_type_alias
{"acknowledged":true}
```

(7) 发出 GET _alias 请求以获取所有索引字首为 fund_type 的索引别名。如果操作成功，则返回的响应状态值为 200（OK）。响应正文中显示它们都有相同的别名，名为 fund_type_alias。fund_type_alias 可以代表该 5 个索引，并同时接受操作。

```
curl -XGET http://localhost:9200/fund_type*/_alias?pretty=true
{
  "fund_type_commodity" : {
    "aliases" : {
      "fund_type_alias" : { }
    }
  },
  "fund_type_stock" : {
    "aliases" : {
      "fund_type_alias" : { }
    }
  },
  "fund_type_bond" : {
    "aliases" : {
      "fund_type_alias" : { }
    }
  },
  "fund_type_hybrid" : {
    "aliases" : {
      "fund_type_alias" : { }
    }
  },
  "fund_type_money_market" : {
    "aliases" : {
      "fund_type_alias" : { }
    }
  }
}
```

3.2 进阶索引管理

掌握了基础的索引管理技术后，接下来开始学习进阶索引管理技术，例如收集信息和减少特定用例的间接开销，并提高整体查询性能，来支撑更加强大和复杂的搜索以及分析应用。

3.2.1 索引存储原理

文档（document）是 Elasticsearch 的基本数据单位。Elasticsearch 通过对文档中的词条进行标记，将文档列表并与可以找到这些词条的位置相关联来创建反向索引。

提示：每个 Elasticsearch 的索引事实上是一个逻辑命名空间，用于组织索引内文档中的数据。

Elasticsearch 索引由一个或多个分片（shard）组成，分片就是 Lucene 索引。Lucene 索引使用反向索引（一种数据结构）存储数据，并由一个或多个不可变的索引分段（segment）组成，这使Lucene索引可以在不重建工作的情况下，将新文档逐步添加到Elasticsearch索引中。

为了保持分段数量的可管理性，Elasticsearch将这些小的分段合并为一个较大的分段，并将新的合并分段储存于磁盘中，并在适当的时候删除旧有的分段。对于每个搜索请求，将在Elasticsearch索引给定的分片内，对所有分段进行搜索。

Elasticsearch在文档索引操作期间，使用事务日志（translog）和临时存储的内存缓冲区（memory buffer）。在适当时，内存缓冲区中的数据将移至新索引分段；最后，这些新索引分段将刷新并储存于硬盘存中，过程如图3-1所示。

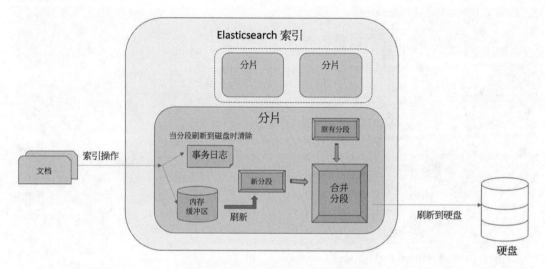

图 3-1　Elasticsearch 索引存储原理

3.2.2　索引的底层信息接口

Elasticsearch支援一些底层信息接口，以提供与索引相关的统计信息、状态信息和在分片中有关Lucene分段的信息。

1. 与索引相关的统计信息

通过发出GET _stats发出请求，可查看索引内有关于分段（segments）、事务日志（translog）、合并分段情况（merges）、刷新内存缓冲区情况（refresh）、刷新到硬盘情况（flush）等的统计信息。

统计结果分两个汇总，主要分片汇总（primaries）信息和总计（total）信息。其中总计信息包括主要分片和分片副本（replicas）。由于返回的响应内容信息太多，因此以下仅显示其主要字段，折叠其他信息。

```
curl -XGET http://localhost:9200/fund_basic/_stats?pretty=true
{
  "_shards" : {
    "total" : 3,
    "successful" : 1,
    "failed" : 0
  },
```

```
"_all" : {
    "primaries" : {
        "docs" : {…},
        "store" : {…},
        "indexing" : {…},
        "get" : {…},
        "search" : {…},
        "merges" : {…},
        "refresh" : {…},
        "flush" : {…},
        "warmer" : {…},
        "query_cache" : {…},
        "fielddata" : {…},
        "completion" : {…},
        "segments" : {…},
        "translog" : {…},
        "request_cache" : {…},
        "recovery" : {…}
    },
    "total" : {…}
},
"indices" : {…}
}
```

2. 获取索引分段信息

以下示例通过 curl 命令发出 GET _segments 请求，来获取索引 fund_basic 内的分段信息。由于还没有操作过文档索引，所以在返回的响应内容中，num_committed_segments（已同步到磁盘的分段）、num_search_segments（搜索分段）和 segments（分段）全部为零或没有值。

```
curl -XGET http://localhost:9200/fund_basic/_segments?pretty=true
HTTP/1.1 200 OK
{
  "_shards" : {
    "total" : 3,
    "successful" : 1,
    "failed" : 0
  },
  "indices" : {
    "fund_basic" : {
      "shards" : {
        "0" : [
          {
            "routing" : {
              "state" : "STARTED",
              "primary" : true,
              "node" : "WS1J2HyJQqeiGWkqby15mA"
            },
            "num_committed_segments" : 0,
            "num_search_segments" : 0,
            "segments" : { }
          }
        ]
      }
    }
  }
}
```

3. 获取索引分片恢复信息

分片恢复是从主分片同步副本分片的过程。在分片恢复完成后，副本分片可用于搜索。以下示例通过 curl 命令发出 GET _recovery 请求，以获取索引 fund_basic 内的分片恢复信息。

在返回的响应内容中，主要显示 target（目标）的信息，以及 index（索引）和 translog（事务日志）在各个恢复阶段的消耗时间、恢复内容的大小和恢复内容的进度。由于返回的响应内容信息太多，因此以下仅显示其中主要信息，折叠其他字段。

```
curl -XGET http://localhost:9200/fund_basic/_recovery?pretty=true
{
  "fund_basic" : {
    "shards" : [
      {
        "id" : 0,
        "type" : "EXISTING_STORE",
        "stage" : "DONE",
        "primary" : true,
        "start_time_in_millis" : 1582913700882,
        "stop_time_in_millis" : 1582913700950,
        "total_time_in_millis" : 67,
        "source" : {…},
        "target" : {
          "id" : "WS1J2HyJQqeiGWkqby15mA",
          "host" : "127.0.0.1",
          …
        },
        "index" : {
          "size" : {…},
          "files" : {…},
          "total_time_in_millis" : 2,
          "source_throttle_time_in_millis" : 0,
          "target_throttle_time_in_millis" : 0
        },
        "translog" : {
          "recovered" : 0,
          "total" : 0,
          "percent" : "100.0%",
          "total_on_start" : 0,
          "total_time_in_millis" : 47
        },
        "verify_index" : {…}
      }
    ]
  }
}
```

4. 获取索引分片存储信息

索引分片存储信息提供了有关在集群中构建索引的分片所有副本的存储信息列表，包括副本的存储（store）节点信息及分配（allocation）信息。以下示例通过 curl 命令发出 GET _shard_stores 请求，以获取索引 fund_basic 内的分片存储的信息。

```
curl -XGET http://localhost:9200/fund_basic/_shard_stores?pretty=true
HTTP/1.1 200 OK
{
  "indices" : {
    "fund_basic" : {
      "shards" : {
        "0" : {
          "stores" : [
            {
              "WS1J2HyJQqeiGWkqby15mA" : {
                "name" : "wai",
                "ephemeral_id" : "L6vCPDQcQcWv3MPpYxGP9A",
                "transport_address" : "127.0.0.1:9300",
```

```
            "attributes" : {
              "ml.machine_memory" : "16668676096",
              "xpack.installed" : "true",
              "ml.max_open_jobs" : "20"
            }
          },
          "allocation_id" : "c9PJP06rT70BoSbDTTE7UQ",
          "allocation" : "primary"
        }
      ]
    }
  }
}
```

5. 用 _cat 接口观看索引内容

以下示例通过 curl 命令发出 GET _cat 请求，以获取索引 fund_basic 的在线或离线模式（status）、健康（health）情况、文件（docs）和存储（store）的统计信息。

```
curl -XGET 'http://localhost:9200/_cat/indices/fund_basic?v&pretty=true'
health status index        uuid                   pri rep docs.count docs.deleted store.size pri.store.size
yellow open   fund_basic   jsACaAWdRx6VtiFZvgD55A   1   2          0            0       283b           283b
```

3.2.3 优化索引

本节将讨论如何用 _split（拆分索引）、_shrink（缩小索引）、_rollover（滚动索引）和 _cache（索引缓存）接口改变索引现状，进而减少特定用例的资源消耗并提高整体查询性能。

1. 拆分索引

可以将现有的索引转换为新的索引，并将原始的主分片拆分为两个或多个主分片。在操作时，新索引的主分片必须是原始主分片的一个分解因子，以下通过两个步骤演示。

提示：索引设置必须为只读模式及索引健康状态为绿色。

（1）通过 curl 命令发出 PUT _settings 请求，以更新索引设置为只读模式（index.blocks.write:true）。如果操作成功，返回的响应状态值为 200（OK）。

```
curl -XPUT -H "Content-Type: application/json" http://localhost:9200/fund_basic/_settings?pretty=true --data $'{"settings": {"index.blocks.write":true}}'
HTTP/1.1 200 OK
{ "acknowledged" : true }
```

（2）通过 curl 命令发出 POST _split 请求，以将 fund_basic 索引拆分为一个名为 fund_basic_split 的新索引，并将分片（index.number_of_shards）设置从原来的 1 拆分为 2。如果操作成功，则返回的响应状态值为 200（OK）。

```
curl -XPOST -H "Content-Type: application/json" http://localhost:9200/fund_basic/_split/fund_basic_split?pretty=true --data $'{"settings": {"index.number_of_shards":2}}'
{
  "acknowledged" : true,
  "shards_acknowledged" : true,
  "index" : "fund_basic_split"
}
```

2. 缩小索引

将现有索引缩小为具有更少主分片的新索引。在操作时，新索引的主分片必须是原来主分片的一个分解因子。此外，索引设置必须为只读模式（index.blocks.write）、索引健康状态为绿色，以及每个分片副本的新位置（index.routing.allocation.require._name）必须是在同一节点，以下通过两个步骤演示。

（1）通过 curl 命令发出 PUT _settings 请求，以更新索引 fund_basic_split 设置。如果操作成功，返回的响应状态值为 200（OK）。

```
curl -XPUT -H "Content-Type: application/json" http://localhost:9200/fund_basic_split/_settings?pretty=true --data $'{"settings": {"index.blocks.write":true, "index.routing.allocation.require._name":"wai"}}'
{ "acknowledged" : true }
```

（2）通过 curl 命令发出 POST _shrink 请求，以将 fund_basic 索引缩小为一个名为 fund_basic_shrink 的新索引，并将分片（index.number_of_shards）设置从原来的 2 缩小为 1。如果操作成功，则返回的响应状态值为 200（OK）。索引的原始设置也将被复制到新索引内。

```
curl -XPOST -H "Content-Type: application/json" "http://localhost:9200/fund_basic_split/_shrink/fund_basic_shrink?pretty=true" --data $'{"settings": {"index.number_of_shards":1}}'
{
  "acknowledged" : true,
  "shards_acknowledged" : true,
  "index" : "fund_basic_shrink"
}
```

3. 滚动索引

在生产环境中经常使用索引别名代替索引，我们可以对索引别名进行条件式滚动操作，条件满足时，新索引将被创建并且现有索引别名将滚动到新索引上。

但是，现有索引必须只具有这单一的别名，而条件（conditions）可对 max_age（索引最大年龄）、max_docs（最大文件数量）和 max_size（最大索引空间）进行设定，以下通过三个步骤演示。

（1）创建一个备用索引 fund_basic_rollover 并具有单一别名 rollover_alias。

```
curl -XPUT -H "Content-Type: application/json" http://localhost:9200/fund_basic_rollover-000001?pretty=true --data $'{"aliases": {"rollover_alias":{}}}'
{
  "acknowledged" : true,
  "shards_acknowledged" : true,
  "index" : "fund_basic_rollover-000001"
}
```

（2）备用索引 fund_basic_rollover-000001 创建后，可以使用 _rollover 接口来切换别名 rollover_alias 指向新索引，而条件设定为索引最大年龄在 5 分钟内。返回的响应内容显示新索引 fund_basic_rollover-000002 被创建。

```
curl -XPOST -H "Content-Type: application/json" http://localhost:9200/rollover_alias/_rollover?pretty=true --data $'{"conditions": {"max_age":"5m"}}'
{
  "acknowledged" : true,
  "shards_acknowledged" : true,
```

```
  "old_index" : "fund_basic_rollover-000001",
  "new_index" : "fund_basic_rollover-000002",
  "rolled_over" : true,
  "dry_run" : false,
  "conditions" : {
    "[max_age: 5m]" : true
  }
}
```

（3）为了证明别名已经从原索引 fund_basic_rollover-000001 滚动到新索引 fund_basic_rollover-000002 上，以下示例通过 curl 命令发出 GET _alias 请求，以获取所有字首为 fund_basic_rollover-00000 的索引的别名。返回的响应内容显示别名 rollover_alias 已从旧索引中删除并添加到新索引上。

```
curl -XGET 'http://localhost:9200/fund_basic_rollover-00000*/_alias?pretty=true'
{
  "fund_basic_rollover-000002" : {
    "aliases" : {
      "rollover_alias" : { }
    }
  },
  "fund_basic_rollover-000001" : {
    "aliases" : { }
  }
}
```

4. 手动清除缓存

当缓存（cache）占用太多，但没有超过设置的阈值，又没法自动清理时，可以发出 POST _cache 请求来强制清理缓存。缓存可以分类为查询（query）缓存、请求（request）缓存和字段数据（fielddata）缓存。默认情况下是清除所有缓存。如果只需要清除特定的缓存，可以设置相关对应参数为 true。以下示例通过 curl 命令发出 POST _cache 请求，以清除 fund_basic 索引的查询缓存。

```
curl -XPOST "http://localhost:9200/fund_basic/_cache/clear?query=true&pretty=true"
{
  "_shards" : {
    "total" : 3,
    "successful" : 1,
    "failed" : 0
  }
}
```

5. 手动刷新内存缓冲区

手动刷新内存缓冲区可使自上次刷新以来对索引执行的所有操作，产生索引分段，变为可搜索的分段。以下示例通过 curl 命令发出 POST _refresh 请求，以手动刷新 fund_basic 索引的内存缓冲区。

```
curl -XPOST http://localhost:9200/fund_basic/_refresh?pretty=true
{
  "_shards" : {
    "total" : 3,
    "successful" : 1,
    "failed" : 0
  }
}
```

6. 手动刷新到硬盘

确保当前仅存储在事务日志中的所有数据也永久存储在硬盘上。以下示例通过 curl 命令发出 POST _flush 请求，以手动刷新 fund_basic 索引到硬盘上。

```
curl -XPOST http://localhost:9200/fund_basic/_flush?pretty=true
{
  "_shards" : {
    "total" : 3,
    "successful" : 1,
    "failed" : 0
  }
}
```

7. 同步刷新

Elasticsearch 跟踪每个分片的索引活动。5 分钟内收到索引操作的分片会自动标记为非活动状态。这为 Elasticsearch 提供了一个减少分片资源的机会，并且执行了一种特殊的同步刷新（synced flush）。同步刷新执行正常刷新，然后将生成的唯一标记（sync_id）添加到所有碎片。同步刷新只对冷索引有效。

```
curl -XPOST http://localhost:9200/fund_basic/_flush/synced?pretty=true
{
  "_shards" : {
    "total" : 2,
    "successful" : 1,
    "failed" : 0
  },
  "fund_basic" : {
    "total" : 2,
    "successful" : 1,
    "failed" : 0
  }
}
```

8. 强制合并分段

当索引不再有写入操作的时候，建议对其进行强制合并分段（force merge）。这个操作可以提升查询速度和减少内存开销。分段是越少越好，最好可以强制合并成一个分段，但是强制合并分段会占用大量的网络、IO 和 CPU 资源。如果不能在业务高峰期之前做完，就需要考虑增大最终的分段数。

```
curl -XPOST http://localhost:9200/fund_basic/_forcemerge?pretty=true
{
  "_shards" : {
    "total" : 2,
    "successful" : 1,
    "failed" : 0
  }
}
```

第 4 章 文档管理

Elasticsearch 是一个面向文档的数据库，这意味着它存储了整个对象或文档。Elasticsearch 中的文档是以 JSON 格式表示，亦即以键值形式编写。键必须是字符串，并且用引号引起来。而值必须是有效的 JSON 数据类型。键和值之间用冒号分隔，各键值对之间用逗号分隔。在文档被索引期间，文档将与元字段（meta-fields）相关联，元字段描述文档本身的字段，包括给定的索引名称、映射类型、文档标识符和路由等。而映射是描述文档结构的数据模型，包括指定字段、字段类型、文档之间的关系与数据转换规则等。本章将同时学习文档管理与文档结构。

4.1 文档管理

索引一个文档包含两种意义，存储与使文档可被检索和搜索。在 3.2.1 节索引存储原理谈到文档是存储在分片中，而搜索请求则在分片内对所有分段进行搜索。以下将深入探讨文档是如何建立索引、如何存储和搜索的原理。

4.1.1 了解 Elasticsearch 文档管理原理

为了帮助读者理解 Elasticsearch 建立索引时会发生什么事件，以下分为两部分内容来说明。

1. 文档路由

无论存储或搜索操作，文档都被转送到特定的分片进行。Elasticsearch 有一个简单的公式来计算路线，而用户可以自定义其路由方法。简单公式如下，routing（路由）默认值为文档标识符：

```
shard = hash(_routing) % number_of_primary_shards
```

基本上就是通过路由值提供定向存储和定向搜索。配合 Elasticsearch 的架构，路由过程说明如图 4-1 所示。在索引文档操作中，我们可以按照以下步骤描述整个过程：

（1）Elasticsearch 客户端将文档发送到 Elasticsearch 集群；第一站是协调节点。

（2）文档转发到由路由公式确定的目标主分片（primary shard）所在的数据节点。例如，从客户端发送的文档将由协调节点转发到数据节点 1 的分片 1。

（3）分析器从文档各键值对的值生成分词（terms）并创建倒排索引，并将其存储在

Lucene 分段中。有关 Lucene 分段的存储过程，请参考 3.2.1 节索引存储原理的 Elasticsearch 索引和 Lucene 索引之间的关系。

（4）主分片负责在适当的时候将文档相关的索引操作，转发到当前同步副本分片（replica shard）的每个副本以进行复制。

（5）当所有副本都成功执行了复制操作并响应了主分片，主分片便会向客户端确认请求成功，而文档标识符 _id 将在响应正文中返回。

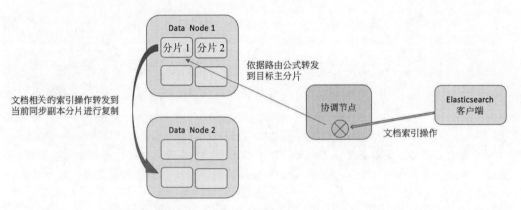

图 4-1　Elasticsearch 文档路由过程说明

2. 文档索引

Elasticsearch 是通过 Lucene 的倒排索引（也称为反向索引）技术，将键值对之值可能出现的每个不同分词（若是文本）对文档（亦即文档标识符）进行标记，产生所有分词的排序列表，并将文档列表与可以在文档中找到这些分词的位置相关联，创建倒排索引，而 Lucene 的索引分段（segment）就是一个具有功能性的倒排索引。由分词到文档标识符的倒排索引关系，说明如图 4-2 所示。

文档 标识符	基金代码	基金管理人 内容
1	159809.SZ	博时基金
2	515130.SH	博时基金
3	515190.SH	中银证券

创建倒排索引 →

分词	计数	文档标识符:标记位置
中银	1	3:1
证券	1	3:2
基金	2	1:2, 2:2
博时	2	1:1, 2:1

图 4-2　由分词到文档标识符的倒排索引关系说明

4.1.2　认识单文档接口功能

基本文档索引功能一般针对单个文档，在开始介绍这些接口之前，让我们先熟悉示例文档。

1. 示例文档

Elasticsearch 中的文档是以 JSON 格式表示，以下是基金代码 159809.SZ 和 515190.SH 公募基金（fund_basic）的 JSON 格式数据，用以操作示例。

```
        {"ts_code":"159809.SZ","name":"恒生湾区","management":"博时基金",
"custodian":"招商银行",
    "fund_type":"股票型","found_date":"20200430","due_date":null,"list_date": "20200521",
    "issue_date":"20200330","delist_date":null,"issue_amount":null,"m_fee":3.9117,
"c_fee":0.15,
    "c_fee":0.05,"duration_year":null,"p_value":1.0,"min_amount":0.100,"exp_
return":null,
    "benchmark":"恒生沪深港通大湾区综合指数收益率","status":"L","invest_type":"被动指数型
","type":"契约型开放式","trustee":null,"purc_startdate":"20200521","redm_startdate":"
20200521","market":"E"}
        {"ts_code":"515190.SH","name":"中银证券500ETF","management":"中银证券
","custodian":"招商银行",
    "fund_type":"股票型","found_date":"20200430","due_date":null,"list_date":"20200527",
    "issue_date":"20200330","delist_date":null,"issue_amount":null,"m_fee":2.5147,
"c_fee":0.15,
    "c_fee":0.05,"duration_year":null,"p_value":1.0,"min_amount":0.100, "exp_return":null,
    "benchmark":"中证500指数收益率","status":"I","invest_type":"被动指数型","type":"
契约型开放式",
    "trustee":null,"purc_startdate":"20200527","redm_startdate":"20200527",
"market":"E"}
```

2．创建（索引）、读取、更新和删除文档等接口

（1）创建文档接口

创建文档可以使用 PUT 或 POST 请求，如果需要自动生成标识符，则只能使用 POST 请求。以下使用索引 fund_basic 为例：

```
PUT /fund_basic/_doc/{_id}
POST /fund_basic/_doc/
PUT /fund_basic/_create/{_id}
POST /fund_basic/_create/{_id}
```

假设索引 fund_basic 已存在，使用 PUT 请求配合文档标识符 _id=1，创建 ts_code=159809.SZ 的文档。如果操作成功，则返回的响应状态值为 201（Created）。在返回的响应主体中，结果 result 为 created，文档标识符 _id 为 1。另外，_version 是一个序列号，用于计数文档更新的次数，而 _seq_no 是一个序列号，用于计算索引上发生操作的次数。

```
curl--request POST http://localhost:9200/fund_basic/_doc/TVFrgnIBzY05muk4f-u5?
pretty=true --header"Content-Type:application/json"--data $'{"ts_code":"515190.
SH",
    "name":"中银证券500ETF","management":"中银证券","custodian":"招商银行",
    "fund_type":"股票型","found_date":"20200430","due_date":null, "list_date":
"20200527","issue_date":"20200330","delist_date":null,"issue_amount":2.5147,
    "m_fee":0.15,"c_fee":0.05,"duration_year":null,"p_value":1.0,"min_
amount":0.100,
    "exp_return":null,"benchmark":"中证500指数收益率","status":"I","invest_type":
"被动指数型","type":"契约型开放式","trustee":null,"purc_startdate":"20200527",
"redm_startdate":"20200527","market":"E"}'
{
  "_index":"fund_basic",
  "_type":"_doc",
  "_id":"TVFrgnIBzY05muk4f-u5",
  "_version":1,
  "result":"created",
  "_shards":{
     "total":2,
     "successful":1,
     "failed":0
  },
  "_seq_no":1,
```

```
    "_primary_term" : 1
}
```

使用 POST 请求，创建 ts_code=515190.SH 的文档并指定标识符（_id）为 TVFrgnIBzY05muk4f-u5。如果操作成功，则返回的响应状态值为 200（OK）。

```
curl --request POST http://localhost:9200/fund_basic/_doc/TVFrgnIBzY05muk4f-u5?pretty=true --header "Content-Type:application/json" --data $'{"ts_code":"515190.SH","name":"中银证券500ETF","management":"中银证券","custodian":"招商银行","fund_type":"股票型","found_date":"20200430","due_date":null,"list_date":"20200527","issue_date":"20200330","delist_date": null,"issue_amount":2.5147, "m_fee":0.15, "c_fee":0.05,"duration_year":null,"p_value":1.0,"min_amount":0.100,"exp_return":null,"benchmark":"中证500指数收益率","status":"I","invest_type":"被动指数型","type":"契约型开放式","trustee":null,"purc_startdate":"20200527","redm_startdate":"20200527","market":"E"}'
{
  "_index" : "fund_basic",
  "_type" : "_doc",
  "_id" : "TVFrgnIBzY05muk4f-u5",
  "_version" : 1,
  "result" : "created",
  "_shards" : {
    "total" : 2,
    "successful" : 1,
    "failed" : 0
  },
  "_seq_no" : 1,
  "_primary_term" : 1
}
```

(2) 读取文档接口

默认情况下，读取文档操作会返回 _source 字段的内容。_source 字段是索引时传递给 Elasticsearch 的原始 JSON 文档主体。如果只需要 _source 中的一个或两个字段，可以使用 _source_includes（简称为 _source）或 _source_excludes 参数包括或过滤出特定字段。若索引的文档具有很多不同数据，检索部分内容，返回较少的字段是可以节省网络开销的。这两个参数可采用逗号分隔的字段或通配符表达式。

- 使用 GET 请求读取整个文档，如果操作成功，则返回的响应状态值为 200（OK）。在返回的响应主体中可以获得整个文档。

```
curl --request GET 'http://localhost:9200/fund_basic/_doc/1?pretty=true'
{
  "_index" : "fund_basic",
  "_type" : "_doc",
  "_id" : "1",
  "_version" : 1,
  "_seq_no" : 0,
  "_primary_term" : 1,
  "found" : true,
  "_source" : {
    "ts_code" : "159809.SZ",
    "name" : "恒生湾区",
    "management" : "博时基金",
    "custodian" : "招商银行",
    "fund_type" : "股票型",
    "found_date" : "20200430",
    "due_date" : null,
    "list_date" : "20200521",
```

```
        "issue_date" : "20200330",
        "delist_date" : null,
        "issue_amount" : 3.9117,
        "m_fee" : 0.15,
        "c_fee" : 0.05,
        "duration_year" : null,
        "p_value" : 1.0,
        "min_amount" : 0.1,
        "exp_return" : null,
        "benchmark" : "恒生沪深港通大湾区综合指数收益率",
        "status" : "L",
        "invest_type" : "被动指数型",
        "type" : "契约型开放式",
        "trustee" : null,
        "purc_startdate" : "20200521",
        "redm_startdate" : "20200521",
        "market" : "E"
    }
}
```

- 使用 GET 请求，配合 _source=ts_code&management 读取文档的部分内容。如果操作成功，则返回的响应状态值为 200（OK）。在返回的响应主体中可以获得文档的基金代码（ts_code）和基金管理人（management）两个字段内容。

```
curl --request GET 'http://localhost:9200/fund_basic/_doc/1?_source=ts_code,management&pretty=true'
{
  "_index" : "fund_basic",
  "_type" : "_doc",
  "_id" : "1",
  "_version" : 1,
  "_seq_no" : 0,
  "_primary_term" : 1,
  "found" : true,
  "_source" : {
    "ts_code" : "159809.SZ",
    "management" : "博时基金"
  }
}
```

（3）更新文档接口

Elasticsearch 提供两种类型的更新操作，即执行补丁操作和使用指定脚本（script）更换文档。更新文档操作用 POST 请求，以下使用 fund_basic 索引为例：

```
POST /fund_basic/_update/{_id}
```

- 补丁操作

如果操作后内容产生变化，则更改的内容和添加的新字段将合并到现有文档中。如果操作后内容没有产生变化，在返回的响应中，result 字段将为 noop。如果文档不存在，则响应为带有 document_missing_exception 的 NOT_FOUND（404）异常。假设需要添加一个新字段 market_cap（市场规模）到标识符 _id=1（恒生湾区）的文档，并设定 market_cap 为"小盘"。使用 POST 请求，配合请求主体设定为 {"doc":{"market_cap":"小盘"}}。如果操作成功，则返回的响应状态值为 200（OK）。在返回的响应主体中，result 为 updated，而文档版本号 _version 已从 1 更改为 2。

```
curl --request POST 'http://localhost:9200/fund_basic/_update/1?pretty=true' --header "Content-Type:application/json" --data $'{"doc":{"market_cap":"小盘"}}'
```

```
{
  "_index" : "fund_basic",
  "_type" : "_doc",
  "_id" : "1",
  "_version" : 2,
  "result" : "updated",
  "_shards" : {
    "total" : 2,
    "successful" : 1,
    "failed" : 0
  },
  "_seq_no" : 2,
  "_primary_term" : 1,
}
```

补丁操作提供一个 boolean 参数 doc_as_upsert。当 doc_as_upsert 设定为 true 时，它允许在文档不存在情况下，与该参数关联的内容将用于创建新文档。

◆ 使用指定脚本更换文档

使用 POST 请求，配合请求主体设定脚本。在操作过程中，Elasticsearch 首先从索引中获取文档，然后运行该脚本，再对运行结果进行重新索引。这项操作将导致整个文档进行替换。假设脚本为从文档中的 _source 删除 market_cap 字段 {"script":"ctx._source.remove('market_cap')"}。使用 POST 请求，配合请求主体的设定。如果操作成功，则返回的响应状态值为 200（OK）。在返回的响应中文档版本号 _version 已从 2 更改为 3。如果读取文档，字段 market_cap 已被删除。

```
curl --request POST 'http://localhost:9200/fund_basic/_update/1?pretty=true'
--header "Content-Type:application/json" --data "{\"script\":\"ctx._source.remove('market_cap')\"}"
{
  "_index" : "fund_basic",
  "_type" : "_doc",
  "_id" : "1",
  "_version" : 3,
  "result" : "updated",
  "_shards" : {
    "total" : 2,
    "successful" : 1,
    "failed" : 0
  },
  "_seq_no" : 3,
  "_primary_term" : 1
}
```

脚本更新文档操作提供两个 boolean 参数 upsert 和 scripted_upsert。如果设定参数 upsert，若文档存在，则更新操作将成功。否则，与该参数关联的内容将用于创建新文档。如果设定参数 scripted_upsert 为 true，无论该文档是否存在，始终运行该脚本。

（4）删除文档接口

删除文档可以使用 DELETE 请求，并设定索引名称和文档标识符 _id。如果操作成功，则返回的响应状态值为 200（OK），在返回的响应主体中 result 为 deleted。当主分片收到删除请求时，可能无法执行该操作。默认情况下，删除操作将在主分片上等待 1 分钟。如果超时，则响应为失败。Elasticsearch 提供超时（timeout）参数设置，提供额外等待时间。

提示：如果已成功删除标识符 _id 为 1 的文档，请再次创建该文档以供后续章节使用。

```
curl --request DELETE 'http://localhost:9200/fund_basic/_doc/1?pretty=true'
{
  "_index" : "fund_basic",
  "_type" : "_doc",
  "_id" : "1",
  "_version" : 7,
  "result" : "deleted",
  "_shards" : {
    "total" : 2,
    "successful" : 1,
    "failed" : 0
  },
  "_seq_no" : 7,
  "_primary_term" : 1
}
```

4.1.3 认识批量多文档接口功能

为避免网络流量开销，Elasticsearch 提供了一组文档接口，用于批量处理多个请求，当处理大量文档时，将获得更好的性能。下面分别介绍各项多文档操作，并用不同示例说明。

1. 同时读取多个文档接口（_mget）

通过给定的多个索引名称和文档标识符对，使用 POST 对 _mget 接口发出请求，可以同时读取多个文档。假设索引 fund_basic 和两个文档（_id=1 和 _id=TVFrgnIBzY05muk4f-u5）已存在。为了避免出现太多不必要的字段，示例使用 _source 参数过滤出基金代码（ts_code）和基金管理人（management）这两个字段内容。如果操作成功，则返回的响应状态值为 200（OK）。在返回的响应主体中仅显示 ts_code 和 management 两个字段。

```
curl --request POST 'http://localhost:9200/_mget?pretty=true'--header
"Content-Type: application/json" --data $'{"docs":[{"_index":"fund_basic", "_id":"1", "_source":["ts_code","management"]}, {"_index":"fund_basic", "_id":"TVFrgnIBzY05muk4f-u5", "_source":["ts_code", "management"]}]}'
{
  "docs" : [
    {
      "_index" : "fund_basic",
      "_type" : "_doc",
      "_id" : "1",
      "_version" : 2,
      "_seq_no" : 9,
      "_primary_term" : 1,
      "found" : true,
      "_source" : {
        "ts_code" : "159809.SZ",
        "management" : "博时基金"
      }
    },
    {
      "_index" : "fund_basic",
      "_type" : "_doc",
      "_id" : "TVFrgnIBzY05muk4f-u5",
      "_version" : 1,
      "_seq_no" : 12,
      "_primary_term" : 1,
      "found" : true,
      "_source" : {
        "ts_code" : "515190.SH",
        "management" : "中银证券"
```

```
            }
        }
    ]
}
```

提示：如果文档来自不同的索引，可以在 docs 字段中每个数组元素指定索引名称。若文档来自相同的索引，可只在 URL 中指定索引名称。

由于只读取相同索引中的多个文档，请求主体内容可以简化为以下示例。

```
curl --request POST 'http://localhost:9200/fund_basic/_mget?pretty=true' --header "Content-Type: application/json" --data $'{"docs":[{"_id":"1", "_source":["ts_code","management"]},{"_id":"TVFrgnIBzY05muk4f-u5", "_source":["ts_code", "management"]}]}'
```

如果读取整个文件，请求主体内容可以再一步简化为以下示例。

```
curl --request POST 'http://localhost:9200/fund_basic/_mget?pretty=true' --header "Content-Type: application/json" --data $'{"ids":["1", "TVFrgnIBzY05muk4f-u5"]}'
```

2. 通过查询更新多个文档接口（_update_by_query）

与之前所描述的指定脚本更新单文档操作类似，_update_by_query 接口对脚本中与给定条件匹配的所有文档执行相同的任务。在操作过程中，Elasticsearch 首先对索引进行快照，然后获取匹配的文档运行脚本，再对运行结果进行重新索引。如果在整个操作过程中有任何文档被更改，则该文档将因版本冲突而失败，将导致暂停整个操作。但是，已执行的更新将不会回滚。默认情况下，在返回的响应主体中显示所有成功的更新和失败的文档。如果想要的结果是不中断执行，接口提供 URL 参数 conflicts 并设定为 proceed（或设定在请求主体内），更新操作会忽略错误而不会中断，继续更新下一个文档。

假设在 fund_basic 索引中，基金管理人（management）为"博时基金"的受托人（trustee）需要全部更新"为世界开发银行"。使用 POST 对 _update_by_query 接口发出请求，配合请求主体设定的脚本，更新通过查询的文档。以下示例在返回的响应主体中，操作成功的文档总数（total）为 2，而更新（updated）文档数为 2，版本冲突（version_of_conflicts）文档数为 0。如果操作成功，返回的响应状态值为 200（OK）。

提示：尽管只有一个文档与指定的查询匹配（文档标识符 _id=1），但根据官网中的描述，当版本匹配时，将更新文档并增加版本号。读者可以发出获取文档请求，以验证两个文档均已更新 _version 及 _seq_no。

```
curl --request POST 'http://localhost:9200/fund_basic/_update_by_query?pretty=true'--header "Content-Type: application/json" --data "{\"script\": \"if (ctx._source['management']=='博时基金') {ctx._source['trustee']='世界开发银行'}\"}"
{
    "took" : 41,
    "timed_out" : false,
    "total" : 2,
    "updated" : 2,
    "deleted" : 0,
    "batches" : 1,
    "version_conflicts" : 0,
    "noops" : 0,
    "retries" : {
      "bulk" : 0,
      "search" : 0
```

```
},
"throttled_millis" : 0,
"requests_per_second" : -1.0,
"throttled_until_millis" : 0,
"failures" : [ ]
}
```

3. 通过查询删除多个文档接口（_delete_by_query）

尽管尚未引入查询表达式(Query DSL)，但仍然可以通过简单的查询作为一个例子，从 fund_basic 索引中删除基金管理人（management）为"博时基金"的文档。接口提供参数 conflicts，会忽略任何错误而不会中断操作。在返回的响应主体中显示，0 个没有发生冲突（version_conflicts=0）的文档，而 1 个文档被删除（deleted=1）。

提示：如果已成功删除基金管理人为"博时基金"的文档，请暂时不要创建该文档。下一节将使用批量处理文档接口重新创建。

```
curl --request POST 'http://localhost:9200/fund_basic/_delete_by_
   query?pretty=true' --header "Content-Type: application/json" --data
$'{"query":{"m
atch":{"management":"博时基金"}}}'
{
  "took" : 103,
  "timed_out" : false,
  "total" : 1,
  "deleted" : 1,
  "batches" : 1,
  "version_conflicts" : 0,
  "noops" : 0,
  "retries" : {
    "bulk" : 0,
    "search" : 0
  },
  "throttled_millis" : 0,
  "requests_per_second" : -1.0,
  "throttled_until_millis" : 0,
  "failures" : [ ]
}
```

4. 批量处理文档接口（_bulk）

批量模式操作包括创建（create）文档，索引（index）文档，更新文档（update）和删除（delete）文档。使用方法是将一系列操作和相对应的请求放入批量接口的请求主体中，批量接口请求语法如下：

```
POST /_bulk
POST /<索引名称>/_bulk
```

请求主体中，一系列操作的请求，需要使用换行符分隔的 JSON 结构，语法如下：

```
{操作类型：{元数据}} \newline
{请求主体} \newline
...
```

以下的示例利用与前面的示例进行相同操作的批量处理，步骤如下：

（1）在 fund_basic 索引中，创建 _id=1 和与具有 ts_code=159809.SZ 相关信息的文档。

（2）在 fund_basic 索引中，添加一个新字段 market_cap（市场规模）到标识符 _id=1（恒生湾区）的文档，并设定 market_cap 为小盘。

由于返回的响应主体过长，内容也类似于单文档操作结果，所以不予显示结果。

```
curl --request POST 'http://localhost:9200/_bulk?pretty=true' --header"Content-
Type: application/json" --data $'{"create":{"_index":"fund_basic","_id":"1"}}\n
{"ts_code":"159809.SZ","name":"恒生湾区","management":"博时基金","custodian":"招
商银行",
    "fund_type":"股 票 型","found_date":"20200430","due_date":null,"list_
date":"20200521","issue_date": "20200330","delist_date": null,"issue_amount":3.9117,
"m_fee":0.15,"c_fee":0.05,"duration_year":null,
    "p_value":1.0,"min_amount":0.100,"exp_return":null,"benchmark":"恒生沪深港通大湾区
综合指数收益率",
    "status":"L","invest_type":"被动指数型","type":"契约型开放式","trustee":null,
"purc_startdate":"20200521",
    "redm_startdate":"20200521","market":"E"}\n {"update":{"_index":"fund_basic","_
id":"1"}}\n
    {"doc":{"market_cap":"小盘"}}\n'
```

5. 重新索引接口（_reindex）

在开发的早期阶段，索引的设计不可能完美，后期需要对现有字段或分析器进行修改，所以重新索引操作是不可避免的。由于接口不会从源索引复制任何设置，因此重建索引之前，必须准备好新的索引和更新后的设置。此外，源索引何以能支持重新索引的功能，在于当它进行索引操作期间启用了映射（_mappings）元数据字段 _source，否则就没有复制的源头。在执行示例之前，必须先创建目标索引。创建新索引 fund_basic_copy 如下：

```
curl --request PUT http://localhost:9200/fund_basic_copy?pretty=true
{
  "acknowledged" : true,
  "shards_acknowledged" : true,
  "index" : "fund_basic_copy"
}
```

重新索引接口就像是将文档从源索引复制到目标索引。请求主体需要指定 source（来源）和 dest（目标）两个字段，语法如下：

```
{来源：{元数据}, 目标：{元数据}}
```

接口提供如下几种复制文档的方法：

（1）从一个索引复制文档

使用 POST 对 _reindex 接口发出请求。以下示例将 fund_basic 重新索引到 fund_basic_copy。在返回的响应主体中显示 created 为 2，意思是使目标索引生成 2 个文档。如果在重新索引操作之前文档已经存在于目标索引，而且其文档标识符与来自源索引的相同，则该文档将被覆盖。

```
curl --request POST 'http://localhost:9200/_reindex?pretty=true' --header
"Content-Type: application/json" --data $'{"source":{"index":"fund_basic"},
"dest":{"index":"fund_basic_copy"}}'
  {
    "took" : 316,
    "timed_out" : false,
    "total" : 2,
    "updated" : 0,
    "created" : 2,
    "deleted" : 0,
    "batches" : 1,
    "version_conflicts" : 0,
    "noops" : 0,
    "retries" : {
      "bulk" : 0,
      "search" : 0
```

```
        },
        "throttled_millis" : 0,
        "requests_per_second" : -1.0,
        "throttled_until_millis" : 0,
        "failures" : [ ]
}
```

Elasticsearch 具有不同的版本控制系统，分别为 internal、external、external_gt 和 external_gte。可以使用 version_type 字段放入目标 dest 字段中指定版本系统，其用法请参阅官方网站。

（2）从多个索引中复制文档

如果有多个索引来源，只需在请求中使用数组列出索引值，如下：

```
{"source":{"index":["fund_basic1", "fund_basic2"]}, "dest":{"index":"fund_basic_copy"}}
```

（3）仅复制丢失的文档

为避免目标索引的文档被覆盖，可以在请求中使用 op_type 字段放入目标 dest 中指定 create 选项，仅复制丢失的文档（即不存在于目标索引的文档标识符）。

```
{"source":{"index":"fund_basic"}, "dest":{"index":"fund_basic_copy", "op_type":"create"}}
```

（4）仅复制与查询匹配的文档

可以通过查询，只对匹配的所有文档执行重新索引操作。例如只对基金管理人（management）为"博时基金"进行重新索引操作，可以在请求中将查询（query）字段放入来源 source 中：

```
{"source":{"index":"fund_basic","query":{"match":{"management":"博时基金"}}}, "dest":{"index":"fund_basic_copy"}}
```

（5）从远程 Elasticsearch 服务器复制文档

若源索引在远程 Elasticsearch 服务器中，将远程主机名称（host）、用户名（username）和密码设置（password）的参数添加到请求中的来源 source，例如主机名称为 remotehost、用户名为 me 和密码为 pass，例子如下：

```
{"source":{"remote": {"host":"http://remotehost:9200/fund_basic", "username":"me", "password":"pass"}}, "dest":{"index":"fund_basic_copy"}}
```

4.2 文档结构

映射是描述索引中文档结构的数据模型，它提供指定字段、选择字段类型、建立文档之间的关系和数据转换规则等功能。一个索引只能有一个数据模型，就是一种映射下定义的规则。以下从映射类型、字段类型和映射参数等方向，深入探讨文档结构。

4.2.1 认识映射类型

Elasticsearch 提供动态映射和显式映射两种类型。动态映射（dynamic）为默认模式，指无需定义字段类型即可对文档建立索引，而显式映射（explicit）首先创建自定义映射。

提示：如果有一个详细的搜索数据模型设计，应该使用显式映射，因为动态映射只提供固定的规则，可能不适用于已有的设计。

1. 动态映射

此模式可以在不预先定义字段映射的情况下，从第一个文档的索引操作，即时根据内置的映射规则，立即检测文档的新字段的数据类型，定义字段并建立其数据类型映射，产生文档结构，然后创建文档索引。在执行文档索引操作期间，新增的字段（从未定义的字段）可以随时添加。

Elasticsearch 文档为 JSON 格式。在 JSON 格式中，有效的数据类型为字符串（string）、数字（number）、JSON 对象（object）、数组（array）、布尔（boolean）和 null。而 JSON 数据值及其映射规则将确定文档字段的最终数据类型。Elasticsearch 支持映射的数据类型有很多，将在 4.2.2 节认识映射的数据类型中介绍。在这里先介绍 text（文本）与 keyword（关键词）的区别。文本数据是根据文本分析器先进行分词，再进行索引。因此，搜索时就可以按照分词，支持部分匹配文本的全文搜索。而关键词数据则按原样索引。表 4-1 描述了内置的映射规则。

表 4-1 内置的映射规则

JSON 数据类型	JSON 数据值	映射设置	映射的数据类型名称
字符串	字符串		text 与 keyword
	日期	date_detection=true（默认值）	date
	整数	numeric_detection=true	long
		numeric_detection=false	text 与 keyword
	浮点数	numeric_detection=true	float
		numeric_detection=false（默认值）	text 与 keyword
数字	整数		long
	浮点数		float
对象			object
数组			取决于数组中的第一个非 null 值
布尔	True 或 false		boolean
null	null		null

为了进一步了解 fund_basic 索引的动态映射定义，可以使用 GET 对 _mapping 接口向 fund_basic 索引发出请求，以获取映射结果。由于返回的响应内容信息太多，因此以下仅显示其中部分的信息。

```
curl --request GET http://localhost:9200/fund_basic/_mappings?pretty=true
{
  "fund_basic" : {
    "mappings" : {
      "properties" : {
        "custodian" : {
          "type" : "text",
          "fields" : {
            "keyword" : {
              "type" : "keyword",
              "ignore_above" : 256
            }
          }
```

```
            },
            "found_date" : {
              "type" : "text",
              "fields" : {
                "keyword" : {
                  "type" : "keyword",
                  "ignore_above" : 256
                }
              }
            },
            ......
        }
```

返回的响应内容，汇总后可以获得表 4-2。在表 4-2 中，所有与 date 数据类型相关的字段没有根据映射的数据类型变换成功。原因是还需要 dynamic_date_formats 这个映射设置来指定日期格式。默认值为 ["strict_date_optional_time", "yyyy/MM/dd HH:mm:ss Z||yyyy/MM/dd Z"]，而日期数据 found_date 的格式为 yyyyMMdd，所以匹配不成功。

表 4-2 fund_basic 索引的动态映射结果

数据名称	数据类型	映射的数据类型
ts_code、name、management、custodian、fund_type、found_date、due_date、list_date、issue_date、delist_date、benchmark、status、invest_type、type、trustee、purc_startdate、redm_startdate、market	str	text 与 keyword
issue_amount、m_fee、c_free、duration_year、p_value、min_amount、exp_return	float	float

提示：我们曾经尝试指定映射设置 dynamic_date_formats 为 ["strict_date_optional_time", "yyyyMMdd"]，但是也匹配不成功。因此，尝试使用 "strict_date_optional_time"、"yyyy-MM-dd"] 和更改数据的日期格式，然后才匹配成功。

以下通过 3 个步骤演示动态映射，并把与日期有关的 JSON 字符串（string）数据类型映射变换为日期（date）数据类型。

（1）以下创建新 fund_basic_copy2 索引并给定相关的映射设置

```
curl -XPUT localhost:9200/fund_basic_copy2 --header "Content-Type: application/json" --data $'{"mappings":{"date_detection":true,"dynamic_date_formats":["strict_date_optional_time","yyyy-MM-dd"]}}'
```

（2）创建 ts_code=159809.SZ 的文档以测试新的动态映射

```
curl --request PUT http://localhost:9200/fund_basic_copy2/_doc/1?pretty=true --header "Content-Type: application/json" --data $'{"ts_code":"159809.SZ","name":"恒生湾区",
"management":"博时基金","custodian":"招商银行","fund_type":"股票型","found_date":"2020-04-30",
"due_date":null,"list_date":"2020-05-21","issue_date":"2020-03-30","delist_date": null,"issue_amount":3.9117,"m_fee":0.15,"c_fee":0.05,"duration_year":null,"p_value":1.0,
"min_amount":0.100,"exp_return":null,"benchmark":"恒生沪深港通大湾区综合指数收益率","status":"L",
"invest_type":"被动指数型","type":"契约型开放式","trustee":null,"purc_startdate":"2020-05-21",
"redm_startdate":"2020-05-21","market":"E"}'
```

（3）获取 fund_basic_copy2 索引动态映射结果

由于返回的响应内容信息太多，因此以下仅显示其中部分的信息。返回的响应内容结果，汇总后可以获得如表 4-3 所示。在表 4-3 中，所有与 date 数据类型相关的字段映射变换成功。

```
curl -XGET localhost:9200/fund_basic_copy2/_mappings?pretty=true
{
  "fund_basic_copy2" : {
    "mappings" : {
      "dynamic_date_formats" : [
        "strict_date_optional_time",
        "yyyy-MM-dd"
      ],
      "date_detection" : true,
      "properties" : {
        "benchmark" : {
          "type" : "text",
          "fields" : {
            "keyword" : {
              "type" : "keyword",
              "ignore_above" : 256
            }
          }
        },
        "c_fee" : {
          "type" : "float"
        },
        "custodian" : {
          "type" : "text",
          "fields" : {
            "keyword" : {
              "type" : "keyword",
              "ignore_above" : 256
            }
          }
        },
        "found_date" : {
          "type" : "date",
          "format" : "strict_date_optional_time"
        },
......
}
```

表 4-3　fund_basic 索引的动态映射结果

数据名称	数据类型	映射的 数据类型
ts_code，name，management，custodian，fund_type，benchmark，status，invest_type，type，trustee，market	str	text 与 keyword
found_date，due_date，list_date，issue_date，delist_date，purc_startdate，redm_startdate	str	date
issue_amount，m_fee，c_free，duration_year，p_value，min_amount，exp_return	float	float

另一方面，Elasticsearch 提供动态模板（dynamic_templates）映射设置，可以基于检测到的数据类型、字段名称和字段的 full dotted path（one.two.three），对动态添加的字段自行定义映射。有关更多详细信息，请参阅官方网站。

2. 显式映射

显式映射也称为静态（static）映射，指自定义映射。可以在创建索引时定义映射，或添加到现有索引。作为一个示例，表 4-2 中的 str 数据类型的映射要进一步细分。fund_ basic 索引字段名称为投资类型（fund_type）、业绩比较基准（benchmark）、存续状态（status）、投资风格（invest_type）、基金类型（type）和场内场外（market）需要重新定义为 keyword。原因它们没有必要分词，整个内容可以用作关键词。下面首先删除 fund_basic 索引，然后再重新创建，并自定义映射。

（1）删除 fund_basic 索引

```
curl -XDELETE localhost:9200/fund_basic
```

（2）重新创建 fund_basic 索引，并自定义映射

```
curl -XPUT localhost:9200/fund_basic?pretty=true --header "Content-Type: application/json" --data $'{"mappings":{"dynamic":false,"properties":{"benchmark":{"type":"keyword"}, "status":{"type":"keyword"}, "fund_type" : {"type":"keyword"}, "invest_type":{"type":"keyword"},"type":{"type":"keyword"}, "market":{"type":"keyword"}, "c_fee":{"type":"float"}, "issue_amount":{"type":"float"}, "m_fee" : {"type":"float"},"min_amount":{"type":"float"}, "p_value":{"type":"float"}, "duration_year":{"type":"float"}, "exp_return":{"type":"float"}, "issue_date":{"type":"date","format":"yyyyMMdd"}, "list_date":{"type":"date", "format":"yyyyMMdd"}, "purc_startdate":{"type":"date", "format":"yyyyMMdd"}, "redm_startdate":{"type":"date","format":"yyyyMMdd"}, "found_date":{"type":"date", "format":"yyyyMMdd"},"due_date":{"type":"date", "format":"yyyyMMdd"}, "delist_date":{"type":"date", "format":"yyyyMMdd"}, "management":{"type":"text", "fields":{"keyword":{"type":"keyword"}}}, "name":{"type":"text", "fields":{"keyword":{"type":"keyword"}}}, "ts_code":{"type":"text", "fields":{"keyword":{"type":"keyword"}}}, "trustee":{"type":"text", "fields":{"keyword":{"type":"keyword"}}},"custodian":{"type" : "text", "fields":{"keyword":{"type":"keyword"}}}}}'
{
  "acknowledged" : true,
  "shards_acknowledged" : true,
  "index" : "fund_basic"
}
```

（3）创建 ts_code=159809.SZ 的文档以测试新的显式映射

```
curl --request PUT http://localhost:9200/fund_basic/_doc/1?pretty=true --header
    "Content-Type: application/json" --data $'{"ts_code":"159809.SZ","name":"恒生湾区",
    "management":"博时基金","custodian":"招商银行","fund_type":"股票型","found_date":
"20200430",
    "due_date":null,"list_date":"20200521","issue_date":"20200330","delist_date":
null,
    "issue_amount":3.9117,"m_fee":0.15,"c_fee":0.05,"duration_year":null,"p_
value":1.0,
    "min_amount":0.100,"exp_return":null,"benchmark":"恒生沪深港通大湾区综合指数收益率",
    "status":"L","invest_type":" 被动指数型 ","type":" 契约型开放式 ","trustee":null,
    "purc_startdate":"20200521","redm_startdate":"20200521","market":"E"}'
```

（4）根据日期类型字段搜索文档

由于字段 found_date 现在是日期类型（date），因此可以使用一个代表 epoch 的长整数来搜索。以下示例搜索日期在 2020 年 5 月 1 日之前成立的公募基金。由于返回的响应内容信息太多，因此以下仅显示其中的部分信息。结果显示搜索的结果成功匹配（hits）一个文档（value=1）。range 查询将在 7.3.1 认识查询表达式一节详细描述。

提示：epoch 的定义为自 1970 年 1 月 1 日格林尼治标准时间起经过的秒数、毫秒数或纳秒数。

```
curl --request POST 'http://localhost:9200/fund_basic/_search?pretty=true'
--header "Content-Type: application/json" --data $'{"query": {"range": {"found_
date": {"lte": "20200501","format": "yyyyMMdd"}}}}'
```

```
{
  "took" : 2,
  "timed_out" : false,
  "_shards" : {
    "total" : 1,
    "successful" : 1,
    "skipped" : 0,
    "failed" : 0
  },
  "hits" : {
    "total" : {
      "value" : 1,
      "relation" : "eq"
    },
    ……
}
```

4.2.2 映射的元字段简介

在映射中可以设置一些元字段,控制如何处理文档索引的操作。一些重要的元字段,其用法描述如表 4-4 所示。

表 4-4 映射的元数据字段介绍

元数据名称	描 述	用 法
_index	索引名称	对文档的索引先进行匹配,然后操作。例如在索引上进行查询(query)、聚合(aggregate)、排序(sort)和访问脚本中的 _index 字段
_type	索引可以包含自定义类型,用以快速按类型名称搜索。	6.0.0 版本开始已被弃用
_id	文档标识符	文档标识符字段的值也可以先进行匹配,然后进行聚合或排序操作。但官方不鼓励这样做
_source	默认为启用并存储源文档。如果被禁用,则不会保留源文档	"mappings":{ "_source":{ "enabled": false}
_size	使用 mapper-size 插件提供 _size 元字段,启用后会索引原始 _source 字段的大小	有关更多插件信息,请参阅官方网站
_field_names	默认为启用,用以查找特定字段具有或不具有任何非空值的文档	禁用选项已被弃用,并将在以后的主要版本中删除
_meta	用户定义的元数据。开发人员可以使用它来定义应用程序的元数据	"mappings":{ "_meta":{ "user-defined-key": "data" }}
_routing	在映射中配置路由值,用以要求所有后续相关操作必须指定路由值。默认为文档标识符	"mappings":{ "_meta":{ "_routing": { "required": "true" }}

4.2.3 映射数据类型简介

在介绍动态映射时,曾经描述一些数据类型。还有许多其他数据类型可以在静态映射中用来定义字段。数据类型的描述和示例用法。在静态映射中设置字段数据类型的语法如下:

```
"mappings":{"properties":{
字段名称 : {"type":类型名称 },
字段名称 : {"type":类型名称 , 参数名称 : 参数值 }}}
```

一些重要的元字段,其用法描述如表 4-5 所示。

表 4-5 映射数据类型简介

数据类型		描 述	用 法
文本		映射成文本，用于全文检索	"名称"：{ "type"："text" }
关键词		映射成字符串，可用于过滤（filter）、排序和聚合	"名称"：{ "type"："keyword" }
数值		映射成带符号的 64 位元整数，其值在 [-2^{63}, 2^{63}-1] 之间	"名称"：{ "type"："long" }
		映射成带符号的 32 位元整数，其值在 [-2^{31}, 2^{31}-1] 之间	"名称"：{ "type"："integer" }
		映射成带符号的 16 位元整数，其值在 [-32168, 32767] 之间	"名称"：{ "type"："short" }
		映射成带符号的 8 位元整数，其值在 [-128, 127] 之间	"名称"：{ "type"："byte" }
		映射成 64 位 IEEE-754 浮点数，其值在 [2^{-1074}, (2-2^{-52})·2^{1023}] 之间	"名称"：{ "type"："double" }
		映射成 32 位 IEEE-754 浮点数，其值在 [2^{-149}, (2-2^{-23})·2^{127}] 之间	"名称"：{ "type"："float" }
		映射成 16 位 IEEE-754 浮点数，其值在 [2^{-24}, 65504] 之间	"名称"：{ "type"："half_float" }
		映射成用比例因子 (scaling_factor) 来描述的 long 类型浮点数	"名称"：{ "type"："scaled_float", "scaling_factor"：比例因子数值 }
日期		映射成日期格式字符串或数字。如果使用整数，则代表自 epoch 后的秒数或毫秒数。默认日期格式为 date_optional_time || epoch_millis。date_optional_time 默认格式参数 (format) 为 yyyy-MM-dd'T'HH:mm:ss.SSSZ 或 yyyy-MM-dd	"名称"：{ "type"："date" }
		映射成数字，代表 epoch 后的纳秒数，默认日期格式和参数同上	"名称"：{ "type"："date_nanos" }
布尔		正确 (true) 或错误 (false) 值	"名称"：{ "type"："boolean" }
数组		相同数据类型的数组，例如整数数组 [1、2、3、4、5]	"名称"：{ "type"："integer" }
IP		IPv4 或 IPv6 网址	"名称"：{ "type"："ip" }
别名		为现有字段定义别名	"名称"：{ "type"："alias", "path"："源字段的完整虚线路径" }
二元值		二进制值，作为 Base64 编码的字符串。默认为不存储且不可搜索	"名称"：{ "type"："binary" }
范围		数值类型 (integer_range、float_range、long_range 和 double_range)，日期类型 (date_range) 或 IP 类型 (ip_range) 的范围	"名称"：{ "type"："integer_range" }
			"名称"：{ "type"："float_range" }
			"名称"：{ "type"："long_range" }
			"名称"：{ "type"："double_range" }
			"名称"：{ "type"："date_range" }
			"名称"：{ "type"："ip_range" }
地理位置		单个位置点如经纬度对	"名称"：{ "type"："geo_point" }
地理形状		适用于 GeoJSON 格式的数据映射成 geo-shapes，例如 point(点)、circle(圆)、polygon(框) 等	"名称"：{ "type"："geo_shape" }
对象		JSON 对象格式值	"对象名称"：{ "properties"：{ "字段名称"：{ "type"："类型名称" }, "字段名称"：{ "type"："类型名称" },…}}

续表

数据类型	描 述	用 法
巢状对象	JSON 对象格式数组	"名称"：{ "type"："nested" }
词元数	配合分析器名称参数，用于计算文本中标记的词元数量	"名称"：{ "type"："token_count"，"analyzer"："分析器名称" }
percolator	映射成查询原生语句并存储	"名称"：{ "type"："percolator" }
文档父子关系	在相同索引的文档中创建 (join) 父子关系，关系描述于 relations 参数中。可分为一对一，一对多及多层次关系，详细用法请参考第 8 章数据建模	"名称"：{ "type"："join"，"relations"：{ "父字段名称"，"子字段名称" }}
rank_feature	为数值编制索引，以使用 rank_feature 查询时的权重提升	"名称"：{ "type"："rank_feature" }
rank_features	为数值数组编制索引，以使用 rank_features 查询时的权重提升	"名称"：{ "type"："rank_features" }
密集数字向量	配合 dims(维度) 参数，映射成浮点值的密集向量 (density_vector)	"名称"：{ "type"："dense_vector"，"dims"：整数值 }
稀疏数字向量	映射成浮点值的稀疏向量 (sparse_vector)	"名称"：{ "type"："sparse_vector" }
Search-as-you-type	映射成类似文本的数据类型，提供开箱即用，自动补全 (auto-complete) 搜索字段的查询	"名称"：{ "type"："search_as_you_type" }
扁平化数据	如果对象类型具有许多子字段，又不想指定其所有映射数据类型，则可以混合使用动态映射方法，自动生成映射。此种替代方法称为扁平化数据 (flatten)。各个子字段的数据类似 keyword	"名称"：{ "type"："flatten" }
直方图	映射成直方图 (histogram) 方式的预聚合数值数据	"名称"：{ "type"："histogram" }
形状	类似 geo-shape 映射类型，适用于 GeoJSON 格式的数据，映射成笛卡尔二维坐标系统	"名称"：{ "type"："shape" }

备注：还有很多尚未讨论的数据类型如 mapper-murmur3、mapper-annotated-text、Constant keyword 等，请参阅官方网站内容。

4.2.4　映射数据类型的参数简介

映射参数定义了存储文档字段的方式、索引的方法和选项、要公开的字段信息、要执行的附加过程，以及在索引和搜索过程中如何分析映射的数据。根据用途，参数可以区分为如何存储、索引、分析和处理四种，我们将在表 4-6 中描述支持的映射参数。

表 4-6　映射数据类型的参数简介

参数名称	描 述	用 法
如何存储		
store	是否存储字段值，默认为 false	"名称"：{ "type"："shape"，"store"：true }
doc_values	与 _source 相同的值，但是以适用于排序和聚合的面向列 (column-oriented) 方式存储。默认值是 true	"名称"：{ "type"："text"，"doc_values"：false }
term_vector	是否存储分析器生成的分词，默认为 no。选项为 no、yes、with_positions、with_offsets、with_positions_offset 和 with_positions_offsets_payloads	"名称"：{ "type"："text"，"termvector"："yes" }
norm	是否存储各种标准化因子以在查询中评分，默认为 true	"名称"：{ "type"："text"，"norm"：true }

续表

参数名称	描 述	用 法
如何索引		
index	是否索引该字段，默认为 true	"名称":{"type":"text","index":false}
index_option	提供搜索和高亮显示(highlight)功能。选项包括文档(docs)、频率(freqs)、位置(positions)和(offsets)。默认为 positions	"名称":{"type":"text","index_options":"offsets"}
index_phrases	是否在文本中提供更快的短语(phrases)搜索。默认为 false	"名称":{"type":"text","index_phrases":true}
index_prefixes	启用文本的前缀搜索。默认为禁用。配合设置 min_chars 和 max_chars 设定前缀长度。min_chars 默认为 2 而 max_chars 默认为 5	"名称":{"type":"text","index_prefixes":{"min_chars":3,"max_chars":6}}
dynamic	是否动态索引该对象格式字段，默认为 true，选项包括 true、false 和 strict。false 选项将忽略新字段。strict 选项为新字段将引发异常	"名称":{"dynamic":false}
enabled	是否跳过分析处理，只需存储该字段而不对其建立索引。只适用于 mappings 或对象字段	"mappings":{"enabled":false}
fielddata	只适用于文本，当字段用于聚合、排序或在脚本中时，使用称为 fielddata 的查询时内存数据结构。默认为 false	"名称":{"type":"text","fielddata":true}
fields	为一个字段，同时产生不同数据类型索引。字段名称用于全文检索，而"字段名称.名称"可以当作关键词检索	"字段名称":{"type":"text","fields":{"名称":{"type":"keyword"}}}
format	日期格式模式，可以自定义格式或使用内置格式	"名称":{"type":"date","format":"yyyyMMdd"}
ignore_above	设置字段最大字符串长度。比设置长的字符串将被忽略，不索引也不存储。适用于文本或关键词类型	"名称":{"type":"text","ignore_above":30}
ignore_malformed	可以忽略无效字段值的文档，并继续下一个文档的索引操作。默认为 false，并引发异常	"名称":{"type":"类型名称","ignore_malformed":true}
null_value	允许使用指定字符串，例如"NULL"，替换空值(null)，用于索引和搜索	"名称":{"type":"类型名称","null_value":"NULL"}
properties	为对象数据类型或嵌套数据类型创建子字段	"字段名称":{"properties":{"子字段名称1":{"type":"类型名称"},"子字段名称2":{"type":"类型名称"}}}
similarity	配合 similarity 参数，将默认评分算法 BM25 更改为布尔(boolean)评分或其他	"名称":{"type":"类型名称","similarity":"boolean"}
如何分析		
analyzer	使用自定义分析器或非默认分析器，只适用于文本类型	"名称":{"type":"text","analyzer":"分析器名称"}
normalizer	使用归一化(normalizer)产生一个分词，只适用于关键词类型	"名称":{"type":"keyword","normalizer":"归一化名称"}
boost	对字段值的相关性进行加权的方法，默认加权值是 1.0	"名称":{"type":"类型名称","boost":2}
position_increment_gap	文本分析器会考虑分词间之间隙，配合此参数，设定较大的间隙，可以防止大多数短语查询在分词之间进行匹配。默认为 100	"名称":{"type":"text","position_increment_gap":120}

续表

参数名称	描 述	用 法
search_analyzer	通常，在索引时和搜索时应使用相同的分析器。配合此参数，可设定在搜索时使用其他文本分析器	"名称":{"type":"text","search_analyzer":"分析器名称"}
如何处理		
coerce	清理脏值以适合字段的数据类型。例如将数字字符串转换为数字，或将浮点数舍入为整数值。默认值为 false	"名称":{"type":"integer","coerce":true}
copy_to	将多个字段中的数据复制到单个字段中，以从该字段中查询数据	"名称1":{"type":"text","copy_to":"新字段名称"}，"名称2":{"type":"text","copy_to":"新字段名称"}
eager_global_ordinals	每个索引段都定义了各自的 doc_values 序数，但是聚合操作收集了整个分片上的数据。为此，Elasticsearch 提供一个称为全局序数的映射参数，默认为 false。至于什么样的聚合操作需要此参数，请参阅官方网址	"名称":{"type":"keyword","eager_global_ordinals":true}

4.2.5 更新显式映射内容并刷新文档索引

如果使用显式映射，映射中未指定的任何字段将不会被索引和搜索，但会存储在 _source 中。若以后需要搜索这些非映射字段，则只需更新显式映射，再通过查询更新多个文档接口并配合 refresh 参数，文档不需要重新索引而进行刷新。

以下按照 4.2.1 节认识映射类型中显式映射示例的步骤，以下为刷新显式映射内容：

（1）删除 fund_basic 索引。

```
curl -XDELETE localhost:9200/fund_basic
```

（2）重新创建 fund_basic 索引，并自定义映射。与显式映射示例区别的是忽略 found_date 字段。

```
curl -XPUT localhost:9200/fund_basic?pretty=true --header "Content-Type: application/json" --data $'{"mappings":{"dynamic":false,"properties": {"benchmark":{"type":"keyword"},"status":{"type":"keyword"},"fund_type":{"type":"keyword"},"invest_type":{"type":"keyword"},"type":{"type":"keyword"}, "market":{"type":"keyword"},"c_fee":{"type":"float"},"issue_amount":{"type":"float"},"m_fee":{"type":"float"}, "min_amount":{"type":"float"},"p_value":{"type":"float"},"duration_year":{"type":"float"},"exp_return":{"type":"float"},"issue_date":{"type":"date","format":"yyyyMMdd"},"list_date":{"type":"date","format":"yyyyMMdd"},"purc_startdate":{"type":"date","format":"yyyyMMdd"},"redm_startdate":{"type":"date","format":"yyyyMMdd"}, "due_date":{"type":"date","format":"yyyyMMdd"},"delist_date":{"type":"date", "format":"yyyyMMdd"},"management":{"type":"text","fields":{"keyword":{"type":"keyword"}}},"name":{"type":"text","fields":{"keyword":{"type":"keyword"}}},"ts_code":{"type":"text","fields":{"keyword":{"type":"keyword"}}},"trustee":{"type":"text","fields":{"keyword":{"type":"keyword"}}},"custodian":{"type":"text","fields":{"keyword":{"type":"keyword"}}}}}}'
{ "acknowledged" : true }
```

（3）在被忽略 found_date 的自定义显式映射下，创建 ts_code=159809.SZ 的文档。

```
curl --request PUT http://localhost:9200/fund_basic/_doc/1?pretty=true --header "Content-Type: application/json" --data $'{"ts_code":"159809.SZ","name":"恒生湾区","management":"博时基金","custodian":"招商银行","fund_type":"股票型","found_date":"20200430",
"due_date":null,"list_date":"20200521","issue_date":"20200330","delist_date":null,
```

```
"issue_amount":3.9117,"m_fee":0.15,"c_fee":0.05,"duration_year":null,"p_value":1.0,
"min_amount":0.100,
"exp_return":null,"benchmark":"恒生沪深港通大湾区综合指数收益率",
"status":"L","invest_type":"被动指数型","type":"契约型开放式","trustee":null,
"purc_startdate":"20200521","redm_startdate":"20200521","market":"E"}
```

（4）根据日期类型字段搜索文档，由于字段 found_date 在自定义显式映射中被忽略，所以没有被索引而无法搜索。

```
curl --request POST 'http://localhost:9200/fund_basic/_search?pretty=true' --header "Content-Type: application/json" --data $'{\"query\": {\"range\": {\"found_date\": {\"lte\": \"20200501\",\"format\": \"yyyyMMdd\"}}}}'
{
  "took" : 0,
  ……
  "hits" : {
    "total" : {
      "value" : 0,
  ……
```

（5）添加 found_date 字段到显式映射定义，使用 PUT 对 fund_basic 索引的 _mapping 接口发出请求，更新映射。

```
curl -XPUT localhost:9200/fund_basic/_mappings?pretty=true --header "Content-Type: application/json" --data $'{"dynamic":false, "properties":{"benchmark":{"type":"keyword"}, "status":{"type":"keyword"}, "fund_type" : {"type":"keyword"}, "invest_type":{"type":"keyword"}, "type":{"type":"keyword"}, "market":{"type":"keyword"}, "c_fee":{"type":"float"}, "issue_amount":{"type":"float"}, "m_fee" : {"type":"float"}, "min_amount":{"type":"float"}, "p_value":{"type":"float"}, "duration_year":{"type":"float"}, "exp_return":{"type":"float"}, "issue_date":{"type":"date", "format":"yyyyMMdd"}, "list_date":{"type":"date", "format":"yyyyMMdd"}, "purc_startdate":{"type":"date", "format":"yyyyMMdd"}, "redm_startdate":{"type":"date", "format":"yyyyMMdd"}, "found_date":{"type":"date", "format":"yyyyMMdd"}, "due_date":{"type":"date", "format":"yyyyMMdd"}, "delist_date":{"type":"date", "format":"yyyyMMdd"}, "management":{"type":"text", "fields":{"keyword":{"type":"keyword"}}}, "name":{"type":"text", "fields":{"keyword":{"type":"keyword"}}}, "ts_code":{"type":"text", "fields":{"keyword":{"type":"keyword"}}}, "trustee":{"type":"text", "fields":{"keyword":{"type":"keyword"}}}, "custodian":{"type" : "text", "fields":{"keyword":{"type":"keyword"}}}}}'
{
  "acknowledged" : true
}
```

（6）使用 POST 对 fund_basic 索引的 _update_by_query 接口发出请求，配合请求 refresh 参数刷新文档索引，可以立刻进行搜索。

```
curl -XPOST 'localhost:9200/fund_basic/_update_by_query?refresh&pretty=true'
{
  "took" : 419,
  "timed_out" : false,
  "total" : 1,
  "updated" : 1,
  "deleted" : 0,
  "batches" : 1,
  "version_conflicts" : 0,
  "noops" : 0,
  "retries" : {
    "bulk" : 0,
    "search" : 0
  },
```

```
    "throttled_millis" : 0,
    "requests_per_second" : -1.0,
    "throttled_until_millis" : 0,
    "failures" : [ ]
}
```

（7）根据日期类型字段搜索文档，由于字段 found_date 现在是日期类型（date），因此可以成功使用一个代表 epoch 的日期来搜索。以下示例为搜索日期在 2020 年 5 月 1 日之前成立的公募基金。

```
    curl --request POST 'http://localhost:9200/fund_basic/_search?pretty=true' --header "Content-Type: application/json" --data $'{"query": {"range": {"found_date": {"lte": "20200501","format": "yyyyMMdd"}}}}'
{
    "took" : 0,
    ……
    "hits" : {
      "total" : {
        "value" : 0,
    ……
    "hits" : [
      {
        "_index" : "fund_basic",
        "_type" : "_doc",
        "_id" : "1",
        "_score" : 1.0,
        "_source" : {
          "ts_code" : "159809.SZ",
          "name" : "恒生湾区",
    ……
}
```

第 5 章 分析文本内容

文档的文本进行索引操作期间,文本的内容由分析器执行,并且生成分词用于构建反向索引。在搜索操作期间,查询的内容也由分析器生成分词,再进行搜索操作。通常,在索引时和搜索时应使用相同的分析器。但也可设定映射数据类型的参数 search_analyzer,在搜索时使用其他文本分析器。分析器可以是内置分析器,也可以是自定义的分析器。本章将学习分析器以及如何分析文本内容。

5.1 分析器的结构

分析器由字符过滤器(character filter)、分词器(tokenizer)和词汇单元过滤器(token filter)组成。字符过滤器用于分词之前对字符进行预处理,分词器根据一组规则将文本拆分为词汇单元,而词汇单元过滤器分别对这些词汇单元流执行进一步操作。最后输出的词汇单元,将构建反向索引,并在以后用于检索结果。Elasticsearch 提供分析接口(_analyze),配合请求主体设定字符过滤器、分词器和 / 或词汇单元过滤器,供测试使用。

5.1.1 分析接口

此接口使用 GET 或 POST 请求,对文本字符串(text)执行分析并返回词汇单元。如果使用 GET 请求,可用 URL 参数方法。如果使用 POST 请求,则在请求主体中指定参数。可配合指定索引名称,使用在索引中的分析器、分词器、词汇单元过滤器、定义的字段映射方式等。接口请求语法如下:

```
GET /_analyze
POST /_analyze
GET /<索引名称>/_analyze
POST /<索引名称>/_analyze
```

一些重要的参数,其用法描述于表 5-1。

表 5-1 映射数据类型简介

参数名称	描述	用法
text	要分析的文本。若是字符串数组,则当作多个字段	"text":"文本内容"
analyzer	分析器名称。默认为 standard 分析器	"analyzer":"分析器名称"

参数名称	描述	用法
char_filter	字符过滤器数组，列出要使用的过滤器名称	"char_filter":["过滤器名称"]
tokenizer	分词器名称	"analyzer":"分词器名称"
filter	词汇单元过滤器数组，列出要使用的过滤器名称	"filter":["过滤器名称"]
normalizer	归一化处理器名称	"normalizer":"归一化处理器名称"
field	要使用此参数，必须指定一个索引。提取基于字段的映射中配置的文本分析器	"field":"字段的路径名称"
explain	使用此参数可获得更多有关于分析的详细信息，默认为 false	"explain":true
attributes	用于过滤 explain 参数输出的属性数组	"attributes":["属性名称"]

5.1.2 字符过滤器

主要功能是将原始输入文本转换为字符流，然后对其进行预处理，再将其传递给分词器作为输入。支持三种内置的字符过滤器：html_strip、mapping 和 pattern_replace。以下将使用 HTML 文本来进行练习。

提示：由于默认分析器 standard 对于中文用户而言并不适合，所以本章中的示例仅提供英文文本。在第 6 章介绍中文文本分析插件后，我们将开始使用中文示例。

```
<h3>Shanghai Stock Exchange's financing balance increased by 566.14 billion yuan, an increase of 0.1%</h3>
```

1. html_strip 字符过滤器

默认情况下，文本内容保持不变，但是所有 HTML 标签将被删除，对内容进行解码，并用 UTF-8 字符替换，例如将 HTML entity（实体）&amp 替换为 &。以下示例文本应用 html_strip 字符过滤器后的输出如下，HTML 标签 <h3> 将被删除：

```
curl --request POST 'http://localhost:9200/_analyze?pretty=true' --header "Content-Type: application/json" --data $'{"text":"<h3>Shanghai Stock Exchange's financing balance increased by 566.14 billion yuan, an increase of 0.1%</h3>", "char_filter":["html_strip"]}}'
{
  "tokens" : [
    {
      "token" : "\nShanghai Stock Exchange's financing balance increased by 566.14 billion yuan, an increase of 0.1%\n",
      "start_offset" : 0,
      "end_offset" : 106,
      "type" : "word",
      "position" : 0
    }
  ]
}
```

2. mapping 字符过滤器

用键值对列表来指定替换值。当输入字符串具有这样的键值时，与原始值中的字符串匹配最长的键值将"获胜"，而原始值将被相应的替换值取代。以下示例使用两个字符过滤器

（html_strip 和 mapping），HTML 标签 <h3> 将被删除，而字符串"SSE's"将取代"Shanghai Stock Exchange's"。

```
curl --request POST 'http://localhost:9200/_analyze?pretty=true' --header 
"Content-Type: application/json" --data $'{"text":"<h3>Shanghai Stock Exchange's 
financing balance increased by 566.14 billion yuan, an increase of 0.1%</h3>", "char_
filter":["html_strip", {"type":"mapping","mappings":["Shanghai Stock Exchange's 
=> SSE's"]}]}'
    {
      "tokens" : [
        {
          "token" : "\nSSE's financing balance increased by 566.14 billion yuan, an 
increase of 0.1%\n",
          "start_offset" : 0,
          "end_offset" : 106,
          "type" : "word",
          "position" : 0
        }
      ]
    }
```

3. pattern_replace 字符过滤器

类似 mapping 字符过滤器，但是用正则表达式以匹配字符序列，并指定替换字符串，也可以使用正则表达式中捕获组（group）表示法。要定义此过滤器，需要指定三个参数：

- pattern（匹配字符序列的 Java 正则表达式）。
- replacement（替换字符串）。
- flags（Java 正则表达式标志）。

以下示例使用三个字符过滤器（html_strip、mapping 和 pattern_replace），其中编写了一个正则表达式 pattern_replace 用于字符过滤器，如果满足正则表达式条件，则使用 percentage 字符串替换百分点的符号（%）。替换字符串使用了捕获组别（group）的表示法。

提示：正则表达式"(\\\\d)+.(\\\\d)+%"会匹配两个数字序列之间有点号分隔，然后跟跟着一个百分号。两个序列号是两个组别，在模式中重复使用。

```
curl --request POST 'http://localhost:9200/_analyze?pretty=true' --header 
"Content-Type: application/json" --data $'{"text": "<h3>Shanghai Stock Exchange's 
financing balance increased by 566.14 billion yuan, an increase of 0.1%</h3>","char_
filter": ["html_strip", {"type": "pattern_replace", "pattern": "(\\\\d)+.
(\\\\d)+%","replacement": "$1.$2 percentage"}]}'
    {
      "tokens" : [
        {
          "token" : "\nShanghai Stock Exchange's financing balance increased by 
566.14 billion yuan, an increase of 0.1 percentage\n",
          "start_offset" : 0,
          "end_offset" : 106,
          "type" : "word",
          "position" : 0
        }
      ]
    }
```

5.1.3 分词器

字符过滤器输出字符流，再传递输入分词器。分词器根据一组规则将文本拆分为词汇单元。Elasticsearch 支持三种分词器，描述如下：

- 面向单词的分词器:将字符流拆分为个别的词汇单元。
- 供文本部分匹配使用:将字符流分成给定长度内的小片段字符串。
- 供结构化文本使用:这会将字符流拆分为已知结构的字符串,例如关键字、电子邮件地址和邮政编码。

Elasticsearch 提供了许多内置分词器,以下为各个分词器提供示例。练习文本,显示如下:

```
192.168.0.1 Mountain View, California 94040. General +1 650-4582-620 info@elastic.co https://www.elastic.co/contact
```

除了 standard 分词器的示例外,其他结果按照分词器种类编排到表 5-2 至表 5-4。由于返回的响应内容信息太多,因此以下仅显示其中部分的信息。

```
curl --request POST 'http://localhost:9200/_analyze?pretty=true' --header "Content-Type: application/json" --data $'{"text": "192.168.0.1Mountain View, California 94040. General +1 650-4582-620 info@elastic.co https://www.elastic.co/contact", "tokenizer":"standard"}'
{
  "tokens" : [
    {
      "token" : "192.168.0.1",
      "start_offset" : 0,
      "end_offset" : 11,
      "type" : "<NUM>",
      "position" : 0
    },
    {
      "token" : "Mountain",
      "start_offset" : 13,
      "end_offset" : 21,
      "type" : "<ALPHANUM>",
      "position" : 1
    },
    ...
}
```

表 5-2 面向单词的分词器

分词器名称		描述及产生词汇单元
standard	描述	基于语法的分词器,支持 max_token_length 参数将输入文本划分为多个字段
	产生词汇单元	\<NUM\> 类型:192.168.0.1、94040、1、650、4582、620 \<ALPHANUM\> 类型:Mountain、View、California、General、info、elastic.co、https、www.elastic.co、contact
letter	描述	使用非字母作为分隔符,将字符流划分为多个词汇单元
	产生词汇单元	\<ALPHANUM\> 类型:Mountain、View、California、General、info、elastic、co、https、www、elastic、co、contact
lowercase	描述	如同 letter 分词器,还将字母从大写转换为小写
	产生词汇单元	\<ALPHANUM\> 类型:mountain、view、california、general、info、elastic、co、https、www、elastic、co、contact
whitespace	描述	使用空格字符作为分隔符,以将字符流分成多个词汇单元
	产生词汇单元	word 类型:192.168.0.1、Mountain、View、California、94040、General、+1、650-4582-620、info@elastic.co、https://www.elastic.co/contact

续表

分词器名称		描述及产生词汇单元
uax_url_email	描述	将字符流分为 URL 格式和电子邮件地址格式
	产生词汇单元	<NUM> 类型：94040、1、650、4582、620 <ALPHANUM> 类型：Mountain、View、California、General <URL> 类型：192.168.0.1、https://www.elastic.co/contact <EMAIL> 类型：info@elastic.co
classic	描述	基于语法的分词器，此外，它使用标点符号作为分隔符，但保留一些特殊的格式，例如非空格字符之间的点、数字、电子邮件地址和互联网主机名之间的连接符
	产生词汇单元	<NUM> 类型：650-4582-620 <ALPHANUM> 类型：Mountain、View、California、94040、General、1、https、contact <EMAIL> 类型：info@elastic.co <HOST> 类型：192.168.0.1、www.elastic.co
thai	请参考官网	

表 5-3 供文本部分匹配使用的分词器

分词器名称		描述、参数用法及产生词汇单元
ngram	描述	沿输入字符流滑动，以指定长度的字符提供词汇单元。指定长度参数 min_gram（默认为 1）和 max_gram（默认为 2），可用参数 token_chars 指定字母、数字、空格、标点符号和符号作为词汇单元的一部分。也可用参数 custom token_chars 允许自定义字符作为词汇单元的一部分
	参数用法	"tokenizer":{ "type": "ngram", "min_gram":8, "max_gram":9, "token_chars": ["letter"]}
	产生词汇单元	word 类型：Mountain、Californ、Californi、aliforni、alifornia、lifornia
edge_ngram	描述	如同 ngram 分词器，区别在于每个项目都锚定到候选单词的起点。提供相同的参数
	参数用法	"tokenizer":{ "type": "edge_gram", "min_gram":8, "max_gram":9, "token_chars": ["letter"]}
	产生词汇单元	word 类型：Mountain、Californ、Californi

表 5-4 供结构化文本使用的分词器

分词器名称		描述、参数用法及产生词汇单元
keyword	描述	与输入字符流相同的文本作为词汇单元。参数 buffer_size 为缓冲区的字符数，默认为 256，官方不建议更改此值
	产生词汇单元	word 类型："192.168.0.1 Mountain View, California 94040. General +1 650- 4582-620 info@elastic.co https://www.elastic.co/contact"
pattern	描述	参数 pattern 值用正则表达式以匹配字符流。匹配的字段，可用作词汇单元输出，也可用作分隔文本。参数 group 默认为 -1，用作分隔符
	参数用法	"tokenizer":{ "type": "pattern", "pattern": "(\d)(\s)" }
	产生词汇单元	word 类型：192.168.0.、Mountain View, California 94040. General +、650-4582-62、info@elastic.co https://www.elastic.co/contact 如果添加参数 "group"：1，则产生 word 类型：1、1、0
char_group	描述	通过使用参数 tokenize_on_chars 定义分隔符列表，将输入字符流拆分为词汇单元
	参数用法	"tokenizer":{ "type": "char_group", "tokenize_on_chars": ["whitespace"]}
	产生词汇单元	word 类型：192.168.0.1、Mountain、View,、California、94040.、General、+1、650-4582-620、info@elastic.co、https://www.elastic.co/contact

续表

分词器名称		描述、参数用法及产生词汇单元
simple_pattern	描述	类似 pattern 分词器，但使用 Lucene 正则表达式。通过使用参数 pattern 和 Lucene 正则表达式，将输入字符流中匹配的字段为词汇单元输出。pattern 默认值为空字符串，即没有输出词汇单元
	参数用法	"tokenizer":{ "type": "simple_pattern", "pattern": "[a-zA-Z]*" }
	产生词汇单元	word 类型：Mountain、View、California、General、info、elastic、co、https、www、elastic、co、contact
simple_patten_split	描述	如同 simple_pattern 分词器，但是将输入字符流中匹配的字段设为分隔字段，用作分隔文本
	参数用法	"tokenizer":{ "type": "simple_pattern_split", "pattern": "[a-zA-Z@,//-:()(.)]" }
	产生词汇单元	word 类型：192、168、0、1、94040、+1、650、4582、620
path_hierarchy	描述	使用路径分隔符将输入字符流拆分为词汇单元，再通过替换字符的连接，从最开始的词汇单元，一个接一个地组成一系列输出的分词。可以设置以下参数：delimiter（分隔符，默认为 /）、replacement（替换分隔符的字符，默认与 delimiter 相同）、buffer_size（批处理的最大长度，默认为 1024）、reverse（操作从拆分后的词汇单元流，反向开始。默认为 false）和 skip（跳过初始生成的词汇单元的数目，默认为 0）
	输入文本	https://www.elastic.co/contact
	参数用法	"tokenizer":{ "type": "path_hierarchy", "delimiter": "#", "replacement": "#", "reverse":true}
	产生词汇单元	word 类型："https:##www.elastic.co#contact"、"#www.elastic.co#contact"、"www.elastic.co#contact"、"contact"

5.1.4 词汇单元过滤器

此过滤器的主要功能是添加、修改或删除分词器最后输出的词汇单元。Elasticsearch 大约提供 50 个内置词汇单元过滤器。用来练习的文本，除了额外在表内列出的，其他与 5.1.3 节分词器所使用的相同。示例都只使用 uax_url_email 分词器和介绍的过滤器，并没有应用字符过滤器。除了 lowercase 过滤器的示例显示如下部分结果外，其他常用的过滤器的结果将编排到表 5-5 中，没有列出的过滤器请参考官网。由于返回的响应内容信息太多，因此以下仅显示其中的部分信息。

```
curl --request POST 'http://localhost:9200/_analyze?pretty=true' --header "Content-Type: application/json" --data $'{"text": "192.168.0.1Mountain View, California 94040. General +1 650-4582-620 info@elastic.co https://www.elastic.co/contact", "tokenizer":"uax_url_email", "filter":{"type": "lowercase"}}'
{
  "tokens" : [
    {
      "token" : "192.168.0.1",
      "start_offset" : 0,
      "end_offset" : 11,
      "type" : "<URL>",
      "position" : 0
    },
    {
      "token" : "mountain",
      "start_offset" : 13,
      "end_offset" : 21,
```

```
            "type" : "<ALPHANUM>",
            "position" : 1
        },
        ......
    }
```

表 5-5　常用的词汇单元过滤器

词汇单元过滤器名称		描述、用法及词汇单元输出
lowercase（uppercase）	描述	将词汇单元的所有字母转换为小写（大写）
	用法	"filter":{"type":"lowercase"}
	词汇单元输出	<NUM> 类型：94040、1、650、4582、620 <ALPHANUM> 类型：mountain、view、california、general <URL> 类型：192.168.0.1、https://www.elastic.co/contact <EMAIL> 类型：info@elastic.co
fingerprint	描述	对分词器最后输出的词汇单元进行排序，删除重复数据后，连接成一个字段。参数 separator 默认为空格，用于连接词汇单元。参数 max_output_size 可以限制字段长度，如果过长，将导致没有字段输出。默认为 255
	用法	"filter":{"type":"fingerprint","separator":"%"}
	词汇单元输出	fingerprint 类型：1%192.168.0.1%4582%620%650%94040%California%General%Mountain%View%https://www.elastic.co/contact%info@elas ticco
keep	描述	仅保留指定列表中定义的词汇单元。提供了三个参数：keep_words 参数允许指定过滤器中的词汇单元；keep_words_path 参数可以指定文件路径 和 keep_words_case 参数将词汇单元转换为小写（默认为 false）
	用法	"filter":{"type":"keep","keep_words":["Mountain","General"]}
	词汇单元输出	<ALPHANUM> 类型：Mountain、General
keep_types	描述	保留或排除列表中参数 types 定义的那些类型的词汇单元。参数 mode 默认为 include（保留）。可以设为 exclude（排除）
	用法	"filter":{"type":"keep_types","types":["<NUM>","<ALPHANUM>"],"mode":"exclude"}
	词汇单元输出	<URL> 类型：192.168.0.1、https://www.elastic.co/contact <EMAIL> 类型：info@elastic.co
reverse	描述	对分词器最后输出的词汇单元进行反转。示例首先使用 kee_typesp 过滤器只选择 <EMAIL> 类型，然后再使用反向词元过滤器（reverse）
	用法	"filter":[{"type":"keep_types","types":["<EMAIL>"],"mode":"include"},{"type":"reverse"}]
	词汇单元输出	<EMAIL> 类型：oc.citsale@ofni
condition	描述	可以使用参数 script 指定脚本，如果条件匹配，则应用过滤器于词汇单元。示例中，若脚本条件匹配，则使用反向词元过滤器（reverse）来对 <ALPHANUM> 类型词汇单元进行反向转换
	用法	"filter":[{"type":"keep_types","types":["<EMAIL>","<ALPHANUM>"]},{"type":"condition","script":{"source":"token.getType()=='<ALPHANUM>'"},"filter":["reverse"]}]
	词汇单元输出	<ALPHANUM> 类型：niatnuoM、weiV、ainrofilaC、lareneG <EMAIL> 类型：info@elastic.co

续表

词汇单元过滤器名称		描述、用法及词汇单元输出
predicate_token_filter	描述	类似 condition 过滤器，可用参数 script 指定脚本，如果条件匹配，保留该词汇单元；否则删除
	用法	"filter"：{"type"："predicate_token_filter"，"script"：{"source"："token.getType() == '<EMAIL>'"}}
	词汇单元输出	<EMAIL> 类型：info@elastic.co
stemmer	描述	允许以指定的语言，只提取词汇单元的词干（stem）
	文本输入	loved loving lovely loveness
	用法	"filter"：[{"type"："keep_types"，"types"：["<EMAIL>"，"<ALPHANUM>"]，{"type"："stemmer"，"name"："English"}}]
	词汇单元输出	<ALPHANUM> 类型：love、love、love、love
unique	描述	只产生唯一的词汇单元。使用 stemmer 示例，在最后阶段添加 unique 过滤器
	文本输入	loved loving lovely loveness
	用法	"filter"：[{"type"："keep_types"，"types"：["<EMAIL>"，"<ALPHANUM>"]，{"type"："stemmer"，"name"："English"}，{"type"："unique"}}]
	词汇单元输出	<ALPHANUM> 类型：love
stop	描述	此过滤器会删除英语停用词，例如冠词、代词、连接词、第三人称代词等。可用参数 stopwords 指定语言、参数 stopwords_path 指定包含停用词的文件路径、参数 ignore_case 指定是否忽略停用词的大小写（默认为 false）及参数 remove_trailing 指定是否删除最后一个词汇单元（默认为 true）
	文本输入	a A and AND the The he there There
	用法	"filter"：{"type"："stop"，"stopwords"："_english_"}
	词汇单元输出	<ALPHANUM> 类型：A、AND、The、he、There
ngram	描述	对分词器最后输出的各个词汇单元，沿单个词汇单元滑动，以指定长度的字符产生更多词汇单元。指定长度参数 min_gram（默认为 1）和 max_gram（默认为 2）
	文本输入	Mountain View California
	用法	"filter"：{"type"："ngram"，"min_gram"：8，"max_gram"：9}
	词汇单元输出	<ALPHANUM> 类型：Mountain、Californ、Californi、aliforni、lifornia、lifornia
edge_ngram	描述	如同 ngram 过滤器，区别在于沿单个词汇单元滑动时都只锚定起点。提供相同的参数
	文本输入	Mountain View California
	用法	"filter"：{"type"："edge_ngram"，"min_gram"：8，"max_gram"：9}
	词汇单元输出	<ALPHANUM> 类型：Mountain、Californ、Californi
asciifolding	描述	当字母，数字和 Unicode 符号不在前 127 个 ASCII 字符中时，词汇单元会转换为 ASCII。reserve_original 参数（默认为 false）将保留原始词汇单元
	文本输入	ÀÇÈÎÒÙ
	用法	"filter"：{"type"："asciifolding"，"preserve"："true}
	词汇单元输出	<ALPHANUM> 类型：ACEIOU、ÀÇÈÎÒÙ

由于 word_delimiter 是一个复杂的词汇单元过滤器，因此我们在这里单独介绍它。默认情况下，这个过滤器使用以下规则进一步将词汇单元拆分再输出：

- 它使用所有非字母数字字符作为分隔符。
- 从每个词汇单元中删除前导或尾随的非字母数字字符。
- 在字母大小写转换时拆分。
- 从每个词汇单元的末尾除去英语所有格（'s）。
- 此过滤器提供许多参数可采用，以下只对其布尔型参数进行更详细的描述，每个示例只使用 standard 分词器和 word_delimiter 过滤器，并没有应用字符过滤器。参数设为 true 或 false 的结果如表 5-6 所示。至于其他类型参数，请参考官网。用来练习的文本显示如下：

```
It's ElasticSearch7.7.0 word_delimiter !
```

表 5-6　word_delimiter 词汇单元过滤器常用的参数

布尔型参数名称	描述、用法及词汇单元输出		
generate_word_parts	描述	保留字母部分	
	用法	"filter":{ "type": "word_delimiter", "generate_word_parts":true|false}	
	词汇单元输出	true（默认）	\<ALPHANUM\> 类型：It、s、Elastic、Search、word、delimiter　\<NUM\> 类型：7、7、0
		false	\<NUM\> 类型：7、7、0
generate_number_parts	描述	保留数字部分	
	用法	"filter":{ "type": "word_delimiter", "generate_number_parts":true|false}	
	词汇单元输出	true（默认）	\<ALPHANUM\> 类型：It、s、Elastic、Search、word、delimiter　\<NUM\> 类型：7、7、0
		false	\<ALPHANUM\> 类型：It、s、Elastic、Search、word、delimiter
catenate_words	描述	包括合并原为一起但拆分了的字母部分	
	用法	"filter":{ "type": "word_delimiter", "catenate_words":true|false}	
	词汇单元输出	true	\<ALPHANUM\> 类型：It、Its、s、Elastic、ElasticSearch、Search、word、worddelimiter、delimiter　\<NUM\> 类型：7、7、0
		false（默认）	\<ALPHANUM\> 类型：It、s、Elastic、Search、word、delimiter　\<NUM\> 类型：7、7、0
catenate_numbers	描述	包括合并原为一起但拆分了的数字部分	
	用法	"filter":{ "type": "word_delimiter", "catenate_numbers":true|false}	
	词汇单元输出	true	\<ALPHANUM\> 类型：It、s、Elastic、Search、word、delimiter　\<NUM\> 类型：7、770、7、0
		false（默认）	\<ALPHANUM\> 类型：It、s、Elastic、Search、word、delimiter　\<NUM\> 类型：7、7、0

续表

布尔型参数名称	描述、用法及词汇单元输出		
catenate_all	描述	包括合并原为一起但拆分了的字母及数字部分	
	用法	"filter":{"type":"word_delimiter","catenate_all":true|false}	
	词汇单元输出	true	<ALPHANUM> 类型：It、Its、s、Elastic、ElasticSearch770、Search、word、worddelimiter、delimiter <NUM> 类型：7、7、0
		false（默认）	<ALPHANUM> 类型：It、s、Elastic、Search、word、delimiter <NUM> 类型：7、7、0
split_on_case_change	描述	在字母大小写转换时拆分	
	用法	"filter":{"type":"word_delimiter","split_on_case_change":true|false}	
	词汇单元输出	true（默认）	<ALPHANUM> 类型：It、Elastic、Search、word、delimiter <NUM> 类型：7、7、0
		false	<ALPHANUM> 类型：It、ElasticSearch、word、delimiter <NUM> 类型：7、7、0
preserve_original	描述	保留原始字体	
	用法	"filter":{"type":"word_delimiter","preserve_original":true|false}	
	词汇单元输出	true	<ALPHANUM> 类型：It's、It、s、ElasticSearch7.7.0、Elastic、Search、word_delimiter、word、delimiter <NUM> 类型：7、7、0
		False（默认）	<ALPHANUM> 类型：It、s、Elastic、Search、word、delimiter <NUM> 类型：7、7、0
split_on_numerics	描述	在字母和数字边界时拆分	
	用法	"filter":{"type":"word_delimiter","split_on_numerics":true|false}	
	词汇单元输出	true（默认）	<ALPHANUM> 类型：It、s、Elastic、Search7、word、delimiter <NUM> 类型：7、0
		false	<ALPHANUM> 类型：It、s、Elastic、Search、word、delimiter <NUM> 类型：7、7、0
stem_english_possessive	描述	这从所有格形容词中删除了撇号	
	用法	"filter":{"type":"word_delimiter","stem_english_possessive":true|false}	
	词汇单元输出	true（默认）	<ALPHANUM> 类型：It、s、Elastic、Search、word、delimiter <NUM> 类型：7、7、0
		false	与 true 相同，这可能是这个版本中一个错误

5.2 利用内置分析器进行分析

每个内置分析器都包含一个分词器和零个一个或多个词汇单元过滤器。用来练习的文本与 5.1.3 节分词器所使用的相同。以下对所有支持的内置分析器进行测试，默认为 standard 分析器。以下示例指定使用 simple 分析器，其他分析器的结果如表 5-7 所示。由于返回的响应内容信息太多，因此以下仅显示其中部分的信息。

```
curl --request POST 'http://localhost:9200/_analyze?pretty=true' --header
 "Content-Type: application/json" --data $'{"text": "Shanghai Stock Exchange's
 financing balance increased by 566.14 billion yuan, an increase of
0.1%","analyzer":
"simple"}'
{
  "tokens" : [
    {
      "token" : "shanghai",
      "start_offset" : 0,
      "end_offset" : 8,
      "type" : "word",
      "position" : 0
    },
    {
      "token" : "stock",
      "start_offset" : 9,
      "end_offset" : 14,
      "type" : "word",
      "position" : 1
    },
    ...
}
```

表 5-7 内置分析器测试结果

内置分析器	组件		词汇单元输出
	分词器	词汇单元过滤器	
standard	standard	lowercase + stop (默认禁用)	<ALPHANUM> 类型：shanghai、stock、exchange's、financing、balance、increased、by、billion、yuan、an、increase、of <NUM> 类型：566.14、0.1
simple	lowercase		word 类型：shanghai、stock、exchange、s、financing、balance、increased、by、billion、yuan、an、increase、of
whitespace	whitespace		word 类型：shanghai、stock、exchange's、financing、balance、increased、by、566.14、billion、yuan、an、increase、of、0.1%
stop	lowercase	stop（默认 _english_）	word 类型：shanghai、stock、exchange、s、financing、balance、increased、billion、yuan、increase
keyword	keyword		token 类型："Shanghai Stock Exchange's financing balance increased by 566.14 billion yuan, an increase of 0.1%"
pattern	pattern	lowercase+stop	word 类型：shanghai、stock、exchange、s、financing、balance、increased、by、566、14、billion、yuan、an、increase、of、0、1
fingerprint	standard	Lowercase + asciifolding + stop (默认禁用)+ fingerprint	token 类型："0.1 566.14 an balance billion by exchange's financing increase increased of shanghai stock yuan"
Language (示例使用 english)	standard	stop(_english_)+ stemmer(englsih, possessive_english)+ keyword_marker	<ALPHANUM> 类型：shanghai、stock、exchang、financ、balanc、increas、billion、increas <NUM> 类型：566.14、0.1

5.3 利用自定义分析器进行分析

Elasticsearch 提供自定义分析器的方法。第一步是在索引设置（settings）内定义分析器，然后在映射中使用它。以下在 fund_basic 索引内创建一个自定义分析器，它的定义与 standard 分析器相同，我们称为 my_standard。组件已列在表 5-7 中。

提示：请先删除 fund_basic 索引，然后使用以下示例再次创建该索引及同时设置 my_standard 分析器。

```
curl -XPUT 'localhost:9200/fund_basic?pretty=true' --header "Content-Type: application/json" --data $'{"settings":{"analysis":{"analyzer":{"my_standard":{"type":"custom", "tokenizer":"standard", "filter":["lowercase"]}}}}}'
{"acknowledged":true,"shards_acknowledged":true,"index":"fund_basic"}
```

用来练习的文本与 5.2 节相同，产生的词汇单元与表 5-7 中的 standard 项目列出的相同。

提示：由于 my_standard 分析器在 fund_basic 创建，使用时需要指定在 fund_basic 索引下的分析接口（_analyze）。

```
curl --request POST 'http://localhost:9200/fund_basic/_analyze?pretty=true' --header "Content-Type: application/json" --data $'{"text": "Shanghai Stock Exchange\'s financing balance increased by 566.14 billion yuan, an increase of 0.1%","analyzer": "my_standard"}'
```

5.4 归一化处理器

归一化处理器（normalizer）类似于分析器，但只生成单个词汇单元。Elasticsearch 7.5.1 没有提供内置的归一化处理器。如果需要使用则需要自定义，而且只允许使用基于字符的字符过滤器和词汇单元过滤器。定义归一化处理器的方法与定义分析器类似，不同之处在于它使用 normalizer 关键字而不是 analyzer。my_normalizer 的定义只包含 lowercase 词汇单元过滤器。

提示：请先删除 fund_basic 索引，然后使用以下示例再次创建该索引及同时设置 my_normalizer 归一化处理器。

```
curl -XPUT 'localhost:9200/fund_basic?pretty=true' --header "Content-Type: application/json" --data $'{"settings":{"analysis":{"normalizer":{"my_normalizer":{"type":"custom","filter":["lowercase"]}}}}}'
```

用来练习的文本与 5.2 节相同，使用定义好的 my_normalizer 归一化处理器，指定在 fund_basic 索引下的分析接口（_analyze）进行操作，示例如下：

```
curl --request POST 'http://localhost:9200/fund_basic/_analyze?pretty=true' --header "Content-Type: application/json" --data $'{"text": "Shanghai Stock Exchange\'s financing balance increased by 566.14 billion yuan, an increase of 0.1%","normalizer": "my_normalizer"}'
{
  "tokens" : [
    {
      "token" : "shanghai stock exchange's financing balance increased by 566.14 billion yuan, an increase of 0.1%",
      "start_offset" : 0,
      "end_offset" : 97,
      "type" : "word",
      "position" : 0
    }
  ]
}
```

第 6 章 文本分析插件

插件（plugin）是一种以定制方式增强 Elasticsearch 功能的方法。Elasticsearch 带有许多内置的插件。此外，还有许多已被开发可用的自定义插件。在本章仅关注中文分析插件。这是因为原版的内置分析器，对于中文环境不太合适，需要额外安装中文分析插件，然后才可以适合在中文文本环境下工作。

6.1 Elasticsearch 插件是什么

插件以定制的方式扩展基本功能。例如通过添加自定义映射类型、自定义分析器和脚本等。插件必须包含一个名为 plugin-descriptor.properties 文件，用于定义插件内的属性。插件可以包含 JAR 文件、脚本和配置文件。插件的来源可以分为两种类型。第一种是内置的核心插件，它是 Elasticsearch 软件包的一部分，由 Elastic 团队维护。例如 ICU Analysis 插件和 Smart Chinese Analysis 插件是产品随附的核心插件。第二种是由个人或不同社区开发的。例如 IK Analysis 插件是由一位名为 Medcl 的工程师开发的。插件可以在不同领域增强 Elasticsearch 的功能。一些常用插件的类型描述于表 6-1，并对每种类型提供简要说明。

表 6-1 常用插件类型简介

插件类型	描 述
延伸接口（API Extension）	添加用于搜索或映射功能的新接口
警报（Alerting）	添加用于发送警报和监视的新功能
分析（Analysis）	添加用于分析不同语言的文本的新字符过滤器、新分词器、新词汇单元过滤器和新分析器
探索（Discovery）	添加用于发现集群中节点的新机制
摄取（Ingest）	添加用于摄取节点的新功能
管理（Management）	添加用于新的 UI 功能以与 Elasticsearch 服务器进行交互
映射（Mapper）	添加用于映射新的字段数据类型
安全（Security）	添加用于满足企业的安全需求
仓库（Repository）	添加用于快照备份和恢复功能的新仓库
储存（Store）	添加用于新的存储而不是默认的 Lucene 存储

安装插件

执行 elasticsearch-plugin 指令提供列出、安装和从服务器中删除已安装的插件功能。该指令位于下载的文件夹的 bin 目录中。

提示：插件安装后，必须重新启动 Elasticsearch 服务器才会生效。

假设当下位于 bin 目录中，以下进行一些指令练习：

（1）指令扶助

```
./elasticsearch-plugin -help
A tool for managing installed elasticsearch plugins
Commands
--------
list - Lists installed elasticsearch plugins
install - Install a plugin
remove - removes a plugin from Elasticsearch
Non-option arguments:
command
Option                    Description
------                    -----------
-E <KeyValuePair>         Configure a setting
-h, --help                Show help
-s, --silent              Show minimal output
-v, --verbose             Show verbose output
```

（2）安装插件（安装 analysis-icu）

```
./elasticsearch-plugin install analysis-icu
-> Installing analysis-icu
-> Downloading analysis-icu from elastic
[=================================================] 100%
-> Installed analysis-icu
```

（3）列出已安装的插件清单

```
./elasticsearch-plugin list
analysis-icu
```

（4）删除插件（删除 analysis-icu）

```
./elasticsearch-plugin remove analysis-icu
-> removing [analysis-icu]...
```

6.2 使用 ICU 分析插件

ICU Analysis 插件是一组将 Lucene ICU 模块集成到 Elasticsearch。ICU 插件的目的本质上是增加对 Unicode 和全球化的支持，对语言提供更好的文本分割和分析能力。从 Elasticsearch 的角度来看，此插件提供了文本分析中的新组件，描述如表 6-2 所示。

表 6-2 映射数据类型简介

组件	组件名称	描述
字符过滤器	icu_normalizer	将文本转换为唯一的等效字符序列。支持三个可选参数 name、mode 和 unicode_set_filter。name 参数可以是 nfc、nfkc 和 nfkc_cf（默认值）。若 mode 参数是 decompose，可将 nfc 转换为 nfd 或将 nfkc 转换为 nfkd。可以通过指定 unicode_set_filter 参数的字母集合来进行归一化处理
分词器	icu_tokenizer	将一段文本分割成边界上的词汇单元。对语言的支持有泰语、老挝语、中文、日语和韩语

组件	组件名称	描述
词汇单元过滤器	icu_normalizer	将词汇单元归一化处理。支持两个可选参数 name 和 unicode_set_filter。name 参数可以是 nfc、nfkc 和 nfkc_cf（默认值）。可将 unicode_set_filter 参数设置为 Unicodeset 中定义的正则表达式值
	icu_folding	执行 Unicode 标准化。支持可选参数 unicode_set_filter 来折叠字符，这意味着将字符转换为等效的 ASCII
	icu_collation_keyword	用于以按特定语言的顺序对词汇单元进行排序，然后将排序好的词汇单元直接编码连接像成关键字。支持参数 doc_value 可以指定原元是否存储于硬盘、参数 index 可以指定原元是否可搜索（默认为 true）、参数 ignore_above 可以指定原词元允许的长度大小、参数 null_value 可以允许空值（默认为 false）等
	icu_transform	提供转换、归一化处理、脚本音译和双向文本处理。支持参数 id，其值可以指定哪一类操作、文本处理方向等
分析器	icu_analyzer	包含 icu_normalizer 字符过滤器、icu_tokenizer 分词器和 icu_normalizer 词汇单元过滤器

使用 icu_analyzer 分析器示例

对于中文文本，默认的 standard 分析器生成的词汇单元是一个单字符的单词。在大多数情况下，由 icu_analyzer 分析器生成的词汇单元有可能是一个双字符的单词。用来练习的中文文本，提取自 TuShare 的公募基金数据列表数据，显示如下：

恒生湾区 招商银行 中银证券 500ETF 中国工商银行 中金公司 沪深 300ETF 博时

以下示例使用 icu_analyzer 所有参数的默认值。由于返回的响应内容信息太多，因此以下仅显示其中部分的信息。如表 6-3 所示。

提示：如果 icu-analysis 插件已删除，请重新安装并重新启动 Elasticsearch 服务器。

```
curl --request POST 'http://localhost:9200/_analyze?pretty=true' --header
"Content-Type: application/json" --data $'{"text":
"恒生湾区 招商银行 中银证券 500ETF 中国工商银行 中金公司 沪深 300ETF 博时 ","analyzer":
"icu_analyzer"}'
{
  "tokens" : [
    {
      "token" : "恒生",
      "start_offset" : 0,
      "end_offset" : 2,
      "type" : "<IDEOGRAPHIC>",
      "position" : 0
    },
    {
      "token" : "湾",
      "start_offset" : 2,
      "end_offset" : 3,
      "type" : "<IDEOGRAPHIC>",
      "position" : 1
    },
    {
      "token" : "区",
      "start_offset" : 3,
      "end_offset" : 4,
      "type" : "<IDEOGRAPHIC>",
      "position" : 2
    },
    ...
}
```

生成的词汇单元，总共 22 个。其中有 8 个双字符的单词。词汇单元类型由 icu_analyzer 自定义，用于识别数据类型。显然，icu_analyzer 分析器只能识别文本中简单的单词边界。

表 6-3　icu_analyzer 分析器示例结果（所有参数使用默认值）

词汇单元类型	产生的词汇单元列表
<IDEOGRAPHIC>	恒生、湾、区、招、商、银行、中、银、证券、中国、工商、银行、中金、公司、沪、深、博、时
<ALPHANUM>	etf、etf
<NUM>	500、300

在以下示例中，字符过滤器使用 nfkd（mode 参数是 decompose，可将 nfkc 转换为 nfkd）。但是，无法在结果中找到任何差异，只生成相同的词汇单元。

```
curl --request POST 'http://localhost:9200/_analyze?pretty=true' --header "Content-Type: application/json" --data $'{"text": "恒生湾区 招商银行 中银证券 500ETF 中国工商银行 中金公司 沪深 300ETF 博时 ","char_filter":[{"type":"icu_normalizer", "name":"nfkc","mode":"decompose"}], "tokenizer":"icu_tokenizer", "filter":["icu_normalizer"]}'
```

6.3　使用 Smart Chinese 分析插件

Smart Chinese 分析插件将 Lucene 的 Smart Chinese 分析模块集成到 Elasticsearch 中，用于分析中文或中英文混合文本。该分析器使用基于隐马尔可夫模型的概率知识，来找到简体中文文本的最佳分词。它使用的策略首先将输入文本分解为句子，然后对句子进行切分以获得词汇单元。该插件提供了一个称为 smartcn 的分析器，以及一个名为 smartcn_tokenizer 的分词器。请注意，两者均不能使用任何参数进行配置。

6.3.1　安装 Smart Chinese 分析插件

在做练习之前，需要安装 Smart Chinese 分析插件。假设其位于下载的文件夹的 bin 目录中，执行以下指令，并重新启动 Elasticsearch 服务器：

```
./elasticsearch-plugin install analysis-smartcn
-> Installing analysis-smartcn
-> Downloading analysis-smartcn from elastic
[=================================================] 100%
-> Installed analysis-smartcn
```

6.3.2　使用 smartcn 分析器示例

用来练习的文本与 6.2 节相同，以下示例使用 smartcn 分析器，无参数可以使用。由于返回的响应内容信息太多，因此以下仅显示其中部分的信息。如表 6-4 所示。

```
curl --request POST 'http://localhost:9200/_analyze?pretty=true' --header "Content-Type: application/json" --data $'{"text":
"恒生湾区 招商银行 中银证券 500ETF 中国工商银行 中金公司 沪深 300ETF 博时 ","analyzer":"smartcn"}'
    {
      "tokens" : [
        {
          "token" : "恒生",
          "start_offset" : 0,
          "end_offset" : 2,
```

```
      "type" : "word",
      "position" : 0
    },
    {
     "token" : "湾",
     "start_offset" : 2,
     "end_offset" : 3,
     "type" : "word",
     "position" : 1
    },
    ...
}
```

生成的词汇单元，总共 21 个。其中有 8 个双字符单词。词汇单元类型由 smartcn 自定义，数据类型只有 word。与 icu_analyzer 分析器之间的不同在于"招商"是个双字符单词，而"中金"则是两个单字符单词，初步测试结果没有证据表明 smartcn 分析器比较好。

表 6-4 smartcn 分析器示例结果

词汇单元类型	产生的词汇单元列表
word	恒生、湾、区、招商、银行、中、银、证券、500、etf、中国、工商、银行、中、金、公司、沪、深、300、etf、博、时

6.4 使用 IK 分析插件

IK 分析插件属于社区贡献插件类别。它是一个开源项目，提供了一个基于 Java 语言的轻量级中文分词工具包。它将 Lucene IK 分析器集成到 Elasticsearch 中，并支持自定义词典。它可以使用英文和中文，还可以兼容韩文和日文字符。支持热插拔以更新字典内容，而无需重新启动 Elasticsearch 服务器。该插件提供两种类型的分析：细粒度和粗粒度。细粒度的分析包括 ik_max_word 分析器和 ik_max_word 分词器。这样可以生成更多词汇单元，适合部分分词的全文检索。粗粒度分析包括 ik_smart 分析器和 ik_smart 分词器，这将生成较少但更精确的词汇单元，因此适用于短语查询。

6.4.1 安装 IK 分析插件

由于 IK 分析插件不是由 Elastic 团队维护的，因此我们需要从作者的 GitHub 站点上获取。可以先下载或是使用 URL 运行安装。假设其位于下载的文件夹的 bin 目录中，执行以下指令，并重新启动 Elasticsearch 服务器：

```
./elasticsearch-plugin install https://github.com/medcl/elasticsearch-analysis-ik/releases/download/v7.5.1/elasticsearch-analysis-ik-7.5.1.zip
-> Downloading https://github.com/medcl/elasticsearch-analysis-ik/releases/download/v7.5.1/elasticsearch-analysis-ik-7.5.1.zip
[=================================================] 100%
@@@@@@@@@@@@@@@@@@@@@@@@@@@@@@@@@@@@@@@@@@@@@@@@@@
@     WARNING: plugin requires additional permissions     @
@@@@@@@@@@@@@@@@@@@@@@@@@@@@@@@@@@@@@@@@@@@@@@@@@@
* java.net.SocketPermission * connect,resolve
See http://docs.oracle.com/javase/8/docs/technotes/guides/security/permissions.html
for descriptions of what these permissions allow and the associated risks.
```

```
Continue with installation? [y/N]y
-> Installed analysis-ik
```

提示：如果在使用插件时 Elasticsearch 产生有关于拒绝存取的报错提示，原因是缺少 plugin-security.policy 文件，请将下面的内容放入文件中。

```
grant {
  permission java.net.SocketPermission "*", "connect,resolve";
};
```

6.4.2 使用 ik_smart 分析器示例

用来练习的文本与 6.2 节相同，以下示例使用 ik_smartcn 分析器及默认参数。由于返回的响应内容信息太多，因此以下仅显示其中部分的信息。如表 6-5 所示。

```
curl --request POST 'http://localhost:9200/_analyze?pretty=true' --header 
"Content-Type: application/json" --data $'{"text":
"恒生湾区 招商银行中银证券 500ETF 中国工商银行 中金公司 沪深 300ETF 博时","analyzer":
"ik_smart"}'
{
  "tokens" : [
    {
      "token" : "恒生",
      "start_offset" : 0,
      "end_offset" : 2,
      "type" : "CN_WORD",
      "position" : 0
    },
    {
      "token" : "湾",
      "start_offset" : 2,
      "end_offset" : 3,
      "type" : "CN_CHAR",
      "position" : 1
    },
    ......
}
```

生成的词汇单元，总共 16 个。其中有个双字符单词 4 个、四字符单词 1 个和六字符单词 1 个，也出现混合数字和英文字母的单词。词汇单元类型由 ik_smartcn 自定义，数据类型有 CN_WORD、CN_CHAR 和 LETTER 等。初步测试结果感觉 ik_smart 分析器比较好，能够产生更少但更精确的词汇单元。但是，对于某些词汇单元无法像 smartcn 分析器一样能够识别，例如中金和沪深。混合数字和英文单词的优点（例如 500etf 和 300etf）也没有很明显。

表 6-5 ik_smart 分析器示例结果

词汇单元类型	产生的词汇单元列表
CN_WORD	恒生、招商银行、中银、证券、中国工商银行、公司
CN_CHAR	湾、区、中、金、沪、深、博、时
LETTER	500etf、300etf

6.5 使用 HanLP 分析插件

HanLP 分析插件是由一系列模型和算法组成的 Java 工具包，目的是提供自然语言处理功能，已经被广泛用于知名软件平台如 Lucene、Elasticsearch、Solr 等。HanLP 支持中文分词、命名实体识别如人名、地名、机构组织等。

杭州市新华智云科技有限公司（KennFalcon），是一家大数据人工智能科技公司，提供 Elasticsearch 非常优秀的中文文本分析插件，称为 elasticsearch-analysis-hanlp，相关详细介绍，请参考官网。在本节中，我们将展示 KennFalcon 的 HanLP 插件。

6.5.1　安装 elasticsearch-analysis-hanlp 分析插件

如同 IK 分析插件一样，不是由 Elastic 团队维护，需要先下载或是使用 URL 运行安装。假设其位于下载的文件夹的 bin 目录中，执行以下指令，并重新启动 Elasticsearch 服务器：

```
./elasticsearch-plugin install
https://github.com/KennFalcon/elasticsearch-analysis-hanlp/releases/download/
v7.5.1/elasticsearch-analysis-hanlp-7.5.1.zip
-> Downloading https://github.com/KennFalcon/elasticsearch-analysis-hanlp/
releases/download/v7.5.1/elasticsearch-analysis-hanlp-7.5.1.zip
[=================================================] 100%
@@@@@@@@@@@@@@@@@@@@@@@@@@@@@@@@@@@@@@@@@@@@@@@@@@
@     WARNING: plugin requires additional permissions     @
@@@@@@@@@@@@@@@@@@@@@@@@@@@@@@@@@@@@@@@@@@@@@@@@@@
* java.io.FilePermission <<ALL FILES>> read,write,delete
* java.lang.RuntimePermission getClassLoader
* java.lang.RuntimePermission setContextClassLoader
* java.net.SocketPermission * connect,resolve
* java.util.PropertyPermission * read,write
See http://docs.oracle.com/javase/8/docs/technotes/guides/security/permissions.html
for descriptions of what these permissions allow and the associated risks.

Continue with installation? [y/N]y
-> Installed analysis-hanlp
```

6.5.2　使用 hanlp 分析器示例

用来练习的文本与 6.2 节相同，以下示例使用 hanlp 分析器及默认参数。由于返回的响应内容信息太多，因此以下仅显示其中部分的信息。如表 6-6 所示。

```
curl --request POST 'http://localhost:9200/_analyze?pretty=true' --header
"Content-Type: application/json" --data $'{"text":
"恒生湾区 招商银行中银证券 500ETF 中国工商银行 中金公司 沪深 300ETF 博时","analyzer": "hanlp"}'
  {
    "tokens" : [
      {
        "token" : "恒生",
        "start_offset" : 0,
        "end_offset" : 2,
        "type" : "nz",
        "position" : 0
      },
      {
        "token" : "湾区",
        "start_offset" : 2,
        "end_offset" : 4,
        "type" : "ns",
        "position" : 1
      },
      ......
  }
```

生成的词汇单元，总共 15 个。其中有个双字符单词 4 个、四字符单词 2 个和六字符单词 1 个。数字和英文单词分离而不会结合，初步测试结果发现 hanlp 分析器能够产生更少也更精确的词汇单元。但是，某些词汇单元也无法识别，例如沪深。数字和英文单词会分别生成词汇单元，这样对搜索显得更有用，例如 300、500 和 ETF。hanlp 分析器可以识别很多词汇单元的主要原因是因为词典文件中有很多已经定义的词汇，相关详细介绍，请参考官网。

表 6-6　hanlp 分析器示例结果

词汇单元类型	产生的词汇单元列表
nz	恒生
ns	湾区
ntcb	招商银行、中国工商银行
nz	中银
nis	证券
m	500、300
nx	ETF、ETF
ntc	中金公司
b	沪
a	深
ag	博
qt	时

6.5.3　使用 hanlp 自定义词典热更新

7.5.1 版本的 HanLP 分析插件提供词典热更新（每个节点都需要安装插件），可以按照以下步骤执行：

（1）假设当下位于下载的文件夹的 plugins/analysis-hanlp/data/dictionary/custom 目录中，添加自定义词典名为 mydic.txt 的文件。文件内容只有两行文字，一行是沪深，另一行是博时。

```
沪深
博时
```

（2）假设当下位于下载的文件夹的 config/analysis-hanlp 目录中，修改 hanlp.properties 文件中 CustomDictionaryPath 的设定，添加自定义词典文件 mydic.txt 路径（没有其他改变）如下：

```
CustomDictionaryPath=data/dictionary/custom/CustomDictionary.txt; mydict.txt;
ModernChineseSupplementaryWord.txt; ChinesePlaceName.txt ns; PersonalName.txt;
OrganizationName.txt; ShanghaiPlaceName.txt ns;data/dictionary/person/nrf.txt nrf;
```

（3）等待 1 分钟后，词典将自动加载提示：如果没有效果，新增词典文件第一次可能需要重新启动 Elasticsearch 服务器。

使用 6.5.2 节相同的示例和文本，如表 6-7 所示。生成的词汇单元，除了双字符单词沪深和博时外，其他与表 6-6 相同。根据目前所有测试结果显示，使用 hanlp 分析器和自定义词典，能够产生更少及更精确的词汇单元，供中文文本分析使用。

表 6-7　hanlp 分析器和自定义词典示例结果

词汇单元类型	产生的词汇单元列表
nz	恒生
ns	湾区
ntcb	招商银行、中国工商银行
nz	中银
nis	证券
m	500、300
nx	ETF、ETF
ntc	中金公司
n	沪深、博时

6.5.4　使用 hanlp 分词器自定义分析器

Elasticsearch-analysis-hanlp 分析插件包括了几种类型的分词器，基于 HanLP 提供大部分的分词方式，例如 hanlp（默认分词器）、hanlp_standard（标准分词器）、hanlp_index（索引分词）、hanlp_nlp（NLP 分词）等。Elasticsearch-analysis-hanlp 分析插件在支持的分词器上，提供了一系列布尔型参数。如果需要采用这些参数，enable_custom_config 参数需要设定为 true。表 6-8 中对这些参数进行简要的描述。

表 6-8　HanLP 支持的分词器可采用的参数

参数名称	简要的描述
enable_custom_config	启用自定义配置
enable_index_mode	启用索引模式
enable_number_quantifier_recognize	启用数字量词识别
enable_custom_dictionary	启用自定义词典
enable_translated_name_recognize	启用翻译名称识别
enable_japanese_name_recognize	启用日本人名名称识别
enable_organization_recognize	启用机构识别
enable_place_recognize	启用地名识别
enable_name_recognize	启用中国人名识别
enable_traditional_chinese_mode	启用繁体中文
enable_stop_dictionary	启用停用词
enable_part_of_speech_tagging	启用语音标签
enable_remote_dict	启用远程词典
enable_normalization	启用归一化处理
enable_offset	启用偏移计算处理

如何创建一个自定义分析器,曾在 5.3 节中描述。以下示例使用 hanlp 分词器及两个参数,enable_custom_config 为 true 及 enable_custom_dictionary 为 true。此外,再添加一个 unique 词汇单元过滤器删除重复项,并将分析器称之为 my_hanlp_analyzer。

提示:本书在后续章节的示例,大部分使用 my_hanlp_analyzer 分析器,用于分析来自基金的信息。

以下在 fund_basic 索引的设置内创建这个 my_hanlp_analyzer 分析器。首先删除 fund_basic 索引,然后使用以下示例在创建 fund_basic 索引时同时在设置(settings)内定义 my_hanlp_tokenizer 分析器。

```
curl -XPUT localhost:9200/fund_basic --header "Content-Type: application/json" --data $'{"settings":{"analysis":{"analyzer":{"my_hanlp_analyzer":{"tokenizer":"my_hanlp_tokenizer", "filter":["unique"]}}, "tokenizer":{"my_hanlp_tokenizer":{"type":"hanlp", "enable_custom_config":true, "enable_custom_dictionary":true}}}}}'
    {"acknowledged":true,"shards_acknowledged":true,"index":"fund_basic"}
```

用来练习的文本与 6.2 节相同,使用 my_hanlp_analyzer 分析器时需要指定在 fund_basic 索引下的分析接口(_analyze)。由于返回的响应内容信息太多,因此以下仅显示其中部分的信息。如表 6-9 所示。

```
curl --request POST 'http://localhost:9200/fund_basic/_analyze?pretty=true'
 --header "Content-Type: application/json" --data $'{"text":
"恒生湾区 招商银行 中银证券 500ETF 中国工商银行 中金公司 } 沪深 300ETF 博时","analyzer":
"my_hanlp_analyzer"}'
{
  "tokens" : [
    {
      "token" : "恒生",
      "start_offset" : 0,
      "end_offset" : 2,
      "type" : "nz",
      "position" : 0
    },
    {
      "token" : "湾",
      "start_offset" : 2,
      "end_offset" : 3,
      "type" : "ng",
      "position" : 1
    },
    ......
}
```

生成的词汇单元,总共 19 个。其中有 8 个双字符单词,和在定制词典内多字符单词,如招商银行、中国工商银行和中金公司。没有其他超过 2 个字符的单词。

表 6-9 my_hanlp_analyzer 分析器示例结果

词汇单元类型	产生的词汇单元列表
nz	恒生
ns	湾区
ntcb	招商银行、中国工商银行
nz	中银
nis	证券

续表

词汇单元类型	产生的词汇单元列表
m	500、300
nx	ETF
ntc	中金公司
n	沪深、博时

6.5.5　简评 hanlp 分词器的对称性

在测试过程中，我们发现当汉字与数字，或是英文字混合时，hanlp 分词器的对称特性不一致。例如中银 ETF（汉字在前）和 ETF 中银（汉字在后）会产生不同的词汇单元。用户必须意识到这一点，否则搜索结果将会与预想的不一致。以下使用 800、中银和 ETF 混合的文本测试几种情况，结果显示其对称特性不一致。汉字在英文词汇之前，和在英文词汇之后会生成不同的词汇单元。

（1）若是汉字在数字之前，中银 800 的文本产生三个词汇单元，中、银和 800。

```
curl --request POST http://localhost:9200/fund_basic/_analyze?pretty=true
--header "Content-Type: application/json" --data $'{"text" : "中银800",
"analyzer":"hanlp"}'
```

（2）若是汉字在数字之后，800 中银的文本产生三个词汇单元，800、中和银。

```
curl --request POST http://localhost:9200/fund_basic/_analyze?pretty=true
--header "Content-Type: application/json" --data $'{"text" : "800中银",
"analyzer":"hanlp"}'
```

（3）若是汉字在英文字之前，中银 ETF 的文本产生两个词汇单元，中银和 ETF。

```
curl --request POST http://localhost:9200/fund_basic/_analyze?pretty=true
--header "Content-Type: application/json" --data $'{"text" : "中银ETF",
"analyzer":"hanlp"}'
```

（4）若是汉字在英文字之后，ETF 中银的文本产生三个词汇单元，ETF、中和银。

```
curl --request POST http://localhost:9200/fund_basic/_analyze?pretty=true
--header "Content-Type: application/json" --data $'{"text" : "ETF中银",
"analyzer":"hanlp"}'
```

（5）若是英文字在数字之前，ETF800 的文本产生两个词汇单元，ETF 和 800。

```
curl --request POST http://localhost:9200/fund_basic/_analyze?pretty=true
--header "Content-Type: application/json" --data $'{"text" : "ETF800",
"analyzer":"hanlp"}'
```

（6）若是英文字在数字之后，800ETF 的文本产生两个词汇单元，800 和 ETF。

```
curl --request POST http://localhost:9200/fund_basic/_analyze?pretty=true
--header "Content-Type: application/json" --data $'{"text" : "800ETF",
"analyzer":"hanlp"}'
```

6.6　使用 Aliws 分析插件

Aliws 分析插件是阿里云 Elasticsearch 的内置插件，不向公众开放。因此，除非用户使用阿里云 Elasticearch 服务，否则无法使用。该分析插件集成了 aliws 分析器和 aliws_ tokenizer 分词器，提供业界非常优秀的 Elasticsearch 中文文本分析插件，相关详细介绍，请参考官网。

依据 Elastic 中文社区深圳 Meetup 一文的报道，文本若为南京市长江大桥一词，aliws_tokenizer 分词器可以产生南京、市、长江和大桥四个词汇单元以供搜索。若使用相同的输入文本，hanlp 分词器则产生南京市和长江大桥两个词汇单元。如果用户使用南京桥一词来检索该文本，那么使用 hanlp 分析器将无法检索到任何结果，而使用 aliws 分析器则搜索成功，因此该报告称 Aliws 分析插件具有很高的查全率。故此若使用免费的 hanlp 分析插件，必须注意到这一点。也许对 hanlp 分析器进行定制，通过增加词汇单元过滤器，对产生的词汇单元衍生更多比较认可的词汇单元来解决这个问题。

然而，若用户使用桥一字来检索该文本，则依据该文报道所产生的四个词汇单元，也将无法检索到任何结果。

第 7 章 搜索数据和查询表达式

从前几章中学到的知识是为搜索做好了准备,包括索引设置、映射和分析器,然后为文档建立索引。学习 Elasticsearch 最终目标之一是执行查询并从索引中获得高质量的搜索结果。Search 接口提供搜索查询并返回与查询匹配的搜索命中结果。运行搜索的基本方法有两种,一种是简单发送带有 URL 参数的搜索请求,另一种是使用查询表达式(Query DSL),在请求主体执行更复杂条件的搜索。在开始研习搜索操作之前,需要按以下用例准备搜索数据。

7.1 索引样本文件

Tushare 提供的公募基金数据,包括约 1130 只场内基金和约 10 785 只场外基金(包括退市的基金)。场外基金是指通过银行柜台、券商或是第三方销售平台购买,而场内基金则指需要开证券交易账户认购。由于 Tushare 需要用户赚取积分才可以调取。因此,本章仅准备了 50 个场内基金样品进行测试。

本章提供了 fund_basic_bulk.json、fund_basic_bulk_index.sh 和 my_hanlp_analyzer.json 三个文件。与 4.2.5 节里的显式映射内容比较,在 my_hanlp_analyzer.json 文件的字段 management、name、ts_code、trustee 和 custodian 原来只用文本映射类型,而在本章则从原来的默认分析器改用了 my_hanlp_analyzer 分析器,以下显示文件内容。提示:这些字段使用 my_hanlp_analyzer 分析器的原因只是为了演示文本分析测试。

```
{
  "settings" : {
    "analysis": {
      "analyzer": {
        "my_hanlp_analyzer": {"tokenizer": "my_hanlp_tokenizer", "filter":["unique"]}
      },
      "tokenizer": {
        "my_hanlp_tokenizer": {
          "type": "hanlp", "enable_custom_config": true, "enable_custom_dictionary": false
        }
      }
    }
  },
  "mappings" : {
    "dynamic": false,
    "properties": {
      "benchmark": {"type": "keyword"},
      "status": {"type": "keyword"},
```

```
            "fund_type": {"type": "keyword"},
            "invest_type": {"type": "keyword"},
            "type": {"type": "keyword"},
            "market": {"type": "keyword"},
            "c_fee": {"type": "float"},
            "issue_amount": {"type": "float"},
            "m_fee": {"type": "float"},
            "min_amount": {"type": "float"},
            "p_value": {"type": "float"},
            "duration_year": {"type": "float"},
            "exp_return": {"type": "float"},
            "issue_date": {"type": "date", "format": "yyyyMMdd"},
            "list_date": {"type": "date", "format": "yyyyMMdd"},
            "purc_startdate": {"type": "date", "format": "yyyyMMdd"},
            "redm_startdate": {"type": "date", "format": "yyyyMMdd"},
            "found_date": {"type": "date", "format": "yyyyMMdd"},
            "due_date": {"type": "date", "format": "yyyyMMdd"},
            "delist_date": {"type": "date", "format": "yyyyMMdd"},
            "management": {"type": "text", "fields": {"keyword": {"type":"keyword"}},
                    "analyzer":"my_hanlp_analyzer"},
            "name": {"type": "text", "fields": {"keyword": {"type": "keyword"}},
                 "analyzer":"my_hanlp_analyzer"},
            "ts_code": {"type": "text", "fields": {"keyword": {"type": "keyword"}},
                  "analyzer":"my_hanlp_analyzer"},
            "trustee": {"type": "text", "fields": {"keyword": {"type": "keyword"}},
                    "analyzer":"my_hanlp_analyzer"},
             "custodian": {"type": "text", "fields": {"keyword": {"type": "keyword"}},
"analyzer":"my_hanlp_analyzer"}
     }}}
```

fund_basic_bulk_index.sh 是个 bash 文件，运行执行以下任务：

（1）删除 fund_basic 索引。

（2）使用在 my_hanlp_analyzer.json 内定义的设置和分析器创建 fund_basic 索引。

（3）使用 fund_basic_bulk.json 内编写好的批量处理文档接口指令，进行指定的 50 个场内基金的文档索引。

执行 fund_basic_bulk_index.sh 的示例显示部分结果如下：

```
./fund_basic_bulk_index.sh
{"acknowledged":true}
{"acknowledged":true,"shards_acknowledged":true,"index":"fund_basic"}
{
  "took" : 243,
  "errors" : false,
  "items" : [
    {
      "index" : {
        "_index" : "fund_basic",
        "_type" : "_doc",
        "_id" : "C9brxHIBE8603ZMK0w4v",
        "_version" : 1,
        "result" : "created",
        "_shards" : {
          "total" : 2,
          "successful" : 1,
          "failed" : 0
        },
        "_seq_no" : 0,
        "_primary_term" : 1,
        "status" : 201
      }
```

```
        },
        ……
}
```

7.2 基础搜索接口

简单的搜索操作，可以使用 POST 或 GET 请求，并将查询字符串用作 URL 参数来执行 _search 接口。可使用逗号分隔多个索引，同时进行搜索操作。接口请求语法如下：

```
GET /<索引名称,索引名称>/_search?查询字符串
GET /_search?查询字符串
POST /<索引名称,索引名称>/_search?查询字符串
POST /_search?查询字符串
```

7.2.1 通过 URI 进行搜索

URI 搜索使用带有 URL 参数的查询字符串执行搜索操作。常用的 URL 参数如下：

- from 表示返回的搜索结果从第几项开始（默认为 0）。
- size 表示返回的搜索结果总项数（默认为 10）。
- sort 表示对搜索结果进行排序（默认为升序）。可指定 sort=field：asc 或 sort=field：desc。也可对得分结果（_score）进行排序。
- _source 表示默认情况下返回文档的元数据字段内容。如果设为 false，则不返回。

以下测试示例使用参数 from=2，将返回的搜索结果控制为从第 2 项开始、使用参数 size=2，来控制搜索结果仅返回 2 项、使用参数 sort=ts_code.keyword：desc 对基金代码进行降序排序、使用参数 _source=false 禁止该字段返回。请求的搜索结果仅返回 2 个基金，基金代码为 515930.SH 和 515890.SH，排序是 515930.SH 在前，515890.SH 在后，并没有返回 _source 字段。读者可以在请求中清除 from 参数以验证 from 的效果。

```
curl --request GET 'http://localhost:9200/fund_basic/_search?pretty=true&from=2
&size=2&sort=ts_code.keyword:desc&_source=false'
{
    ……
    "hits" : {
      "total" : {
        "value" : 50,
        "relation" : "eq"
      },
      "max_score" : null,
      "hits" : [
        {
          "_index" : "fund_basic",
          "_type" : "_doc",
          "_id" : "EdbrxHIBE8603ZMK0w4v",
          "_score" : null,
          "sort" : [
            "515930.SH"
          ]
        },
        {
          "_index" : "fund_basic",
          "_type" : "_doc",
          "_id" : "G9brxHIBE8603ZMK0w4v",
          "_score" : null,
```

```
      "sort" : [
        "515890.SH"
      ]
    }
    ......
}
```

返回的响应内容，简要描述如下：

- took 表示 Elasticsearch 执行请求所花费的毫秒数，但不包括发送请求给 Elasticsearch、序列化 JSON 响应及发送给客户端所需的时间。
- timeout 表示等待响应的时间，默认为无超时。
- hits 表示返回的搜索结果。
- hits.total.value 表示返回的文档总数。
- hits.total.relation 表示返回的文档是否准确。eq 代表准确而 gte 代表仅仅匹配。
- hits.max_score 表示返回的文档其中的最高得分。
- hits.hits 表示返回的搜索结果的文档数组。对于每个文档，默认包括以下字段：
  ① _index 表示文档的索引名称。
  ② _type 表示文档的索引类型。
  ③ _id 表示文档的标识符。
  ④ _score 表示文档的得分。
  ⑤ _source 表示文档的元数据字段。

以下继续对常用的 URL 搜索参数进行说明。

1. 参数 q 及 default_operator

参数 q 可指定遵循 DSL 语法的查询句子。当查询中有多个条件时，可用 default_operator 参数指定布尔运算符（默认为 or 或 |）对条件进行分组。以下使用参数 q 和参数 default_operator=AND 的示例。该查询用于限制返回的文档为字段 name 必须包含 50 和 ETF 两个词，而且使用参数 _source 只显示 ts_code 和 name 两个字段。请求的搜索结果从 fund_basic 中索引返回 3 个基金，基金简称分别为 5G50ETF、医药 50ETF 基金和红利低波 50ETF。

```
curl --request GET 'http://localhost:9200/fund_basic/_search?pretty=true&q=name
:50ETF&default_operator=AND&_source_includes=name,ts_code'
{
  ......
  "hits" : {
    "total" : {
      "value" : 3,
      "relation" : "eq"
    },
    "max_score" : 2.6059227,
    "hits" : [
      {
        "_index" : "fund_basic",
        "_type" : "_doc",
        "_id" : "fdaixXIBE8603ZMklQ7G",
        "_score" : 2.6059227,
        "_source" : {
          "ts_code" : "159811.SZ",
          "name" : "5G50ETF"
        }
      },
```

```
        {
          "_index" : "fund_basic",
          "_type" : "_doc",
          "_id" : "hdaixXIBE8603ZMKlQ7G",
          "_score" : 2.6059227,
          "_source" : {
            "ts_code" : "515950.SH",
            "name" : "医药50ETF基金"
          }
        },
        {
          "_index" : "fund_basic",
          "_type" : "_doc",
          "_id" : "nNaixXIBE8603ZMKlQ7G",
          "_score" : 2.3196278,
          "_source" : {
            "ts_code" : "515450.SH",
            "name" : "红利低波50ETF"
          }
        ......
}
```

2. 参数 explain

参数 explain 提供了详细的描述来解释相关性得分的计算。以下使用参数 q 查询必须包含字符串 88 的基金示例。请求的搜索结果仅返回 1 个基金，基金简称为创精选 88。显示得分（_score）的计算方法非常详细，有兴趣的读者可以参考官网。简而言之，得分的计算主要根据 4 个指标：

- tf 表示文档中要搜索的词元出现的频率。
- idf 表示档中要搜索的词元出现的逆向文档频率。
- boost 指提升权重的词元在该文档的得分。
- norm 表示字段长度归一值，字段中词汇单元的总数平方根的倒数。

提示：返回的响应只包含 tf、idf 和 boost。原因是没有使用归一化（normalizer）处理。由于返回的响应内容信息太多，因此以下仅显示其中部分的信息。

```
curl --request GET 'http://localhost:9200/fund_basic/_search?pretty=true&q=88&_source_includes=name,ts_code&explain=true'
{
  ......
  "_score" : 3.446918,
"_source" : {
    "ts_code" : "159804.SZ",
    "name" : "创精选88"
},
  "_explanation" : {
    "value" : 3.446918,
    "description" : "max of:",
    "details" : [
      {
        "value" : 3.446918,
        "description" : "weight(name:88 in 28) [PerFieldSimilarity], result of:",
        "details" : [
          {
            "value" : 3.446918,
            "description" : "score(freq=1.0), product of:",
            "details" : [
              {
                "value" : 2.2,
```

```
              "description" : "boost",
              "details" : [ ]
            },
            {
              "value" : 3.5263605,
              "description" : "idf, computed as log(1 + (N - n + 0.5) / (n + 0.5)) from:",
              ......
            },
            {
              "value" : 0.44430536,
              "description" : "tf, computed as freq / (freq + k1 * (1 - b + b * dl / avgdl)) from:",
              ......
            },
            ......
          }
```

3. 参数 analyzer

默认情况下，在索引时和搜索时会使用相同的分析器，但也可于特殊情况下，在搜索时指定参数 analyzer，使用其他文本分析器。这参数是用于分析搜索时输入的查询字符串的。以下测试使用 7.2.1 节项目 1 相同的示例，但改用关键字（keyword）分析器以代替 my_hanlp_analyzer 分析器。当使用关键字分析器用于输入查询字符串后，会找不到任何搜索结果。这是由于关键字分析器使用 50ETF 进行整个词的搜索比对，而不是全文搜索做部分的比对。

提示：搜索时使用其他分析器，可能会对结果产生负面影响，从而导致出乎意料意外的搜索结果。

```
    curl --request GET 'http://localhost:9200/fund_basic/_search?pretty=true&q=50ETF&
    default_operator=AND&_source_includes=name,ts_code&analyzer=keyword'
    {
      "took" : 2,
      "timed_out" : false,
      ......
      "hits" : {
        "total" : {
          "value" : 0,
          "relation" : "eq"
        },
        "max_score" : null,
        "hits" : [ ]
      }
    }
```

关于 URI 其他常用参数可如表 7-1 所示，留给有兴趣的读者进行实践。

表 7-1　URI 其他常用参数

参数名称	描述
allow_partial_search_results	是否在失败的情况下返回部分结果，默认为 true
analyzer	使用分析器名称
batched_reduce_size	减少了在协调节点中收集的临时结果的数量，以减少内存使用量
df	指定要搜索的默认字段
docvalue_fields	在搜索结果中，返回以逗号分隔的指定字段，以列表方式显示
lenient	忽略搜索时和索引时的数据类型不匹配错误。默认为 false
max_concurrent_shard_requests	限制并发分片请求的数量，默认为 5

续表

参数名称	描 述
preference	指定节点或分片对其执行操作，默认情况下为随机
routing	执行搜索时，可以通过提供路由参数来控制要搜索哪些分片
search_type	定义搜索操作的方式。可选 query_then_fetch 或 dfs_query_then_fetch
sort	返回的搜索结果的字段排序，可用逗号分隔的指定键值对列表＜字段：asc|desc＞
_source, _source_include, _source_exclude	输出文档的原始内容，可选全部包括（_source=true）、部分包括（_source_includes=a,b,c），或过滤出特定字段（_source_excludes=a,b,c）。其中 a、b、c 为字段
stats	用于日志记录和统计目的
stored_fields	检索在映射中标记为 stored 的那些字段。可用逗号分隔的指定字段列表
suggest_field	指定用于建议的字段
suggest_mode	指定建议模式。可用选项 always、missing（默认）和 popular
suggest_size	允许返回的建议总数
suggest_text	是否返回建议的源文本
terminate_after	指定在分片中允许收集的文档数，默认为无限制
timeout	指定完成搜索操作所允许的时间。默认值为无超时
track_scores	追踪评分，即使它们不用于排序。默认为 false
track_total_hits	追踪匹配文档的总数。可以使用 false 值禁用它。默认为 10000
typed_keys	是否应在聚合名称或建议名称的前面加上各自的类型作为前缀
version	是否返回文档版本，默认为 true

7.2.2 通过请求主体（request body）进行搜索

对于复杂的查询，可以使用查询表达式编写 JSON 查询句子，通过 GET 或 POST 请求执行搜索。如果必须使用 GET 请求，可将请求主体内容设置为 source 的 URL 参数。以下测试使用 7.2.1 节项目 1 相同的示例，但改用 POST 请求而查询表达式在请求主体中。请求的搜索结果同前例。仅返回 2 个基金，代码为 515930.SH 和 515890.SH，并没有返回 _source 字段。

```
curl --request POST http://localhost:9200/fund_basic/_search?pretty=true
  --header "Content-Type: application/json" --data $'{"from":2, "size":2,
"sort":[{"ts_code.keyword":"desc"}], "_source":false}'
```

除了参数 search_type、request_cache 和 allow_partial_search_results 只可用于 URI 参数外，大部分 URI 参数可用于搜索的请求主体，另外可用参数 query 代替 q，及用参数 default_field 代替 df。以下将讨论属于通过请求主体进行搜索使用的一些重要参数。

1. 参数 collapse

参数 collapse（折叠）提供了一种根据不同字段值对搜索结果进行分组的方法，仅收集每个字段值的第一个文档，并将其返回到响应正文中。这种显示方式看起来像折叠。以下使用此参数对 50 个场内基金样品进行折叠投资风格字段测试，结果是只有 3 种投资风格。分别是被动指数型、增强指数型和灵活配置型。由于返回的响应内容信息太多，因此以下仅显示其中部分的信息。

```
curl --request POST http://localhost:9200/fund_basic/_search?pretty=true
  --header "Content-Type: application/json" --data $'{"collapse": {"field":"invest_
type"}, "_source":false}'
  {
```

```
......
    "hits" : {
      "total" : {
        "value" : 50,
        "relation" : "eq"
      },
      "max_score" : null,
      "hits" : [
        {
          "_index" : "fund_basic",
          "_type" : "_doc",
          "_id" : "cNaixXIBE8603ZMKlQ7G",
          "_score" : 1.0,
          "fields" : {
            "invest_type" : [
              "被动指数型"
            ]
          }
        },
        {
......
"fields" : {
            "invest_type" : [
              "增强指数型"
            ]
          }
        },
        {
          ......
          "fields" : {
            "invest_type" : [
              "灵活配置型"
            ]
          }
        }
      ......
}
```

2. 参数query

参数query（查询）提供使用查询表达式（DSL）定义查询的功能。此特定语言非常广泛和复杂，在7.3节进阶搜索中将作详细介绍。在这一小节中，仅演示一个简单的查询。示例与在7.2.1节项目1类似，区别是使用参数query下的query_string方式，通过请求主体进行搜索包含50和ETF两个词的基金简称。由于返回的响应内容信息太多，因此以下仅显示其中部分的信息。搜索结果返回3个基金，基金简称为5G50ETF、医药50ETF基金和红利低波50ETF。

```
curl --request POST http://localhost:9200/fund_basic/_search?pretty=true
--header "Content-Type: application/json" --data $'{"query": {"query_string" :
{"query" : "50ETF","default_field" : "name", "default_operator":"AND"}}, "_
source":{"includes":["name","ts_code"]}}'
{
    ......
    "hits" : {
      "total" : {
        "value" : 3,
        "relation" : "eq"
      },
      "max_score" : 2.6059227,
      "hits" : [
```

```
        {
          ......
          "_score" : 2.6059227,
          "_source" : {
            "ts_code" : "159811.SZ",
            "name" : "5G50ETF"
          }
        },
        {
          ......
          "_score" : 2.6059227,
          "_source" : {
            "ts_code" : "515950.SH",
            "name" : "医药50ETF基金"
          }
        },
        {
          ......
          "_score" : 2.3196278,
          "_source" : {
            "ts_code" : "515450.SH",
            "name" : "红利低波50ETF"
          }
        }
      ......
}
```

3. 参数 rescore

此参数提供了一种除默认评分方案以外的其他评分策略。基本上，根据基于 window_size 参数（默认为 10）的顶部匹配文档进行重新排序，重新计算评分结果并提供另一个分数。除对 _score 字段进行反向排序外，不支持其他排序。以下测试使用项目 2 的示例，并添加一个评分策略 rescore_query，将参数 window_size 设置为 3，对医药这个词在基金简称重新进行评分。由于返回的响应内容与项目 2 相似，但是由于重新评分而使排序不同，而医药 50ETF 基金的得分（_score）从 2.605 922 7 增加到 5.627 412。

```
curl --include --request POST http://localhost:9200/fund_basic/_search?pretty=true --header "Content-Type: application/json" --data
$'{"query": {"query_string" : {"query" : "50ETF","default_field" : "name","default_operator":"AND"}}, "rescore":{"window_size":5, "query": {"rescore_query": {"query_string":{"query" : 医药,"default_field" :"name"}}}}, "_source":{"includes":["name","ts_code"]}}'
{
  ......
  "max_score" : 5.627412,
  "hits" : [
    {
      ......
      "_score" : 5.627412,
      "_source" : {
        "ts_code" : "515950.SH",
        "name" : "医药50ETF基金"
      }
    },
    {
      ......
      "_score" : 2.6059227,
      "_source" : {
        "ts_code" : "159811.SZ",
        "name" : "5G50ETF"
      }
```

```
        },
        {
          ......
            "_score" : 2.3196278,
            "_source" : {
              "ts_code" : "515450.SH",
              "name" : "红利低波 50ETF"
            }
          ......
        }
```

4. 参数 sort

此参数提供了许多不同的方式进行排序,可按一个字段、多个字段、一个数组、一个对象或一个嵌套对象进行排序。可以使用原始数据类型或在 _geo_distance(地理距离)中对字段进行排序。并且还可以对元数据 _score 及 _doc 进行排序。

(1)排序顺序选项如下:
- asc 表示升序(默认)。
- desc 表示降序。

对字段 field1 进行降序排序,示例如下:

```
"sort":[{"field1":{"order":"desc"}}]
```

(2)对数组或多值字段排序,可按模式选项控制所属的文档进行排序。模式选项如下:
- min 表示从数组或多值字段中选取最小值进行排序。
- max 表示从数组或多值字段中选取最大值进行排序。
- sum 表示选取总和进行排序。
- avg 表示选取平均数进行排序。
- median 表示选取中位数进行排序。

假设字段 field1 为数字类型数组,需要对其中位数进行降序排序,示例如下:

```
"sort":[{"field1":{"order":"desc", "mode":"median"}}]
```

(3)如何处理在不同索引中同一字段,但不同数字类型的冲突。假设字段 field1 在索引 index1 中为浮点数类型,而 field1 在索引 index2 中为 long 类型。进行搜索多个索引和排序操作时,可设定 field1 为 long 类型,强制转换浮点数类型。示例如下:

```
"sort":[{"field1":{"order":"desc", "numeric_type":"long"}}]
```

(4)如何处理缺少排序字段的文档?如果文档缺少排序字段,可按模式选项控制所属的文档进行排序。 模式选项如下:
- _last 表示选择排序列表的最后一个值(默认)。
- _first 表示选择排序列表的第一个值。
- 自订值。

假设字段起点金额(min_amount)有可能在文档中缺少,而空值将替换为第一个值。示例如下:

```
"sort":[{"min_amount":{"order":"asc","missing":"_first"}}]
```

(5)如何处理排序文档中有未映射的字段。默认情况下,搜索没有映射的字段导致请求将失败。假设字段 field1 的类型为 long,但是在多个索引搜索操作中,可能某个索引没有映射时需要忽略。为避免这个问题,可用 unmapped_type 选项忽略该字段。

```
"sort":[{"field1":{"order":"asc","unmapped_type":"long"}}]
```

5. 参数 scroll

使用 scroll(游标查询)参数,可使搜索上下文保持活性,就像与给定时间戳对应的快照一样,并提供前向游标操作。初始请求的结果中给定了一个标识符 _scroll_id,以便获取下一批结果。如果没有更多结果,则将返回一个空数组。可用参数 size 来限制批次中返回的匹配结果总数。以下示例执行了三次滚动搜索以获取所有文档。参数 scroll 设定为 10 分钟(活性时间)和 size 为 45(fund_basic 有 50 条记录)。

(1)第一次滚动搜索,使用 fund_basic 索引下 _search 接口和参数 scroll 为 10m 发出请求,获取第一批次搜索结果和 scroll_id。

```
curl --request POST 'http://localhost:9200/fund_basic/_search?pretty=true&scroll=10m' --header "Content-Type: application/json" --data $'{"size":45, "_source":{"includes":["name","ts_code"]}}'
{
  "_scroll_id" : "DXF1ZXJ5QW5kRmV0Y2gBAAAAAAAAJcWS1E4TVhoSnpTZ3ltaVhXVWpNcU1xZw==",
  ……
}
```

(2)第二次滚动搜索,使用 _search/scroll 接口、参数 scroll_id 为 DXF1ZXJ5QW5kRmV0Y2gBAAAAAAAAJcWS1E4TVhoSnpTZ3ltaVhXVWpNcU1xZw== 和 scroll 为 10m 发出请求,获取第二批次搜索结果和 scroll_id。

```
curl --request POST 'http://localhost:9200/_search/scroll?pretty=true' --header "Content-Type: application/json" --data $'{"scroll":"10m", "scroll_id":"DXF1ZXJ5QW5kRmV0Y2gBAAAAAAAAJcWS1E4TVhoSnpTZ3ltaVhXVWpNcU1xZw=="}'
{
  "_scroll_id" : "DXF1ZXJ5QW5kRmV0Y2gBAAAAAAAAJcWS1E4TVhoSnpTZ3ltaVhXVWpNcU1xZw==",
  ……
}
```

(3)第三次滚动搜索,使用 _search/scroll 接口、参数 scroll_id 为和 scroll 为 10m 发出请求。返回的 hits.hits 数组为空,表示没有更多可用数据。

```
curl --request POST 'http://localhost:9200/_search/scroll?pretty=true' --header "Content-Type: application/json" --data $'{"scroll":"10m", "scroll_id":"DXF1ZXJ5QW5kRmV0Y2gBAAAAAAAAJcWS1E4TVhoSnpTZ3ltaVhXVWpNcU1xZw=="}'
{
  "_scroll_id" : " "DXF1ZXJ5QW5kRmV0Y2gBAAAAAAAAJcWS1E4TVhoSnpTZ3ltaVhXVWpNcU1xZw==",
  ……
  "hits" : {
    "total" : {
      "value" : 50,
      "relation" : "eq"
    },
    "max_score" : 1.0,
    "hits" : [ ]
  }
}
```

6. 参数 search_after

参数 search_after 提供类似游标查询的操作,主要是借用上一页的结果以帮助下一页的检索,所以使用较少的资源。由于不保留快照,因此不能保证后续跳转在相同的背景下。此参数的操作原理基于排序和次序。而且还需要在 search_after 参数中指定与 sort 参数中相同的值，并且排序方式必须相同，而且不应使用 from 参数。以下示例使用两个步骤进行说明：

（1）从 fund_basic 索引对字段基金代码降序查询，并限制只返回一条记录。因文本类型无法排序，故必须使用基金代码的关键词数据类型（ts_code.keyword）排序。

```
curl --request POST http://localhost:9200/fund_basic/_search?pretty=true
--header "Content-Type: application/json" --data $'{"size":1,
"sort":[{"ts_code.keyword":{"order":"desc"}}], "_source":false}'
{
  "took" : 1,
  "timed_out" : false,
  ……
  "hits" : {
  "total" : {
      "value" : 50,
      "relation" : "eq"
  },
  "max_score" : null,
  "hits" : [
      {
        "_index" : "fund_basic",
        "_type" : "_doc",
        "_id" : "odaixXIBE8603ZMKlQ7G",
        "_score" : null,
        "sort" : [
          "515980.SH"
        ]
      }
  ……
}
```

（2）上一页最后一个搜索结果为 515980.SH，可使用参数 search_after 继续搜索下一批。下一个搜索结果为 515950.SH。

```
curl --request POST http://localhost:9200/fund_basic/_search?pretty=true
--header "Content-Type: application/json" --data $'{"size":1, "sort":[{ "ts_code.keyword":{"order":"desc"}}],"search_after":["515980.SH"], "_source":false}'
{
  "took" : 0,
  "timed_out" : false,
  ……
  "hits" : {
    "total" : {
      "value" : 50,
      "relation" : "eq"
    },
    "max_score" : null,
    "hits" : [
      {
        "_index" : "fund_basic",
        "_type" : "_doc",
        "_id" : "hdaixXIBE8603ZMKlQ7G",
        "_score" : null,
        "sort" : [
          "515950.SH"
        ]
      }
    ……
}
```

7. 参数 highlight

Elasticsearch 支持高亮显示（highlight）功能来强调查询中的匹配词，以改善用户体验。如果没有在映射中的相关字段指定 store 参数，则将从 _source 中提取匹配的单词，在响应主体的原始文本中高亮显示匹配的单词。支持三种高亮类型，统一（unified）、普通（plain）和快速矢量（fvh）。

提示：普通型最适合在单一字段的简单查询匹配。快速矢量型可以处理比较复杂的情况，例如索引的映射带有参数 term_vector 设置为 with_positions_offsets，可以为特定文档检索之用。统一型将文本分成句子进行评分，支持短语和多项匹配。默认为统一型。

以下示例使用参数 query 下的 query_string 方式，通过请求主体进行搜索包含 50ETF 其中一个单词的基金简称。同时在搜索结果中，对匹配的单词以默认高亮显示方式，配合参数 boundary_scanner（边界扫描器）选项为单词，在文本中高亮显示匹配的单词。在搜索结果中高亮显示部分为 <em>50</em><em>ETF</em>，如第 6 章所述，查询字符串"50ETF"被搜索分词器分割为两个单词。默认标签 <em></em> 是一种以 HTML 方式标记强调的文本。

```
curl --request POST http://localhost:9200/fund_basic/_search?pretty=true
 --header "Content-Type: application/json" --data $'{"query": {"query_string" :
 {"query" : "50ETF", "default_field" : "name", "default_operator":"AND"}},
"highlight":{
   "fields":{"name":{"boundary_scanner":"word"}}}, "_source":{"includes":
["name","ts_code"]}}'
{
  "took" : 7,
  "timed_out" : false,
  ......
  "hits" : {
    "total" : {
      "value" : 35,
      "relation" : "eq"
    },
    "max_score" : 2.6059227,
    "hits" : [
      {
        "_index" : "fund_basic",
        "_type" : "_doc",
        "_id" : "fdaixXIBE8603ZMKlQ7G",
        "_score" : 2.6059227,
        "_source" : {
          "ts_code" : "159811.SZ",
          "name" : "5G50ETF"
        },
        "highlight" : {
          "name" : [
            "5G<em>50</em><em>ETF</em>"
          ]
        }
      },
      ......
}
```

以下示例在前例基础上，配合参数 pre_tags 及 post_tags，使用自定义标签 <mark></mark> 替换默认的标签 <em></em>，以高亮显示匹配的单词。

```
curl --request POST http://localhost:9200/fund_basic/_search?pretty=true
 --header "Content-Type: application/json" --data $'{"query": {"query_string" :
```

```
{"query" : "50ETF","default_field" : "name", "default_operator":"AND"}}, "highl
ight":{"fields":{"name":{"boundary_scanner":"word"}}, "pre_tags":"<mark>", "post_
tags":"</mark>"}, "_source":{"includes":["name","ts_code"]}}'
{
  "took" : 7,
  "timed_out" : false,
  ……
  "hits" : {
    "total" : {
      "value" : 35,
      "relation" : "eq"
    },
    "max_score" : 2.6059227,
    "hits" : [
      {
        "_index" : "fund_basic",
        "_type" : "_doc",
        "_id" : "fdaixXIBE8603ZMK1Q7G",
        "_score" : 2.6059227,
        "_source" : {
          "ts_code" : "159811.SZ",
          "name" : "5G50ETF"
        },
        "highlight" : {
          "name" : [
            " 5G<mark>50</mark><mark>ETF</mark>"
          ]
        }
      },
      ……
  }
```

配合 highlight 同时使用的参数描述，如表 7-2 所示。这些设置可以在全局级别或局部级别（字段级别）中声明，局部级别将覆盖全局级别声明。

表 7-2 配合参数 highlight 的其他常用参数

参数名称	描 述
有关如何拆分文本以找到匹配的词汇单元或短语	
boundary_chars	定义高亮显示的单词或短语的边框字符。默认字符是。，！ \t\n。仅支持 fvh 类型
boundary_max_scan	控制参数 border_chars 扫描的距离。默认为 20 个字符
boundary_scanner	指定拆分文本字段的策略。共有三种策略：chars、sentence 和 word
boundary_scanner_locale	对于 sentence 或 word 策略，可选择语言环境，例如 en-us、zh-cn 或 ja-jp，搜索句子和单词的边界
fields	指定高亮显示的字段。若字段数据类型为关键字，则可用通配符指定。若为 JSON 对象类型，则指定其子对象。若为数组类型，则指定其元素
highlight_query	用以匹配重新评分查询
type	指定高亮显示类型：plain、fvh 和 unified。默认为 unified
有关如何适当分割字段	
fragmenter	仅在 fvh 类型中支持的 chars 策略，涉及使用由 border_chars 定义的一组字符，作为分隔符。句子策略旨在在句子边界处实现拆分。统一边界扫描器的默认策略是句子。当使用统一的荧光笔并且句子的长度大于 fragment_size 参数时，将在刚好超过大小的单词边界处进行拆分。单词策略旨在在单词边界处实现拆分
fragment_size	此参数将包含匹配单词 / 短语的原始文本部分，剪切到要呈现的给定 fragment_size。默认为 100 个字符。当此参数设定为 0 时，将忽略此值
fragment_offset	在文本字段中指定起始偏移置

续表

参数名称	描述
number_of_fragments	指定要返回的分段数。默认为 5。如果将其设置为 0，则显示整个字段
matched_fields	指定的那些字段必须使用 term_vector 映射参数和 with_positions_offsets 选项；此设置只适合 fvh 高亮显示类型
phrase_limit	限制了匹配短语的数量，以避免资源的过度使用，默认为 256。此设置只适合 fvh 高亮显示类型
有关如何呈现高亮显示样式	
encoder	指定高亮显示字段的编码类型，默认为无编码。可设为 HTML
no_match_size	若搜索结果中没有匹配的高亮显示字段，当然没有高亮显示结果。可以设置此参数以返回原始文本的字符个数，默认返回 0 个字符
order	默认情况下，高亮显示的部分会按照 fields 参数的顺序。当高亮显示类型为 unified 时，提供 order 选项为 score 分数的顺序
pre_tags/post_tags	指定高亮显示部分的 HTML 标签，默认为 <em> ... </em>
require_field_match	允许同时指定多个高亮显示字段，但只有匹配的字段才会显示高亮。默认为 true。如果将其设为 false，则全部指定字段将以高亮显示
tags_schema	指定要使用的标签的方式。内置的样式为将 pre_tags 和 post_tags 分别定义为 <em class = "hltx"> 和 </ em class = "hltx">，其中 x 为 1 到 10

8. 其他参数

通过请求主体进行搜索的其他参数将编排到表 7-3，并提供简短说明和示例。

表 7-3　通过请求主体进行搜索的其他常用参数

参数名称		描述及示例
allow_partial_search_results	描述	是否在失败的情况下返回部分结果，默认为 true。参数仅可用于 URI
ccs_minimize_roundtrip	描述	跨集群搜索请求执行过程中，网络往返行程是否应最小化，默认为 true。参数仅可用于 URI
docvalue_fields	描述	在搜索结果中，是否返回指定的 doc_values 字段，以列表方式显示
	示例	"docvalue_fields":["name.keyword"，"fund_type"]
explain	描述	是否提供详细的描述来解释相关性得分的计算
	示例	"explain":true
min_score	描述	仅返回分数大于或等于 min_score 设定值的那些文档
	示例	"query" : { "query_string" : { "query" : "50ETF"，"default_field" : "name" }}, "min_score" :2
preference	描述	指定节点或分片以执行搜索请求。参数仅可用于 URI，有许多不同的设置可供选择，详情请参考官网
request_cache	描述	缓存是否应用于此请求，默认为索引级别设置。参数仅可用于 URI
script_fields	描述	指定测试脚本（脚本涉及源字段）以评估结果，并在响应主体中显示结果。示例脚本为计算当天与上市时间之间的日距离
	示例	"script_fields":{ "my_script":{ "script" : "((new Date()).getTime()-doc['list_date'].value.getMillis()) / (24×60×60×1000)" }}
search_type	描述	搜索操作类型，选项为 query_then_fetch 或 dfs_query_then_fetch。参数仅可用于 URI
seq_no_primary_term	描述	用于返回文档的 _seq_no 和 _primary_term 字段
	示例	"seq_no_primary_term":true

续表

参数名称	描述及示例	
stored_fields	描述	检索在映射中标记为 stored 的那些字段。可用逗号分隔的指定字段列表
	示例1	"stored_fields" : "_none"
	示例2	"stored_fields" : "field1"
	示例3	"stored_fields" :["field1" , "field2"]
version	描述	返回文档版本号（_version）字段
	示例	"version" : true
_source	描述	输出文档的原始内容，字段名称可使用通配符
	示例1	"_source" :false
	示例2	"_source" : "name"
	示例3	"_source" :["name" , "ts_code"]
	示例4	"_source" :{ "includes" :["name" , "ts_code"]}
	示例5	"_source" :{ "excludes" :["name" , "ts_code"]}
inner_hits	描述	检查哪个嵌套对象导致实际匹配，细节可以参考 8.1.3 节
post_filter	描述	后置过滤器，细节可以参考 9.7 节

7.3 进阶搜索

通常，为了解决复杂的搜索范围，可用查询表达式（Query DSL）构建搜索查询，其中包括通配符来匹配多个字段的搜索等。符合给定条件的文档会按相似性进行排名，以显示查询结果的匹配程度。

认识查询表达式

查询表达式使用基于 JSON 的搜索语言来查询特定数据集和分析数据集。基本上，查询句子可以分为两部分：查询操作和过滤操作。过滤操作是基于搜索结果，可以继续对其进行条件匹配过滤。默认情况下，Elasticsearch 通过相关性进行评分（_score），测量每个文档与查询的匹配程度，对匹配的搜索结果的得分进行排序。

1. 全文搜索

全文搜索（full text query）模式首先使用分析器（或给定的搜索分析器）对查询字符串进行分析，并获得词汇单元组。然后，使用每个词汇单元来匹配索引下的文档内给定的文本。并为每个搜索结果提供评分。可根据所用的查询方法含义，分为三种类型全文查询：包括匹配查询、查询语句查询和间隔查询。以下逐一介绍：

（1）匹配查询（match）：可以根据匹配的方式进一步细分为以下五种匹配查询类型：

① match

标准匹配查询以匹配提供的文本、数字、日期或布尔值。对查询字符串进行分析，并获取词汇单元，然后将各个词汇单元根据参数 operator（默认为 OR）进行匹配及布尔运算，获得最终匹配结果。以下示例在索引 fund_basic 中的基金简称文本，匹配查询字符串中银 ETF。查询字符通过 my_hanlp_analyzer 分析器后，获取两个词汇单元：中银及 ETF。

示例使用布尔运算符 AND（意味并存），获得匹配结果的基金简称为中银证券 500ETF 和中银中证 100ETF。这两个基金包含所有要求的词汇单元。由于返回的响应内容信息太多，以下只显示请求命令。

```
curl --request POST http://localhost:9200/fund_basic/_search?pretty=true
--header "Content-Type: application/json" --data $'{"query": {"match" : {"name":
{"query" : "中银ETF", "operator":"AND"}}}, "_source":["name","ts_code"]}'
```

可配合 query 参数（指定查询字符串）同时使用的常用参数，如表 7-4 所示。

表 7-4 可配合 query 参数同时使用的常用参数

参数名称		描述及示例
analyzer	描述	指定用来分析查询字符串的分析器
	示例	{"query":{"match":{"name":{"query":"中银","analyzer":"ik_smart"}}}}
fuzziness	描述	通过使用模糊匹配，来匹配查询字符串所产生的词汇单元
	示例	{"query":{"match":{"name":{"query":"500","funzziness":1}}}}
prefix_length	描述	模糊匹配时，对于查询字符串所产生的词汇单元，需要固定开始字符数的个数。默认为 0
	示例	{"query":{"match":{"name":{"query":"中银","funzziness":1,"prefix_length":1}}}}
fuzzy_transpositions	描述	模糊匹配时，对于查询字符串所产生的词汇单元，允许两个相邻字符换位。默认为 true
	示例	{"query":{"match":{"name":{"query":"05","funzziness":1,"fuzzy_transpositions":false}}}}
max_expansions	描述	模糊匹配时，对于查询字符串所产生的词汇单元最大个数，默认为 50
	示例	{"query":{"match":{"name":{"query":"中银证券","funzziness":1,"max_expansions":2}}}}
lenient	描述	忽略基于格式的错误，例如为数字字段提供文本查询值。默认为 false
	示例	{"query":{"match":{"list_date":{"query":5,"lenient":true}}}}
operator	描述	对于查询字符串所产生的词汇单元，使用布尔运算符合并条件。默认为 OR
	示例	{"query":{"match":{"name":{"query":"中证100","operator":"AND"}}}}
minimum_should_match	描述	对于查询字符串所产生的词汇单元，最小应该匹配的个数
	示例	{"query":{"match":{"name":{"query":"中银100","minimum_should_match":1}}}}
zero_terms_query	描述	如果查询字符通过分析器后产生 0 个词汇单元，那应该怎么返回结果。选项为 none 或 all，默认为 none，意思是不返回数据
	示例	{"query":{"match":{"name":{"query":"","zero_terms_query":"all"}}}}

② 布尔与前缀匹配（match_bool_prefix）

布尔与前缀匹配对查询字符串进行分析，并获取词汇单元，并根据这些词汇单元构造一个布尔查询（bool query）。最后一项为前缀（prefix）查询，其他项则为词条（term）查询，每个子句查询（包含前缀查询）为应（should）出现在匹配的文档中，相当于逻辑或（OR）。

以下示例在文本基金简称中，匹配查询字符串 800 证券。获得最终匹配结果的基金简称为 800ETF、香港证券 ETF、证券 ETF 富国、中证 800ETF 基金和中银证券 500ETF。观察前缀"证券"这个词汇单元可以不需要出现（例如 800ETF 和中证 800ETF 基金），也可以在任何位置出现（例如香港证券 ETF 和中银证券 500ETF，证券这个前缀可以出现在中间的词汇单元）。由于返回的响应内容信息太多，以下只显示请求命令。

提示：可以使用 5.3 利用自定义分析器进行分析一节的 _analyze 接口方法，分析香港证券 ETF 和中银证券 500ETF 输出的词汇单元。香港证券 ETF 文本输出香港、证券和 ETF。中银证券 500ETF 文本输出中银、证券、500 和 ETF。

```
curl --request POST http://localhost:9200/fund_basic/_search?pretty=true
--header "Content-Type: application/json" --data $'{"query": {"match_bool_prefix" :
{"name" : "800证券"}}, "_source":["name","ts_code"]}'
```

可同时使用的常用参数包括 analyzer、minimum_should_match 和 operator，其用法如表 7-4 所示。

③ 短语匹配（match_phrase）

短语匹配对查询字符串进行分析，并获取词汇单元，然后将各个词汇单元连接，生成一个短语以匹配搜索文本字段。以下示例在文本基金简称中，匹配查询字符串 800ETF。

请注意，介于 800 和 ETF 之间是否具有空格字符是不会改变搜索结果。这是因为两者都生成两个词汇单元并连接成同一个单词"800ETF"。获得最终匹配结果的基金简称为 800ETF 和中证 800ETF 基金。这两个基金包含前述的短语。由于返回的响应内容信息太多，以下只显示请求命令。

```
curl --request POST http://localhost:9200/fund_basic/_search?pretty=true
--header "Content-Type: application/json" --data $'{"query": {"match_phrase" :
{"name": {"query" : "800ETF"}}}, "_source":["name","ts_code"]}'
```

match_phrase 查询支持 slop 参数（斜率默认为 0），以指定的查询字符串中，允许匹配项的字符之间的间隔差，间隔指的是文本中可以忽略的词汇单元数目。以下示例显示斜率值的含义。在文本基金简称中，匹配查询字符串"5ETF"。当斜率等于 0 或 1 时，不存在匹配结果。当斜率等于 2 时，有一个匹配结果 5G50ETF。另外可使用参数 analyzer 及 zero_terms_query，其用法如表 7-4 所示。

提示：文本 5G50ETF 输出 5、G、50 和 ETF 词汇单元，而 5 和 ETF 的间隔差为 2。

```
curl --request POST http://localhost:9200/fund_basic/_search?pretty=true
--header "Content-Type: application/json" --data $'{"query": {"match_phrase" :
{"name": {"query" : "5ETF", "slop":2}}}, "_source":["name","ts_code"]}'
{
  ……
  "hits" : {
    "total" : {
      "value" : 1,
      "relation" : "eq"
```

```
        },
        "max_score" : 1.303079,
        "hits" : [
          {
            ……
            "_source" : {
              "ts_code" : "159811.SZ",
              "name" : "5G50ETF"
            }
            ……
          }
```

④ 短语前缀（match_phrase_prefix）

短语前缀匹配与 match_phrase 查询类似，只是生成的短语以前缀匹配搜索文本字段。前缀可以在文本字段输出的词汇单元的任何位置。以下示例在文本基金简称中，匹配查询字符串"证券ETF"。获得最终匹配结果的基金简称为香港证券ETF和证券ETF富国。查询字符"证券 ETF"出现在"香港证券 ETF"，通过分析器后输出的词汇单元的中间位置。由于返回的响应内容信息太多，以下只显示请求命令。

```
curl --request POST http://localhost:9200/fund_basic/_search?pretty=true
--header "Content-Type: application/json" --data $'{"query": {"match_phrase_prefix"
: {"name": {"query" : "证券ETF"}}}, "_source":["name","ts_code"]}'
```

⑤ mutli_match

多字段匹配将查询字符串生成的词汇单元，匹配到多个字段的文本查询。字段可以使用通配符表示。在多个字段中可指定参数 type 的选项来定义字段间的匹配类型，如表 7-5 所示。

表 7-5 多个字段查询的参数 type 的选项

参数名称	描述
best_fields	根据最佳得分对结果进行排名（默认值）
most_fields	按总得分对结果进行排名
cross_fields	匹配到多个字段的文本查询结果，必须存在至少一个字段才算文档为匹配
phrase	与 best_fields 参数选项类似，但使用 match_phrase 查询代替
phrase_prefix	与 best_fields 参数选项类似，但使用 match_phrase_prefix 查询代替
bool_prefix	与 most_fields 参数选项类似，但使用 match_bool_prefix 查询代替

以下示例在基金管理人和基金托管人文本中，匹配查询字符串"中信建投 招商"。配合表 7-5 所示的参数，获得最终匹配结果，将前两个最高分数搜索结果如表 7-6 所示。由于返回的响应内容信息太多，以下只显示请求命令。

```
curl --request POST http://localhost:9200/fund_basic/_search?pretty=true
--header "Content-Type: application/json" --data $'{"query": {"multi_match" :
{"query" : "中信建投 招商", "fields":["management","custodian"],"type":"best_fields"}},
"_source":["management","custodian"]}'
```

提示：使用参数 phrase 或是 phrase_prefix 没有得到匹配结果的原因，是因为在基金管理人或是基金托管人的文本中没有产生这个"中信建投招商"的词汇单元。使用参数 bool_prefix 有匹配结果是因为查询字符串"中信建投招商"会产生"中信""建""投"和"招商"这四个词汇单元去匹配。

表 7-6 多个字段查询配合各个参数的匹配结果

参数名称	最高得分	第二高得分
使用文本"中信建投 招商"		
best_fields	5.546761 "custodian":"中信建投", "management":"博时基金"	5.546761 "custodian":"中信建投", "management":"招商基金"
most_fields	8.622208 "custodian":"中信建投", "management":"招商基金"	5.546761 "custodian":"中信建投", "management":"博时基金"
cross_fields	8.622208 "custodian":"中信建投", "management":"招商基金"	5.546761 "custodian":"中信建投", "management":"博时基金"
phrase	没有匹配结果	没有匹配结果
phrase_prefix	没有匹配结果	没有匹配结果
bool_prefix	8.622208 "custodian":"中信建投", "management":"招商基金"	5.546761 "custodian":"中信建投", "management":"博时基金"

（2）查询语句查询（query string query）

可以细分为两种子类型：query_string 和 simple_query_string。它们使用不同的解析器来处理查询字符串。其解析语法完全不同。如果查询字符串不适合语法，则 query_string 的解析器将引发异常，而 simple_query_string 的解析器仅丢弃无效部分并继续执行。两个解析器都使用运算符先拆分查询字符串为组别，然后再针对每个组别拆分如下字段：

- simple_query_string

简单查询字符串类型将参数 query 指定的查询字符串，按照基于保留字符的简单语法，拆分为词汇单元，然后查询并返回匹配文档。如果需要使用保留字符作为一般字符，需要使用反斜杠字符"\"作为前缀。表 7-7 描述了查询字符串的语法。如果查询字符串中未指定任何字段，则使用 index.query.default_field 的设置。

表 7-7 simple_query_string 查询字符串的语法

保留字符	描述	示例
+	布尔运算符 AND	{"query":{"simple_query_string":{"query":"中银 +100","fields":["name"]}}}
|	布尔运算符 OR	{"query":{"simple_query_string":{"query":"中银|民生","fields":["name"]}}}
-	不得与以下查询字符串匹配	{"query":{"simple_query_string":{"query":"-ETF","fields":["name"]}}}
" "	匹配短语	{"query":{"simple_query_string":{"query":"\\ 创新药 \\"","fields":["name"]}}}
W~N、P~N	如果在单词 W 之后出现，则表示模糊。如果在短语 P 之后，则表示斜率（slop）。其中 N 是最大跨度字符数	{"query":{"simple_query_string":{"query":"中银 ~1","fields":["name"]}}}
W*	如果在单词 W 后面则表示前缀匹配	{"query":{"simple_query_string":{"query":"中银 *","fields":["name"]}}}
()	用于表示优先权	{"query":{"simple_query_string":{"query":"(中银 +100)|民生","fields":["name"]}}}

以下示例在基金管理人和基金托管人文本中，测试匹配短语天弘基金，由于返回的响应内容信息太多，以下只显示请求命令。

```
curl --request POST http://localhost:9200/fund_basic/_search?pretty=true
--header "Content-Type: application/json" --data $'{"query": {"simple_query_string"
: {"query" : "\\"天弘基金\\"", "fields":["management","custodian"]}}, "_source":
["management","custodian"]}'
```

可配合 query 参数（指定查询字符串）同时使用的常用参数，如表 7-8 所示。

表 7-8　可配合 query 参数同时使用的常用参数

参数名称	描述及示例	
analyzer	描述	指定用来分析查询字符串的分析器
	示例	{"query":{"simple_query_string":{"query":"中*","analyzer":"hanlp","fields":["name"]}}}
analyze_wildcard	描述	是否分析查询字符串中的通配符，默认为 false
	示例	{"query":{"simple_query_string":{"query":"中*","analyze_wildcard":true,"fields":["name"]}}}
auto_generate_synonyms_phrase_query	描述	是否自动为短语查询产生多项同义词（从 synonym_graph 词汇单元过滤器产生）供匹配，默认为 true。用法请参考官网
default_operator	描述	指定布尔运算符（默认为 or 或 |）对条件进行分组
	示例	{"query":{"simple_query_string":{"query":"500ETF","default_operator":"AND","fields":["name"]}}}
fields	描述	要搜索的查询字符串数组，字符串接受通配符表达式
	示例	{"query":{"simple_query_string":{"query":"\\"天弘基金\\"","fields":["management","custodian"]}}}
flag	描述	如果只想启用查询字符串内部分的保留字符，可以使用表 7-9 所示的参数选项来启用
fuzzy_max_expansions	描述	模糊匹配时，对于查询字符串所产生的词汇单元的最大个数，默认为 50
	示例	{"query":{"simple_query_string":{"query":"\\"天弘基金\\"","fields":["management","custodian"]}}}
fuzzy_prefix_length	描述	模糊匹配时，对于查询字符串所产生的词汇单元，需要固定开始字符数的个数。默认为 0
	示例	{"query":{"simple_query_string":{"query":"中银证~2","fuzzy_prefix_length":2,"fields":["name"]}}}
fuzzy_transpositions	描述	模糊匹配时，对于查询字符串所产生的词汇单元，允许两个相邻字符换位。默认为 true
	示例	{"query":{"simple_query_string":{"query":"05~1","fuzzy_transpositions":false,"fields":["name"]}}}
lenient	描述	忽略基于格式的错误，例如为数字字段提供文本查询值。默认为 false
	示例	{"query":{"simple_query_string":{"query":5,"fields":["list_date"],"lenient":true}}}
minimum_should_match	描述	对于查询字符串所产生的词汇单元，最小应该匹配的个数
	示例	{"query":{"simple_query_string":{"query":"中银 100","minimum_should_match":1,"fields":["name"]}}}

启用查询字符串内部分的保留字符的参数选项，如表 7-9 所示。

表 7-9　启用查询字符串内部分的保留字符的参数选项

flag 参数选项	保留字符	描述
ALL		默认使用所有保留字符
AND	+	启用布尔运算符 AND 保留字符
ESCAPE	\	启用反斜杠字符
FUZZY	~N	启用模糊匹配

续表

flag 参数选项	保留字符	描述
NONE		禁用所有保留字符
NOT	-	启用否定语
OR	OR	启用布尔运算符 OR 保留字符
PHRASE	W~	启用短语匹配
PRECEDENCE	()	启用优先权保留字符
PREFIX	W*	启用前缀匹配
SLOP	~N	启用斜率保留字符
NEAR	~N	启用斜率保留字符
WHITESPACE		启用空白字符作为分隔符

◆ query_string

查询字符串类型使用其特殊语法根据运算符（例如 AND 或 NOT）解析和拆分提供的查询字符串。可用于创建复杂的查询搜索，其中包括通配符，跨多个字段的搜索等。query_string 查询也有保留字符，若要作为一般字符，需要使用反斜杠字符 \ 作为前缀。以下首先使用与 simple_query_string 相同的示例，在基金管理人和基金托管人文本中，测试匹配短语天弘基金，由于返回的响应内容信息太多，以下只显示请求命令。

```
curl --request POST http://localhost:9200/fund_basic/_search?pretty=true
--header "Content-Type: application/json" --data $'{"query": {"query_string" :
{"query" : "\\"天弘基金\\"", "fields":["management","custodian"]}}, "_source":["m
anagement","custodian"]}'
```

上面的示例的语法类似于 simple_query_string 查询，但可以有很多变化，表 7-10 描述了查询字符串的语法。如果查询字符串中未指定任何字段，则使用 index.query.default_field 的设置。

提示：由于 query_string 查询会针对任何无效的语法返回错误，因此官网不建议使用于查询框式用法。

表 7-10 query_string 查询字符串的语法

保留字符	描述	示例
:	用于字段和值之间的分隔符	
()	用于表示优先权	
&&、||、!	用于逻辑运算符，注意在两个操作数之间使用空格	{"query":{"query_string":{"query": "name:(中银 && +100)| 民生" }}}
+、-	运算符 + 用于表示必须匹配，运算符 - 用于不得与以下查询字符串匹配，注意紧靠在跟随的字符	
>、<、=	用于比较运算符，注意紧靠在跟随的字符	{"query":{"query_string":{"query": "m_fee:(>=0.15 AND <=0.2)" }}}
*、?	通配符 * 表示匹配零个或多个字符，而? 表示匹配单个字符	{"query":{"query_string":{"query": "name: 中 *" }}}
" "	匹配短语	{"query":{"query_string":{"query": "name: \\ "创新药 \\ " " }}}
~	模糊匹配	{"query":{"query_string":{"query": "name: 中银 ~1" }}}

续表

保留字符	描述	示例
/	将正则表达式装在正斜杠中用作查询字符串。Elasticsearch 使用 Apache Lucene 的正则表达式引擎来解析这些查询	{"query":{"query_string":{"query":"name:/.* 证券 .*/"}}}
^	提升词汇单元权重	{"query":{"query_string":{"query":"(中银 +100)|民生 ^2","fields":["name"]}}}
[]、{}	用于范围查询。{} 符号对表示排除，而 [] 符号对表示包含	{"query":{"query_string":{"query":"list_date:{20200525 TO *}"}}}

可配合 query 参数（指定查询字符串）同时使用的常用参数，如表 7-11 所示。

表 7-11　可配合 query 参数同时使用的常用参数

参数名称		
analyzer	描述	指定用来分析查询字符串的分析器
	示例	{"query":{"query_string":{"query":"name: 中银证券","analyzer":"ik_smart"}}}
analyze_wildcard	描述	是否分析查询字符串中的通配符，默认为 false
	示例	{"query":{"query_string":{"query":"name: 中 *","analyze_wildcard":true}}}
allow_leading_wildcard	描述	是否允许通配符作为查询字符串的第一个字符。默认为 true
	示例	{"query":{"query_string":{"query":"name:* 证","allow_leading_wildcard":true}}}
auto_generate_synonyms_phrase_query	描述	是否自动为短语查询产生多项同义词（从 synonym_graph 词汇单元过滤器产生）供匹配，默认为 true。用法请参考官网
boost	描述	用于降低或增加查询的相关性分数。默认为 1.0 倍
	示例	"query":{"query_string":{"query":"name:800ETF","boost":2}}}
default_operator	描述	指定布尔运算符（默认为 or 或 |）对条件进行分组
	示例	"query":{"query_string":{"query":"name:800ETF","default_operator":"AND"}}}
fields	描述	要搜索的查询字符串数组，字符串接受通配符表达式
	示例	{"query":{"query_string":{"query":"中银","fields":["name"]}}}
fuzziness	描述	通过使用模糊匹配，来匹配查询字符串所产生的词汇单元。其值表示经过多少个字符的编译
	示例	{"query":{"query_string":{"query":"name: 500~","fuzziness":1}}}
fuzzy_max_expansions	描述	模糊匹配时，对于查询字符串所产生的词汇单元的最大个数，默认为 50
	示例	"query":{"query_string":{"query":"name: 中银 ~","fuzziness":1,"fuzzy_max_expansions":2}}}
fuzzy_prefix_length	描述	模糊匹配时，对于查询字符串所产生的词汇单元，需要固定开始字符数的个数。默认为 0
	示例	{"query":{"query_string":{"query":"name: 中银证 ~2","fuzzy_prefix_length":2}}}

续表

参数名称		
fuzzy_transpositions	描述	模糊匹配时,对于查询字符串所产生的词汇单元,允许两个相邻字符换位。默认为 true
	示例	{"query":{"query_string":{"query":"name:05~1","fuzzy_transpositions":false}}}
lenient	描述	忽略基于格式的错误,例如为数字字段提供文本查询值。默认为 false
	示例	{"query":{"query_string":{"query": 5, "fields":["list_date"] "lenient":true }}}
minimum_should_match	描述	对于查询字符串所产生的词汇单元,最小应该匹配的个数
	示例	{"query":{"query_string":{"query":"name: 中银 100",minimum_should_match":2}}}

(3) 间隔查询 (interval query)

间隔查询与跨度查询(span query)类似,根据定义的匹配规则,返回文档。根据所用的查询规则或对象,可分为五个类型全文查询。以下介绍各个类型全文查询:

• match

可根据对查询文本单词的匹配项的顺序(ordered)和间隔接近程度(max_gaps)返回文档。常用参数如表 7-12 所示。

表 7-12 可在 match 参数内使用的常用参数

参数名称	描述
analyzer	指定用来分析查询字符串的分析器
filter	可选的间隔过滤器规则参数,如表 7-15 所示
max_gaps	匹配时,词与词之间的最大允许偏移数,默认为 -1 表示没有限制。如果超过则不视为匹配项
ordered	匹配字词必须按指定查询字符串分析后所获得词汇单元组并按照顺序,默认为 false
query	指定查询字符串
use_field	如果指定此参数值,则匹配此参数值的字段而不是跟随的 interval 参数的字段

以下示例在基金简称文本中,匹配查询字符串中银 ETF。查询字符通过 my_hanlp_analyzer 分析器后,获取两个词汇单元,中银及 ETF。max_gaps 设为 2,获得最终匹配结果的基金简称为中银证券 500ETF。这是因为文本"中银证券 500ETF"被分割成 4 个词汇单元,中银、证券、500 和 ETF。而中银和 ETF 的间隔为两个词汇单元。若将 max_gaps 设为 1,则没有搜索结果。

```
curl --request POST http://localhost:9200/fund_basic/_search?pretty=true
--header "Content-Type: application/json" --data $'{"query": {"intervals": {"name":
{"match": {"query":"中银ETF", "max_gaps":2}}}}}'
```

• prefix

前缀规则匹配以 prefix 参数指定的查询字符串,查询字符串不再分词,该前缀可以在文本中最多产生 128 个匹配。如果超过,算作错误。可在 prefix 参数内使用的常用参数,如表 7-13 所示。

表 7-13　可在 prefix 参数内使用的常用参数

参数名称	描述
analyzer	指定用来分析查询字符串的分析器
prefix	指定前缀字符串，和 match_phrase_prefix 一样，前缀可以在文本字段任何位置匹配
use_field	如果指定此参数值，则匹配此参数值的字段而不是跟随的 interval 参数的字段

以下示例在文本基金简称中，匹配查询前缀字符串为 MSCI 获得最终匹配结果的基金简称为 MSCI 中国招商 ETF。

提示：根据测试结果，prefix 参数只对英文和数字有效，对汉字无效。例如参数 prefix 设置为"中银"会没有匹配结果。

```
curl --request POST http://localhost:9200/fund_basic/_search?pretty=true
--header "Content-Type: application/json" --data $'{"query": {"intervals": {"name":
{"prefix": {"prefix": "MSCI"}}}}}'
```

- wildcard

使用通配符规则匹配，可以在文本中最多产生 128 个匹配。如果超过，算作错误。参数 wildcard 内可使用的常用参数，如表 7-14 所示。

表 7-14　可在 wildcard 参数内使用的常用参数

参数名称	描述
analyzer	指定用来分析查询字符串的分析器
pattern	指定通配符规则，此参数支持两个通配符，？字符匹配任何单个字符，而 * 字符匹配零个或多个字符，注意避免以通配符为开头的模式
use_field	如果指定此参数值，则匹配此参数值的字段而不是跟随的 interval 参数的字段

以下示例在基金简称文本中，匹配查询字符串"中国*"。匹配结果的基金简称为 MSCI 中国招商 ETF。

```
curl --request POST http://localhost:9200/fund_basic/_search?pretty=true
--header "Content-Type: application/json" --data $'{"query": {"intervals": {"name":
{"wildcard": {"pattern": "中国*"}}}}}'
```

- all_of

all_of 规则返回能够匹配所有子规则（match、prefix 和 wildcard 规则）的搜索结果。此参数内可使用的常用参数有 intervals、max_gaps、ordered 和 filter。除了 intervals 参数，其他参数用法如前所述。intervals 参数定义一组规则供匹配。以下示例在文本基金简称中，匹配准则为同时包含"500"和"LOF"两个字符串。搜索结果只有"500 增强 LOF"。

```
curl --request POST http://localhost:9200/fund_basic/_search?pretty=true
--header "Content-Type: application/json" --data $'{"query": {"intervals":
{"name": {"all_of": {"intervals": [{"match" : {"query" : "500"}},
{"match":{"query":"LOF"}}]}}}}, "_source":["name"]}'
```

- any_of

any_of 规则返回匹配任一或者更多子规则（match、prefix 和 wildcard 规则）产生的搜索结果。可在此参数内使用的常用参数有 intervals 和 filter。除了 intervals 参数，其他参数用法如前所述。intervals 参数定义一组规则供匹配。以下示例在文本基金简称中，匹配准则为包含"500"和"LOF"两个字符串其中之一。搜索结果有"500 增强 LOF"和"中银证券

500ETF"两个基金(表7-15)。

```
curl --request POST http://localhost:9200/fund_basic/_search?pretty=true
--header "Content-Type: application/json" --data $'{"query": {"intervals":
{"name": {"any_of": {"intervals": [{"match" : {"query" : "500"}},
{"match":{"query":"LOF"}}]}}}}, "_source":["name"]}'
```

表7-15 可选的间隔过滤器规则选项

filter 参数选项	描 述
after	返回遵循过滤规则间隔的结果
before	返回还没经过过滤规则间隔的结果
contained_by	返回经过过滤规则间隔包含的结果
containing	返回经过过滤规则间隔符合其中之一个间隔的结果
not_contained_by	返回经过过滤规则间隔不包含的结果
not_containing	返回经过过滤规则间隔不符合任何一个间隔的结果
not_overlapping	返回经过过滤规则间隔与其间隔不重叠的结果
overlapping	返回经过过滤规则间隔与其间隔重叠的结果
script	返回经过符合脚本执行的结果,脚本执行的结果是一个布尔值

以下示例在文本基金简称中,匹配查询字符串"500",但是会过滤"500增强LOF"这个基金。获得最终匹配结果的基金简称为中银证券500ETF。500增强LOF这个基金没有出现在搜索结果中。

提示:过滤规则的字符串不再分词,例如使用LOF替代500增强LOF进行过滤,则不会过滤任何搜索结果。

```
curl --request POST http://localhost:9200/fund_basic/_search?pretty=true  --header
"Content-Type: application/json" --data $'{"query": {"intervals": {"name": {"match":
{"query":"500", "filter":{"not_containing":{"match":{"query":"500增强LOF"}}}}}}}}'
```

2. 词条级别搜索

词条级别搜索使用查询中指定的整个查询字符串(不进行分析)来匹配。通常使用数字、日期和关键字数据类型的字段,而不是文本(text)字段。以下介绍词条级别查询的不同子类型:

(1) exists

存在(exists)查询通过查询字符串进行匹配字段名称是否存在该字符串,若是存在则返回文档。查询字符串不存在文档的字段内会有多种原因,例如文档不包括这个字段、源JSON中这个字段为null、没有这个字段的映射、超出字段长度允许值及格式错误。以下示例查询是否有文档存在退市日期(delist_date)数据,搜索结果为没有,因为所有文档在这个字段的值全为null,以下只显示请求命令。

```
curl --request POST http://localhost:9200/fund_basic/_search?pretty=true
--header "Content-Type: application/json" --data $'{"query": {"exists": {"field":
"delist_date"}}}'
```

(2) fuzzy

模糊(fuzzy)查询通过与查询字符串相似的字段值来匹配。以下示例在基金托管人(使用关键字数据类型)中模糊匹配查询字符串中信建投,并允许查询字符串进行2个字符编辑,可以匹配更多词汇单元。

提示：使用模糊查询并设置参数 fuzziness 为 2 之后，中信建投可以模糊匹配中信证券。fuzziness 的意义请参考表 7-11。

```
curl --request POST http://localhost:9200/fund_basic/_search?pretty=true
--header "Content-Type: application/json" --data $'{"query": {"fuzzy": {
"custodian.keyword": {"value":"中信建投", "fuzziness":2}}}}'
```

可在参数 fuzzy 内使用的常用参数，除了参数 value 指定查询字符串外，其他参数如 fuzziness、max_expansions、prefix_length 和 fuzzy_transpositions 如表 7-4 所示。

（3）ids

查询字符串为一个或多个文档标识符（_id），使用参数 values 指定文档标识符数组。以下示例使用基金简称"5G50ETF"和"医药 50ETF 基金"的文档标识符查询。请注意要使用运行环境中正确的文档标识符。

```
curl --request POST http://localhost:9200/fund_basic/_search?pretty=true
--header "Content-Type: application/json" --data $'{"query": {"ids": {
"values":["fdaixXIBE8603ZMKlQ7G", "hdaixXIBE8603ZMKlQ7G"]}}}'
```

（4）prefix

前缀（prefix）查询指定前缀字符串对分词的前缀进行匹配。以下示例使用参数 values 指定前缀为沪深对基金简称（关键字数据类型）进行前缀匹配查询。搜索结果仅为"沪深 300ETF 博时"。

```
curl --request POST http://localhost:9200/fund_basic/_search?pretty=true
--header "Content-Type: application/json" --data $'{"query": {"prefix": {
"name.keyword":{"value":"沪深"}}}}'
```

（5）range

范围（range）查询返回范围内包含的匹配字段值的文档。值的比较可以使用 gt 表示大于，gte 表示大于或等于，lt 表示少于，以及 lte 表示少于或等于。支持参数 format，可以用来转换日期格式，请参考 1.6 节接口用法约定说明中的"日期字符串"。

- 查询管理费范围在 0.15 和 0.2 之间（不含 0.15，含 0.2）的基金。搜索结果为"创大盘"和"股息龙头 ETF"。

```
curl --request POST http://localhost:9200/fund_basic/_search?pretty=true
--header "Content-Type: application/json" --data $'{"query": {"range": {
"m_fee":{"gt":0.15, "lte":0.2}}}}'
```

- 查询上市时间范围在 2020 年 5 月 25 日至今的基金。搜索结果为中银证券 500ETF 及浙江 ETF。上市时间为 20200526 及 20200527。

```
curl --request POST http://localhost:9200/fund_basic/_search?pretty=true
--header "Content-Type: application/json" --data $'{"query": {"range": {
"list_date":{"gt":20200525}}}}'
```

（6）regexp

正则（regexp）查询，返回与以参数 value 指定的正则表达式匹配的分词的文档。以下示例在基金简称（使用关键字数据类型）中，匹配查询字符串"证券"一词。匹配结果的基金简称为中银证券 500ETF、上证券商 ETF、香港证券 ETF 和证券 ETF 富国。可与参数 value 同时使用的参数 flags，请参考官网的正则表达式语法。

```
curl --request POST http://localhost:9200/fund_basic/_search?pretty=true
--header "Content-Type: application/json" --data $'{"query": {"regexp": {
"name.keyword":{"value":".*证券.*"}}}}'
```

（7）term

词条（term）查询返回包含与查询字符串一样的分词的文档。以下示例在基金简称（使用关键字数据类型）中，匹配查询字符串"5G50ETF"一词。匹配结果当然只有该基金。可与参数 value 同时使用的参数 boost，如表 7-11 所示。

```
curl --request POST http://localhost:9200/fund_basic/_search?pretty=true
--header "Content-Type: application/json" --data $'{"query": {"term": {
"name.keyword":{"value":"5G50ETF"}}}}'
```

（8）terms

多个词条（terms）查询返回与查询字符串数组内一样的分词的文档。以下示例在基金简称（关键字数据类型）中，匹配查询字符串 5G50ETF 和上证券商 ETF 两个词，匹配结果当然只有该两个基金。可与查询字段同时使用的参数有 boost，用于增加或减少相关分数，如表 7-11 所示。

```
curl --request POST http://localhost:9200/fund_basic/_search?pretty=true
--header "Content-Type: application/json" --data $'{"query": {"terms": {
"name.keyword":["5G50ETF", "上证券商 ETF"]}}}'
```

（9）terms_set

词条集（terms_set）查询类似词条查询，但是可以定义所需的匹配词条最小数目，适用于词条数组类型数据里面搜索。由于当前索引中没有数组类型数据，因此随意选用业绩比较基准字段并将 minimum_should_match_script 设为返回值 1。

以下示例在业绩比较基准中，查询使用上证证券行业指数或是中证全指证券公司指数收益率的基金，匹配结果有上证券商 ETF 和证券 ETF 富国。另一可以设定最低匹配个数的参数 minimum_should_match_field，因为在文档建立索引期间需要设置 required_matches 字段，用法请参考官网。

```
curl --request POST http://localhost:9200/fund_basic/_search?pretty=true
--header "Content-Type: application/json" --data $'{"query": {"terms_set":
{"benchmark":{"terms":["上证证券行业指数", "中证全指证券公司指数收益率"],
"minimum_should_match_script":{"source":"1"}}}}}'
```

（10）wildcard

通配符（wildcard）查询使用通配符规则匹配，使用参数 value 来指定通配符规则。支持两种通配符规则，？字符匹配任何单个字符，而 * 字符匹配零个或多个字符，注意避免以通配符为开头的模式，原因是比较消耗资源。

以下示例在文本基金简称中，匹配查询字符串"中银*"。最终匹配结果的基金简称为中银证券 500ETF 和中银中证 100ETF。可与参数 value 同时使用的参数 boost，请参考表 7-11。

```
curl --request POST http://localhost:9200/fund_basic/_search?pretty=true
--header "Content-Type: application/json" --data $'{"query": {"wildcard": {
"name.keyword": {"value": "中银*"}}}}'
```

3. 复合查询

复合查询（Compound query）用于组合多个子句以构建复杂的查询。以下介绍复合查询的不同子类型：

（1）bool

布尔（bool）查询使用等效于布尔值运算符 must（AND）、should（OR）和 must_not（NOT）组合查询子句，并提供组合的结果。以下示例查询管理费范围在 0.15 和 0.2 之间（含 0.15 和 0.2），同时上市时间范围不在 2020 年 5 月 2 日至 2020 年 5 月 25 日之间（含首末两天）。最终匹配结果的基金简称为中银证券 500ETF 和民生加银 300ETF。其管理费均为 0.15，上市时间范围不在指定范围其间。另外可使用参数 minimum_should_match，指定最小应该匹配的子句个数。

```
curl --request POST http://localhost:9200/fund_basic/_search?pretty=true
--header "Content-Type: application/json" --data $'{"query": {"bool": {"must":
{"range": {"m_fee":{"gte":0.15, "lte":0.2}}}, "must_not":{"range":{
"list_date":{"gte":20200210, "lte":20200525}}}}}}'
```

（2）boost

权重提升（boost）查询使用 term 查询搜索多个字段，可以使用参数 positive 指定必须匹配的查询或是使用参数 negative 指定负面查询。另外指定参数 negative_boost 以减少匹配文档的得分权重。negative_boost 数值需要设置在 0 和 1.0 之间。以下示例查询管理费大于 0.8，但当它大于 1 时，将得分权重比减小 0.5。匹配结果有国富 100、科创建信和 500 增强 LOF。得分数分别是 1.0、0.5 和 0.5。科创建信和 500 增强 LOF 的分数与范围查询结果比较减少了 0.5。

```
curl --request POST http://localhost:9200/fund_basic/_search?pretty=true
--header "Content-Type: application/json" --data $'{"query": {"boosting":
{"positive": {"range": {"m_fee":{"gte":0.8}}}, "negative":{"range":{
"m_fee":{"gte":1}}}, "negative_boost":0.5}}}'
```

（3）constant_score

固定分数（constant_score）查询使用过滤器查询，返回的每个匹配文档相关分数重设为等于参数 boost 的值。以下示例通过滤器执行 constant_score 查询成立日期为 20200219，返回搜索结果并使其得分等于 boost 参数值 1.2。搜索结果为 500 增强 LOF，得分为 1.2。

```
curl --request POST http://localhost:9200/fund_basic/_search?pretty=true
--header "Content-Type: application/json" --data $'{"query": {"constant_score":
{"filter": {"term":{"found_date":20200219}}, "boost":1.2}}}'
```

（4）dis_max

分离最大化（dis_max）查询使用参数 queries 收集要匹配的子查询。一个文档的得分为该文档所有匹配字段的最佳匹配评分作为结果返回。以下示例在基金简称（使用关键字数据类型）中，在两个子查询中，一个子查询要匹配字符串中 *，而另一个子查询要匹配 ETF。匹配结果显示如表 7-16 所示。

```
curl --request POST http://localhost:9200/fund_basic/_search?pretty=true
--header "Content-Type: application/json" --data $'{"query": {"dis_max": {"queries":
[{"wildcard":{"name.keyword":{"value":"中证*"}}}, {"term": {"name":"800"}}]}}}'
```

可使用参数 tie-breaker，并设为 0 到 1.0 之间的某值，若有更多子查询匹配，则得到更高的得分，以下示例同上并设 tie-breaker 为 1，匹配结果显示如表 7-16 所示。

提示：中证 800ETF 基金在启用 tie_breaker 后多了 1 分，原因是两个子查询都匹配。

```
curl --request POST http://localhost:9200/fund_basic/_search?pretty=true
--header "Content-Type: application/json" --data $'{"query": {"dis_max": {"queries":
[{"wildcard":{"name.keyword":{"value":"中证*"}}}, {"term":
{"name":"800"}}],"tie_breaker":1}}}'
```

表 7-16 分离最大化查询测试结果

匹配基金简称	匹配结果得分	匹配结果得分（tie-breaker=1）
800ETF	3.430 637 4	3.430 637 4
中证 800ETF 基金	2.299 934 4	3.299 934 4
中证红利指数 ETF	1.0	1.0

（5）function_score

函数评分（function_score）查询提供了一种通过用户定义的函数，对匹配文档重新计算评分。用户可以先定义查询，然后再定义一个或多个评分函数。如果使用多个函数，则另外再指定参数 score_mode 以选择如何计算综合得分。综合方式有 multiply（相乘）、sum（相加）、avg（平均数）、first（第一个具有过滤器的函数）、max（最大值）和 min（最小值）等，支持的评分函数简介如下：

- script_score

提供脚本评分（script_score），评分计算结果来自脚本。查询基金简称包含中银两个字，而搜索结果以管理费和托管费相加后降序排序。以下示例以其总和 sum 的倒数为得分。搜索结果为中银证券 500ETF 和中银中证 100ETF。费用总和为 0.2 和 0.54，其得分为 5.0 和 1.851 851 8。

```
curl --request POST http://localhost:9200/fund_basic/_search?pretty=true
--header "Content-Type: application/json" --data $'{"query": {"function_score":
{"query":{"wildcard":{"name.keyword":"中银*"}}, "script_score":{"script":
{"source":
"1/(doc[\'c_fee\'].value + doc[\'m_fee\'].value)"}}}}'
```

- weight

评分计算结果来自评分分数乘以所提供的权重（weight）。以下示例查询基金简称包含中银两个字，而搜索结果以得分乘以 2 作为评分计算结果。搜索结果为中银证券 500ETF 和中银中证 100ETF，原始得分为是 1.0，现在其得分均为 2.0。

```
curl --request POST http://localhost:9200/fund_basic/_search?pretty=true
--header "Content-Type: application/json" --data $'{"query": {"function_score":
{"query":{"wildcard":{"name.keyword":"中银*"}}, "weight":2}}}'
```

- random_score

随机生成（random_score）均匀分布从 0 到 1（但不包括 1）的评分分数。以下示例查询基金简称包含中银两个字而搜索结果以随机数为得分。参数 seed 的数据类型可为 int、long 或 string。以下示例设置参数 seed 为文档标识符 $(id)。

提示：可以检查搜索结果随机生成的评分分数是否小于 1。

```
curl --request POST http://localhost:9200/fund_basic/_search?pretty=true
--header "Content-Type: application/json" --data $'{"query": {"function_score":
{"query": {"wildcard": {"name.keyword":"中银*"}}, "random_score": {"seed":"$id"}}}}'
```

- field_value_factor

类似于脚本评分,但可用文档中的字段值(field_value_factor)来影响得分。参数 field 指定文档中的字段名称,参数 factor 指定相乘的因子,而 modifier 指定一个数学函数,例如 sqrt、square、reciprocal 和 log 等。以下示例查询基金简称包含中银两个字,而搜索结果以管理费降序排序,以其管理费 m_fee 的倒数为得分。搜索结果为中银证券 500ETF 和中银中证 100ETF。其得分为 6.666 666 5 及 2.222 222 3,亦即其管理费 0.15 及的 0.45 倒数值。可使用另一参数 missing 指定如果文档不存在管理费字段时的替代数值。

```
curl --request POST http://localhost:9200/fund_basic/_search?pretty=true
--header "Content-Type: application/json" --data $'{"query": {"function_score":
{"query": {"wildcard": {"name.keyword":"中银*"}}, "field_value_factor": {"field":
"m_fee", "modifier":"reciprocal", "missing":1}}}}'
```

- gauss、linear 或 exp 等衰减函数

对用户给定文档的数字字段,通过衰减函数对文档评分。支持的衰减函数有 gauss、linear 及 exp。这三个函数的计算方法会涉及以下四个参数,用户可指定其参数值。

  ○ origin

  原点(origin)为给定的起始值,文档的数字字段数值等于这个值将评分为 1.0。

  ○ offset

  偏移量(offset)指用户给定原点的正负偏移范围。当文档的数字字段数值落在这范围内,将评分为 1.0。超出范围的数值则按照衰减函数计算其值。偏移量默认为 0。

  ○ scale

  距离尺度(scale)为给定的值。当文档的数字字段数值远离范围边界(origin±offset)为 scale 数值时,评分便衰减至 decay 值。

  ○ decay

  衰减(decay)值指当文档的数字字段数值远离范围边界(origin±offset)为 scale 数值时,评分便衰减至此指定值。默认为 0.5。

若通过衰减函数对文档评分的数字字段为数组类型,可以指定参数 multi_value_mode 决定如何计算最终值,选项有 avg(数组平均值)、min(数组内最小值)、max(数组内最大值)和 sum(数组合计值)。以下示例查询基金简称包含 "50" 这个两个字符,并分别对管理费和托管费使用 linear 衰减函数,给定的原点为 0 和距离尺度为 0.2,另外再指定参数 score_mode 以选择 sum(相加)计算综合得分。搜索结果以降序排序,匹配结果显示如表 7-17 所示。

```
curl --request POST http://localhost:9200/fund_basic/_search?pretty=true
--header "Content-Type: application/json" --data $'{"query": {"function_score":
{"functions":[{"linear":{"m_fee":{"origin":"0","scale":"0.25"}}},{"linear":{
"c_fee":{"origin":"0","scale":"0.2"}}}],"query":{"wildcard":{"name.keyword":"*50*"}},
"score_mode":"sum"}}, "_source":["name", "m_fee", "c_fee"]}'
```

表 7-17 分别对管理费和托管费使用 linear 衰减函数查询结果

匹配基金简称	管理费	托管费	综合得分
中银证券 500ETF	0.15	0.05	1.574 999 9
5G50ETF	0.45	0.1	0.85

匹配基金简称	管理费	托管费	综合得分
医药 50ETF 基金	0.5	0.1	0.75
红利低波 50ETF	0.5	0.1	0.75
500 增强 LOF	1.0	0.2	0.5

4. 脚本查询

脚本（script）查询使用自定义脚本，通常脚本查询的子句被放在布尔（bool）查询之后的过滤器（filter）子句中。以下示例查询基金管理人为易方达基金（must 查询子句），并且基金的管理费和托管费总和不大于 0.25（filter 查询子句）。在 fund_basic 索引中易方达基金有三支，其中之一的"浙江 ETF"，其费用总和大于 0.25 而被过滤，搜索结果为科技 ETF 和香港证券 ETF，费用总和都是 0.2。

```
curl --request POST http://localhost:9200/fund_basic/_search?pretty=true
--header "Content-Type: application/json" --data $'{"query": {"bool":
{"must":{"term":{"management.keyword":"易方达基金"}},"filter":{"script":{"script":{"source":
"if ((doc[\'m_fee\'].value + doc[\'c_fee\'].value) <= 1) return true; else return
false;"}}}}}, "_source":["name", "m_fee", "c_fee", "management"]}'
```

7.4 其他相关功能

Elasticsearch 还支持一些与搜索相关的接口，例如对搜索问题进行故障排除和协助复杂的搜索查询以获得更精确的搜索结果等，以下介绍常用的接口。

7.4.1 搜索多重目标接口

搜索多重目标（_msearch）接口与批量处理文档接口相似，使用方法是将多个搜索请求放入接口的请求主体中，收集在一起以批量的方式发送。请求语法如下：

```
POST /_msearch
POST /<索引名称>/_msearch
```

请求主体中，多个搜索的请求，需要使用换行符分隔的 JSON 结构，语法如下：

```
{} \newline
{搜索请求主体} \newline
{"index": "index name"} \newline
{搜索请求主体} \newline
```

搜索请求主体中可以使用适当的参数，包括 query（查询句子）、search_type（搜索操作方式）、preference（偏好）、routing（路由）、from（返回的搜索结果从第几项开始）、size（返回的搜索结果总项数）和 max_concurrent_searches（最大并发搜索数）。如果使用档案文件作为数据输入，标头内容类型必须设置为以换行符分隔的 JSON 格式（application/x-ndjson）。以下示例发出一个 _msearch 请求的，该请求包含两个子请求，查询业绩比较基准为中证 100 指数收益率和中证 500 指数收益率的基金。以下内容将被写入名为 msearch.json 的文件，注意最后一行是空行。返回的源字段设置仅为 name。

```
{}
{"query":{"term":{"benchmark":"中证100指数收益率"}}, "_source":["name"]}
{}
{"query":{"term":{"benchmark":"中证500指数收益率"}}, "_source":["name"]}
```

使用 curl 命令行参数 --data-binary 引用文件，并用符号 @ 标示文件名称 msearch.json。响应主体具有一个响应字段，其数据类型为数组，包含与两个子请求相对应的两个搜索结果，中银中证 100ETF 及中银证券 500ETF。

```
curl --request POST http://localhost:9200/fund_basic/_msearch?pretty=true
--header "Content-Type: application/x-ndjson" --data-binary @msearch.json
{
  "took" : 1,
  "responses" : [
    {
      ......
      "hits" : [
        {
          ......
          "_source" : {
            "name" : "中银中证100ETF"
          }
        },
        ......
      ]
    },
    {
      ......
      "hits" : [
        {
          ......
          "_source" : {
            "name" : "中银证券500ETF""
          }
        },
        ......
      ]
    }
  ]
}
```

7.4.2　搜索结果试算接口

搜索结果试算接口（_count）提供对搜索结果进行计数，但不提供详细信息。可以通过 URI 搜索或通过请求主体（request body）进行搜索。在表 7-18 所示列出可用的参数，并在表 7-19 中提供其简要说明，用法与 _search 接口相同。以下使用与 7.3.1 节中的项目 3 复合查询中的布尔查询相同的示例。示例查询管理费范围在 0.15 和 0.2 之间（含 0.15 和 0.2），同时上市时间范围不在 2020 年 5 月 2 日至 2020 年 5 月 25 日之间（含首末两天）。试算结果为两个基金。

```
curl --request POST http://localhost:9200/fund_basic/_count?pretty=true
--header "Content-Type: application/json" --data $'{"query": {"bool": {"must":
{"range": {"m_fee":{"gte":0.15, "lte":0.2}}}, "must_not":{"range":{"list_
date":{"gte":20200210, "lte":20200525}}}}}}'
{
  "count" : 2,
  "_shards" : {
    "total" : 1,
```

```
    "successful" : 1,
    "skipped" : 0,
    "failed" : 0
  }
}
```

表 7-18 其他搜索接口可用的参数

参数名称	_count	_explain	_field_caps	_validate
all_shards				√
allow_no_indices	√		√	√
analyzer	√	√		√
analyze_wildcard	√	√		√
default_operator	√	√		√
df	√	√		√
explain				√
expand_wildcards	√		√	√
fields			√	
ignore_throttled	√			
ignore_unavailable	√		√	√
include_unmapped			√	
lenient	√	√		√
min_score	√			
preference	√	√		
q	√	√		√
query	√	√		
rewrite				√
routing		√		
stored_fields		√		
_source		√		
_source_excludes		√		
_source_includes		√		

表 7-19 URI 其他常用参数

参数名称	描述
all_shards	是否对所有分片执行验证。默认为 false
allow_no_indices	是否当请求的索引不可用（例如关闭、不存在）时，请求不会返回错误
analyzer	使用的分析器名称
analyze_wildcard	是否分析查询字符串中的通配符，默认为 false
default_operator	指定布尔运算符（默认为 or 或 |）对条件进行分组
df	指定要搜索的默认字段
explain	是否在发生错误时响应将返回详细信息。默认为 false
expand_wildcards	通配符表达式可以扩展到什么状态的索引，选项为 all、open、closed 和 none
fields	以逗号分隔或通配符表达式，以包括的字段列表在统计信息中。
ignore_throttled	是否忽略线程的限速配置
ignore_unavailable	是否在响应中忽略丢失或闭合的索引。默认为 false
Include_unmapped	是否在响应中包括未映射的字段。默认为 false。

续表

参数名称	描述
lenient	是否忽略搜索时和索引时的数据类型不匹配错误。默认为 false
min_score	仅返回分数大于或等于 min_score 设定值的那些文档
preference	指定节点或分片以执行搜索请求
q	遵循 DSL 语法的查询句子，仅用于 URI 搜索
query	遵循 DSL 语法的查询句子，仅用于搜索请求主体中
rewrite	是否在重写 Lucene 查询返回更详细的说明，默认为 false
routing	执行搜索时，可以通过提供路由参数来控制要搜索哪些分片
stored_fields	检索在映射中标记为 stored 的那些字段。可用逗号分隔的指定字段列表
_source	输出文档的原始内容，字段名称可使用通配符
_source_excludes	过滤出文档的特定字段后再输出原始内容（_source_excludes=a,b,c）
_source_includes	输出文档的部分原始内容（_source_includes=a,b,c）

7.4.3 评分说明接口

评分说明（_explain）接口相当于 7.2.1 节通过 URI 进行搜索中的项目 2 参数 explain，提供了详细的描述来解释相关性得分的计算。请求语法如下：

```
GET /<索引名称>/_explain/<id>
POST /<索引名称>/_explain/<id>
```

表 7-18 列出可用的参数，并在表 7-19 中提供其简要说明，用法与 _search 接口相同。以下使用与 7.2.1 节相同的示例。示例使用参数 q 查询必须包含字符串 88 的基金，并使用搜索结果的文档标识符 jNaixXIBE8603ZMKlQ7G 获得该基金的得分说明。

请注意要使用运行环境中正确的文档标识符。

```
curl --request GET 'http://localhost:9200/fund_basic/_explain/jNaixXIBE8603ZMKlQ7G?pretty=true&q=88&_source_includes=name,ts_code'
{
  "_index" : "fund_basic",
  "_type" : "_doc",
  "_id" : "jNaixXIBE8603ZMKlQ7G",
  "matched" : true,
  "explanation" : {
    "value" : 3.446918,
    "description" : "max of:",
    "details" : [
      {
        "value" : 3.446918,
        "description" : "weight(name:88 in 28) [PerFieldSimilarity], result of:",
        "details" : [
          ]
      },
      ]
      ......
}
```

7.4.4 字段功能接口

字段功能（_field_caps）接口提供字段在索引中的检索功能。列举五个属性，包括是否在全部索引中可搜索（searchable），是否在全部索引中可聚合（aggregatable），是否在全部索引中具有相同的数据类型（indices），不支持这字段搜索的索引清单（non_searchable_indices）和不支持这字段聚合的索引清单（non_aggregatable_indices）。在表 7-18 所示列出可用的参数，并可在 7-19 中提供其简要说明，用法与 _search 接口相同。请求语法如下：

```
GET /_field_caps
POST /_field_caps
GET /<索引名称>/_field_caps
POST /<索引名称>/_field_caps
```

以下示例发出一个 _field_caps 请求，查询字段基金简称（name）和管理费（m_fee）的检索功能。结果显示基金简称字段是可搜索（searchable），但不可聚合（aggregatable），原因是被设定为文本数据类型。而管理费两者都支持。

提示：如果对索引字段的属性不太了解，可以使用此接口熟悉字段的使用功能。

```
curl --request GET "http://localhost:9200/_field_caps?fields=name,m_fee&pretty=true"
{
  "indices" : [
    "fund_basic"
  ],
  "fields" : {
    "m_fee" : {
      "float" : {
        "type" : "float",
        "searchable" : true,
        "aggregatable" : true
      }
    },
    "name" : {
      "text" : {
        "type" : "text",
        "searchable" : true,
        "aggregatable" : false
      }
    }
  }
}
```

7.4.5 搜索查询评估接口

搜索查询评估（_validate）接口提供了一种无需耗时真实执行查询，即可对查询进行评估的方法。可针对复杂的查询，缩短其测试时间。在表 7-18 列出可用的参数，并可在 7-19 中提供其简要说明，用法与 _search 接口相同。请求语法如下：

```
GET /<index>/_validate/<query>
```

以下示例验证无效的查询，该查询将字段管理费（数字数据类型）与文字匹配。同时可以使用 URL 参数 explain=true 以获得有关失败原因的详细说明。示例结果（"valid"：false）显示为无效的查询，原因是数字格式异常（NumberFormatException）。

```
curl --request GET "http://localhost:9200/fund_basic/_validate/query?pretty=true&explain=true" --header "Content-Type: application/json" --data $'{"query":{"term":{"m_fee":{"value":"abc"}}}}'
{
  "_shards" : {
    "total" : 1,
    "successful" : 1,
```

```
          "failed" : 0
      },
      "valid" : false,
      "explanations" : [
        {
          "index" : "fund_basic",
          "valid" : false,
          "error" : "[fund_basic/H0jaCw-kS5-kjmMigPm79w] QueryShardException[failed to create
query: {\n                   \"term\" : {\n\"m_fee\" : {\n\"value\" : \"abc\",\n\"boost\"
 : 1.0\n}\n}\n}]; nested: NumberFormatException[For input string:\"abc\"];; java.lang.
NumberFormatException: For input string: \"abc\""
        },
        ……
      ]
    }
```

7.4.6 性能分析设置参数

在搜索请求中,可通过性能分析参数 profile 设置为 true 打开性能分析,搜索结果附带有关查询过程中各个步骤非常具体的信息。Elasticsearch 为了获得所有这些详细信息,许多低阶方法被使用来提供分析每个步骤所用时间的明细。在完全深入了解查询结果的结构之后,有助于改进查询方式。以下使用与 7.4.2 节相同的示例,并开启参数 profile 设置。示例查询管理费范围在 0.15 和 0.2 之间(含 0.15 和 0.2),同时上市时间范围不在 2020 年 5 月 2 日至 2020 年 5 月 25 日之间(含首末两天)。结果显示查询时间为 354 749 纳秒并分解描述每一小步所花费的时间(breakdown),此外有更多说明两个条件中每个条件(children)所需的时间。

```
    curl --request POST http://localhost:9200/fund_basic/_search?pretty=true
--header "Content-Type: application/json" --data $'{"query": {"bool": {"must":
{"range": {"m_fee":{"gte":0.15, "lte":0.2}}}, "must_not":{"range":{"list_
date":{"gte":20200210, "lte":20200525}}}}}, "profile":true}'
{
    "took" : 2,
    "timed_out" : false,
    ……
    "hits" : {
    ……
    },
    "profile" : {
      "shards" : [
        {
          "id" : "[KQ8MXhJzSgymiXWUjMqMqg][fund_basic][0]",
          "searches" : [
            {
              "query" : [
                {
                  "type" : "BooleanQuery",
                  "description" : "+m_fee:[0.15 TO 0.2] -list_date:[1581292800000
 TO 1590451199999]",
                  "time_in_nanos" : 354749,
                  "breakdown" : {
                  ……
                  },
                  "children" : [
                    {
                      "type" : "IndexOrDocValuesQuery",
                      "description" : "m_fee:[0.15 TO 0.2]",
                      "time_in_nanos" : 117608,
                      "breakdown" : {
```

```
            ......
         }
      },
      {
         "type" : "IndexOrDocValuesQuery",
         "description" : "list_date:[1581292800000 TO 1590451199999]",
         "time_in_nanos" : 97117,
         "breakdown" : {
            ......
         }
      },
   ......
}
```

7.4.7 查询建议器

查询建议器（suggester）被广泛用于改善用户体验的搜索解决方案中。建议器功能通过参数 suggest、提供的文本和目标字段来建议查询字符串。查询建议器有以下四种类型：

1. 词条（term）建议器

此建议器先对提供的文本进行分析，再返回可能正确的词条列表提供搜索使用。以下示例在文本中银证券 500ETF 对基金简称字段寻求查询字符串的建议。结果显示四个建议词条，分别为中银、证券、500 和 ETF，可用于搜索操作。

```
curl --request POST http://localhost:9200/fund_basic/_search?pretty=true
--header "Content-Type: application/json" --data $'{"suggest": {"text":"中银证券500ETF", "suggester1":{"term":{"field":"name"}}}}'
{
......
         "suggest" : {
            "suggester1" : [
               {
                  "text" : "中银",
                  ......
               },
               {
                  "text" : "证券",
                  ......
               },
               {
                  "text" : "500",
                  ......
               },
               {
                  "text" : "ETF",
                  ......
               }
            ]
         …..
}
```

2. 短语（phrase）建议器

此建议器返回可能正确的短语列表提供搜索使用。提供的建议是根据词条建议器改进发展而来。生成的短语是基于词汇单元的频率和并发出现的选择方法。以下示例在文本中银证券 500ETF 对基金简称字段寻求查询字符串短语建议。结果显示建议短语为原来文本。

```
curl --request POST http://localhost:9200/fund_basic/_search?pretty=true
--header "Content-Type: application/json" --data $'{"suggest": {"text":"中银证券500ETF", "suggester1":{"phrase":{"field":"name"}}}}'
```

3. 完成（completion）建议器

此建议器提供了一种以完成用户键入自动补全的建议词条。要提供此功能的成本比较高，必须首先在映射中指定具有建议完成的字段。以下示例将基金简称和基金代码这两个字段从原来的文本类型更改为完成（completion）数据类型，原来要更改的内容如下：

```
"name": {"type": "text", "fields": {"keyword": {"type": "keyword"}},
"analyzer":"my_hanlp_analyzer"}, "ts_code": {"type": "text", "fields": {"keyword":
{"type": "keyword"}}, "analyzer":"my_hanlp_analyzer"},
```

更改后的内容如下："name":｛"type"："completion"，"analyzer"："my_hanlp_analyzer"｝，"ts_code"：｛"type"："completion"，"analyzer"："my_hanlp_analyzer"｝，

文件 my_suggester_mappings.json 包含上述映射，步骤是通过运行 bash 文件 fund_basic_suggester_bulk_index.sh。这 bash 文件首先删除 fund_basic_suggester 索引（如果存在）。然后使用 my_suggester_mappings.json 内的映射和设置创建 fund_basic_suggester 索引。最后，对文件 fund_basic_bulk.json 内 50 只基金在索引 fund_basic_suggester 下进行了文件索引。执行步骤以及相应的结果是如下：

（1）执行 fund_basic_suggester_bulk_index.sh bash 文件

```
./fund_basic_suggester_bulk_index.sh
{"acknowledged":true,"shards_acknowledged":true,"index":"fund_basic_
suggester"}{
  "took" : 109,
  "errors" : false,
  "items" : [
    {
      "index" : {
        "_index" : "fund_basic_suggester",
        "_type" : "_doc",
        "_id" : "OJ5URHMBpfHDzdHU1_zu",
        "_version" : 1,
        "result" : "created",
        "_shards" : {
          "total" : 2,
          "successful" : 1,
          "failed" : 0
        },
        "_seq_no" : 0,
        "_primary_term" : 1,
        "status" : 201
      }
      ......
    ]
    ......
}
```

（2）对基金简称这个字段使用前缀为"中银"来测试完成建议器，执行命令以及相应的结果如下，建议查询的基金为中银中证 100ETF 和中银证券 500ETF：

```
curl --request POST http://localhost:9200/fund_basic_suggester/_
search?pretty=true--header "Content-Type: application/json" --data $'{"suggest":
{"suggest_name" :{"prefix" : 中银, "completion": {"field":"name"}}},
"_source":["name", "ts_code"]}'
{
  ......
  "suggest" : {
    "suggest_name" : [
```

```
      {
        "text" : "中银",
        "offset" : 0,
        "length" : 2,
        "options" : [
          {
            "text" : "中银中证100ETF",
            "_index" : "fund_basic_suggester",
            "_type" : "_doc",
            "_id" : "PJ5URHMBpfHDzdHU1_zu",
            "_score" : 1.0,
            "_source" : {
              "ts_code" : "515670.SH",
              "name" : "中银中证100ETF"
            }
          },
          {
            "text" : "中银证券500ETF",
            "_index" : "fund_basic_suggester",
            "_type" : "_doc",
            "_id" : "OZ5URHMBpfHDzdHU1_zu",
            "_score" : 1.0,
            "_source" : {
              "ts_code" : "515190.SH",
              "name" : "中银证券500ETF"
            }
          }
      ……
}
```

（3）对基金代码这个字段使用前缀为"5150"的完成建议器，执行命令以及相应的结果如下，建议查询的基金为515030.SH 和 515090.SH：

```
curl --request POST http://localhost:9200/fund_basic_suggester/_search?pretty=true--header "Content-Type: application/json" --data $'{"suggest":{"suggest_ts_code" :{"prefix" : "5150", "completion": {"field":"ts_code"}}}, "_source":["name", "ts_code"]}'
{
……
  "suggest" : {
    "suggest_ts_code" : [
      {
        "text" : "5150",
        "offset" : 0,
        "length" : 4,
        "options" : [
          {
            "text" : "515030.SH",
            "_index" : "fund_basic_suggester",
            "_type" : "_doc",
            "_id" : "kZ7hRHMBpfHDzdHUCfyQ",
            "_score" : 1.0,
            "_source" : {
              "ts_code" : "515030.SH",
              "name" : "新能源车 ETF"
            }
          },
          {
            "text" : "515090.SH",
            "_index" : "fund_basic_suggester",
            "_type" : "_doc",
            "_id" : "mp7hRHMBpfHDzdHUCfyQ",
```

```
            "_score" : 1.0,
            "_source" : {
              "ts_code" : "515090.SH",
              "name" : "可持续发展ETF"
            }
          }
          ......
        }
```

(4) 对基金代码这个字段使用正则（regex）查询，前缀为 5150 再加向上数字 0 至 5 其中之一的完成建议器。执行命令以及相应的结果如下，建议查询的基金为 515030.SH。

提示：前缀为 5150 有两个基金，515030.SH 和 515090.SH。因为 515090.SH 不满足接着是 0 至 5 的条件，因此没有被建议。

```
curl --request POST http://localhost:9200/fund_basic_suggester/_search?pretty=true --header "Content-Type: application/json" --data $'{"suggest":
{"suggest_name" :{"regex" : "5150[0-5]+", "completion": {"field":"ts_code"}}},
"_source":["name", "ts_code"]}'
{
  ......
  "suggest" : {
    "suggest_name" : [
      {
        "text" : "5150[0-5]+",
        "offset" : 0,
        "length" : 10,
        "options" : [
          {
            "text" : "515030.SH",
            ......
            "_source" : {
              "ts_code" : "515030.SH",
              "name" : "新能源车ETF"
            }
          }
          ......
        }
```

4. 上下文（context）建议器

此建议器提供了一种与上下文（语境）相关联方式给出自动补全完成的建议词条。

（1）在映射中指定上下文列表（亦即不同的上下文名称），并按照上下文索引文档，以便可以在指定的上下文进行搜索。假设字段 suggest_with_context 与风险评测相关，可以设置语境 risk_type 为 1 至 5 以获得不同语境下的建议。按照以下步骤测试上下文建议器：

① 在定义字段 suggest_with_context 的映射时，也定义上下文 contexts。contexts 在当前支持的数据类型 type 有 category 和 geo。contexts 可以同时定义多个上下文。以下对字段 suggest_with_context 定义了 risk_type 为 category。

```
    "suggest_with_context": {"type":"completion", "contexts":[{"name":"risk_type",
"type":"category"}]},
```

② 在文档索引时，字段 suggest_with_context 使用参数 input 指定字段内容，并使用参数 contexts.risk_type 指定上下文。

```
    "suggest_with_context":{"input":["ETF"], "contexts":{"risk_type":["1","2"]}}
```

（2）本测试提供了 fund_basic_suggester_bulk_index.sh、fund_basic_context_bulk.json 和 fund_basic_context_suggester_bulk_index.sh 三个文件。执行 bash 文件将创建索引 fund_basic_ context_suggester 和对四个文档进行索引。这四个文档包括基金简称、字段 suggest_with_context 的内容和随意给予的上下文 risk_type 显示如表 7-20 所示。

（3）定义名因为 risk_suggestion 的上下文建议器。使用上下文 contexts.risk_type 为 3，获取字段 suggest_with_context 的字首为 "e" 的查询建议。结果显示自动补全完成建议的基金只有中银证券 500ETF。其字段 suggest_with_context 的内容为 ETF，以及其 contexts.risk_type 为 3 和 4。

```
curl --request POST http://localhost:9200/fund_basic_context_suggester/_search?pretty=true--header "Content-Type: application/json" --data
$'{"suggest":{"risk_suggestion":{"prefix": "e","completion":{"field": "suggest_with_context","contexts":{"risk_type":["3"]}}}}}'
{
  "suggest" : {
    "risk_suggestion" : [
      {
        "text" : "ETF",
        "index" : "fund_basic_context_suggester",
        "_source" : {
          "suggest_with_context" : {
            "input" : [
              "ETF"
            ],
            "contexts" : {
              "risk_type" : [
                "2",
                "3"
              ]
            }
          },
          "ts_code" : "515190.SH",
          "name" : "中银证券500ETF",
          ......
}
```

表 7-20 四个文档的基金简称和上下文的 risk_type

基金简称	input	上下文的 risk_type
家电 ETF	ETF	["1","2"]
中银证券 500ETF	ETF	["2","3"]
浙江 ETF	ETF	["4","5"]
银行 ETF 富国	ETF	["5"]

上下文 contexts 的数据类型 type 为 geo 的测试，读者可以参考官方网站深入研究。

第二篇 数据建模、聚合框架、管道处理和探索性数据分析

第 8 章

数据建模

Elasticsearch 是面向文档的分布式 NoSQL 数据库。旨在存储，检索和管理面向文档或半结构化的数据。基本上，存储文档时仅对文档之间的关系提供最小的支持。本章重点介绍数据建模技术，以及一些最常见的问题及其使用不同技术的解决方案，其中包括使用嵌套关系和父类子类关系处理关系数据。

8.1 数据建模及方法

数据建模的常用方法是通过使用关系数据库的规范化过程，来减少数据冗余并提高数据完整性。但是 Elasticsearch 不是为了要建立关系型数据库而产生的，因此这种数据建模类型并不适用。与大多数 NoSQL 数据库一样，Elasticsearch 一般将现实世界视为一个平面世界，每个文档都是独立并包含所有信息。为了弥合两个世界之间的"鸿沟"，Elasticsearch 支持几种常用的数据建模技术，包括数据非规范化、内部对象、嵌套对象和连接数据类型。在某种意义上，数据建模的目的通常是使搜索操作尽可能轻便。无论使用哪种技术，都需要预测用户查询的方式，然后才能做出良好的数据建模决策。

8.1.1 使用非规范化方法

对数据进行非规范化的真正含义，是通过在两种不同类型的文档中保留一个或多个冗余字段，来保持一种扁平结构来建立关系。由于每种类型的文档都满足查询所需的所有信息，不需要关联两种不同类型的文档，因此可以避免索引之间文档的其他联接，加快搜索速度。

在第二章已经介绍 Tushare 提供 fund_company 接口，可以获得公募基金管理公司的相关信息。数据之一名为简称（shortname），该数据是相对应于公募基金数据的基金管理人（name），这意味着在搜索系统中是个冗余字段。如果用户想查看某基金的基金管理人的详细信息，可以建立 fund_company 索引，再从 fund_basic 索引中引用基金管理人名称来搜索 fund_company 索引的简称字段。使用数据非规范化方法建模说明如图 8-1 所示。

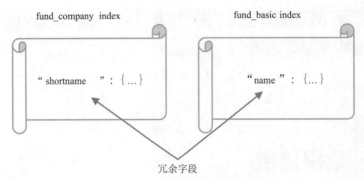

图 8-1 使用数据非规范化方法建模

1. 准备测试环境

本章提供了 fund_company_bulk.json、fund_company_bulk_index.sh 和 fund_company_mappings.json 3 个文件，fund_company_bulk_index.sh 是个 bash 文件，运行执行以下任务：

（1）使用在 fund_company_mappings.json 内定义的设定和分析器创建 fund_company 索引。除了 location 字段外，其他字段与表 2-2 所示的公募基金公司数据列表一一对应。location 字段为基金公司注册地址所在城市的经度和纬度，数据类型是 geo_point。采用的显式映射内容如下：

```
{
    "settings" : {"analysis": {"analyzer": {
            "my_hanlp_analyzer": {"tokenizer": "my_hanlp_tokenizer", "filter":
["unique"]},
            "comma_analyzer": {"tokenizer":"comma_tokenizer"}},
            "tokenizer": {"my_hanlp_tokenizer": {"type": "hanlp",
                "enable_custom_config": true, " enable_custom_dictionary ":
false},
                "comma_tokenizer":{"type":"pattern", "pattern":"[,]"}}}},
    "mappings" : {"dynamic": false, "properties": {
        "name": {"type": "keyword"},
        "shortname": {"type": "keyword"},
        "province": {"type": "keyword"},
        "city": {"type": "keyword"},
        "address": {"type": "keyword"},
        "phone": {"type": "text", "analyzer":"classic"},
        "office": {"type": "keyword"},
        "website": {"type": "keyword"},
        "chairman": {"type": "keyword"},
        "manager": {"type": "keyword"},
        "setup_date": {"type": "date", "format": "yyyyMMdd"},
        "end_date": {"type": "date", "format": "yyyyMMdd"},
        "main_business": {"type": "text", "fields": {"keyword":
{"type": "keyword"}}, "analyzer":"my_hanlp_analyzer"},
        "org_code": {"type": "keyword"},
        "credit_code": {"type": "keyword"},
        "reg_capital": {"type": "integer"},
        "employees": {"type": "integer"},
        "location": { "type": "geo_point" }}}}
```

（2）使用 fund_company_bulk.json 内编写好的批量处理文档接口指令，只对那 50 个场内基金的所有基金管理人进行文档索引，总共 26 家基金公司。执行 fund_company_bulk_index.sh 后的部分索引结果显示如下：

```
./fund_company_bulk_index.sh
{"acknowledged":true,"shards_acknowledged":true,"index":"fund_company"}{
  "took" : 40,
  "errors" : false,
  "items" : [
    {
      "index" : {
        "_index" : "fund_company",
        "_type" : "_doc",
        "_id" : "xp6-UHMBpfHDzdHUlvwB",
        "_version" : 1,
        "result" : "created",
        "_shards" : {
          "total" : 2,
          "successful" : 1,
          "failed" : 0
        },
        "_seq_no" : 0,
        "_primary_term" : 1,
        "status" : 201
      }
    },
    ……
}
```

2. 搜索简称为博时基金的基金公司

基金简称为恒生湾区的基金管理人为博时基金，以下示例在 fund_company 索引下对简称（shortname）字段进行匹配查询（match query），查询字符串为博时基金。搜索结果如下：

```
curl --request POST http://localhost:9200/fund_company/_search?pretty=true
--header "Content-Type: application/json" --data $'{"query": {"match" : {"shortname":
{"query" : "博时基金"}}}}'
{
  ……
  "hits" : [
    {
      "_index" : "fund_company",
      "_type" : "_doc",
      "_id" : "lawAtnUBIhDpB0g9Zh87",
      "_score" : 2.8903718,
      "_source" : {
        "name" : "博时基金管理有限公司",
        "shortname" : "博时基金",
        "province" : "广东",
        "city" : "深圳市",
        "address" : "广东省深圳市福田区深南大道 7088 号招商银行大厦 29 层",
        "phone" : "86-755-83169999,86-95105568",
        "office" : "广东省深圳市福田区益田路 5999 号基金大厦 21 层",
        "website" : "www.bosera.com",
        "chairman" : "张光华",
        "manager" : "江向阳",
        "reg_capital" : 25000,
        "setup_date" : "19980713",
        "end_date" : null,
        "employees" : 323,
        "main_business" : null,
        "org_code" : "710922202",
        "credit_code" : "91440300710922202N",
        "location" : {
          "lat" : 22.5431,
          "lon" : 114.0579
        }
```

```
        ……
}
```

使用数据非规范化方法来解决关系很简单。但是若需要执行涉及多个相关索引（例如 fund_company 和 fund_basic）的查询，则仅使用数据非规范化方法就会无法实现，必须使用编程方式来解决问题。

8.1.2 使用对象数据类型方法

若需要时常执行涉及多个相关索引的查询，使用对象数组将公募基金数据嵌入到公募基金公司数据中是一种解决方法。通过这种方法，无需为单个公募基金数据索引很多文档。只有当该公募基金公司成为另一个新基金的管理人时，才需要更新相应的文档。由于要成为一个新基金的管理人的机会不高，因此这种方法似乎是可行的。

在 fund_company 数据结构中，添加了一个名为 fund_list 的对象字段、以表示由该基金公司管理的基金清单，从而合并 fund_basic 数据到 fund_company 数据。而公募基金数据中的基金管理人为冗余字段，所以删除。正如我们所知 Elasticsearch 没有数组数据类型，任何字段都可以包含零个或多个值（表示数组）。使用对象数据类型方法建模说明如图 8-2 所示。

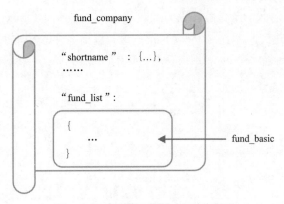

图 8-2　使用对象数据类型方法建模

1. fund_company_fund_basic 索引的显式映射内容

采用的显式映射内容如下：

```
{
        "settings" : {"analysis": {"analyzer": {
            "my_hanlp_analyzer": {"tokenizer": "my_hanlp_tokenizer", "filter": ["unique"]},
            "comma_analyzer": {"tokenizer":"comma_tokenizer"}},
            "tokenizer": {"my_hanlp_tokenizer": {"type": "hanlp",
            "enable_custom_config": true, " enable_custom_dictionary ": false},
            "comma_tokenizer":{"type":"pattern", "pattern":"[,]"}}}},
        "mappings" : {"dynamic": false, "properties": {
            "name": {"type": "keyword"},
            "shortname": {"type": "keyword"},
            "province": {"type": "keyword"},
            "city": {"type": "keyword"},
            "address": {"type": "keyword"},
            "phone": {"type": "text", "analyzer":"classic"},
            "office": {"type": "keyword"},
            "website": {"type": "keyword"},
```

```
                "chairman": {"type": "keyword"},
                "manager": {"type": "keyword"},
                "setup_date": {"type": "date", "format": "yyyyMMdd"},
                "end_date": {"type": "date", "format": "yyyyMMdd"},
                "main_business": {"type": "text", "fields": {"keyword": {"type": "keyword"}}, "analyzer":"my_hanlp_analyzer"},
                "org_code": {"type": "keyword"},
                "credit_code": {"type": "keyword"},
                "reg_capital": {"type": "integer"},
                "employees": {"type": "integer"},
                "location": { "type": "geo_point" },
                "fund_list":{ "properties": {
                    "benchmark": {"type": "keyword"},
                    "status": {"type": "keyword"},
                    "fund_type": {"type": "keyword"},
                    "invest_type": {"type": "keyword"},
                    "type": {"type": "keyword"},
                    "market": {"type": "keyword"},
                    "c_fee": {"type": "float"},
                    "issue_amount": {"type": "float"},
                    "m_fee": {"type": "float"},
                    "min_amount": {"type": "float"},
                    "p_value": {"type": "float"},
                    "duration_year": {"type": "float"},
                    "exp_return": {"type": "float"},
                    "issue_date": {"type": "date", "format": "yyyyMMdd"},
                    "list_date": {"type": "date", "format": "yyyyMMdd"},
                    "purc_startdate": {"type": "date", "format": "yyyyMMdd"},
                    "redm_startdate": {"type": "date", "format": "yyyyMMdd"},
                    "found_date": {"type": "date", "format": "yyyyMMdd"},
                    "due_date": {"type": "date", "format": "yyyyMMdd"},
                    "delist_date": {"type": "date", "format": "yyyyMMdd"},
                    "management": {"type": "text", "fields": {"keyword": {"type": "keyword"}}, "analyzer":"my_hanlp_analyzer"},
                    "name": {"type": "text", "fields": {"keyword": {"type": "keyword"}}, "analyzer":"my_hanlp_analyzer"},
                    "ts_code": {"type": "text", "fields": {"keyword": {"type": "keyword"}}, "analyzer":"my_hanlp_analyzer"},
                    "trustee": {"type": "text", "fields": {"keyword": {"type": "keyword"}}, "analyzer":"my_hanlp_analyzer"},
                    "custodian": {"type": "text", "fields": {"keyword": {"type": "keyword"}}, "analyzer":"my_hanlp_analyzer"}}}}
    }
```

2. 准备测试环境

本测试需求 3 个文件，分别为 fund_company_fund_basic_bulk_index.sh、fund_company_fund_basic_bulk.json 和 fund_company_fund_basic_object_mappings.json。其中 fund_company_fund_basic_bulk_index.sh 是个 bash 文件，运行执行以下任务：

（1）使用在 fund_company_fund_basic_object_mappings.json 内定义的设定和分析器创建 fund_company_fund_basic 索引。

（2）使用 fund_company_fund_basic_bulk.json 内编写好的批量处理文档接口指令，对 26 家基金管理公司管理的 50 个基金进行文档索引。而 fund_basic 数据合并到 fund_company 数据中。部分索引结果显示如下：

```
./fund_company_fund_basic_bulk_index.sh
{"acknowledged":true,"shards_acknowledged":true,"index":"fund_company_fund_basic"}{
  "took" : 56,
  "errors" : false,
```

```
    "items" : [
      {
        "index" : {
          "_index" : "fund_company_fund_basic",
          "_type" : "_doc",
          "_id" : "1J6MVHMBpfHDzdHUJv29",
          "_version" : 1,
          "result" : "created",
          "_shards" : {
            "total" : 2,
            "successful" : 1,
            "failed" : 0
          },
          "_seq_no" : 0,
          "_primary_term" : 1,
          "status" : 201
        }
      },
      ......
}
```

（3）获取索引 fund_company_fund_basic 的文档总数，结果显示如下：

```
curl --request POST http://localhost:9200/fund_company_fund_basic/_count?pretty=true
{
  "count" : 26,
  ......
}
```

3. 搜索与聚合

以下示例对注册地址所在城市为"珠海市"的基金管理公司进行搜索与值计数聚合。该示例涉及聚合操作将在第 9 中介绍，并关联公募基金公司数据和公募基金数据，结果显示有 5 个公募基金，包括浙江 ETF、科技 ETF、香港证券 ETF、800ETF 和芯片基金。

提示：搜索结果返回两个文档总共 5 个基金。在易方达基金管理有限公司旗下有浙江 ETF、科技 ETF 和香港证券 ETF。在广发基金管理有限公司旗下有 800ETF 和芯片基金。

```
curl --request POST http://localhost:9200/fund_company_fund_basic/_search?pretty=true   --header "Content-Type: application/json" --data $'{"query":{"match" : {"city": {"query" :
"珠海市"}}}, "aggs":{"types_count":{"value_count":{"field":"fund_list.ts_code.keyword"}}},
"_source":["name", "city","fund_list.name"]}'
{
  ......
  "hits" : {
    "total" : {
      "value" : 2,
      "relation" : "eq"
    },
    "hits" : [
      {
        ......
        "_id" : "sawHtnUBIhDpB0g93R-D",
        "_source" : {
          "city" : "珠海市",
          "fund_list" : [
            {
              "name" : "浙江ETF"
            },
            {
              "name" : "科技ETF"
```

```
            },
            {
              "name" : "香港证券ETF"
            }
          ],
          "name" : "易方达基金管理有限公司"
        }
      },
      {
        ……
        "_id" : "tqwHtnUBIhDpB0g93R-D",
        "_source" : {
          "city" : "珠海市",
          "fund_list" : [
            {
              "name" : "800ETF"
            },
            {
              "name" : "芯片基金"
            }
          ],
          "name" : "广发基金管理有限公司"
        }
      }
      ……
    },
    "aggregations" : {
      "types_count" : {
        "value" : 5
      }
    ……
}
```

4. 搜索数组内对象

以下示例对注册地址所在城市为"珠海市"的基金管理公司，并且其管理的基金中，其托管人为"中国银行"进行搜索。结果显示有1个基金管理公司及其所有管理的基金，而其中一些资基金并没有满足搜索条件。

提示：Elasticsearch没有内部对象的概念，当使用对象数组时，它将对象层次结构简化为字段名称和值的简单列表，而数组内各个对象没有保留其独立属性，所以结果是个整体。

```
curl --request POST http://localhost:9200/fund_company_fund_basic/_search?
pretty=true --header "Content-Type: application/json" --data $'{"query":
{"bool":{"must":[{"term" : {"city" : "珠海市"}},
{"term":{"fund_list.custodian.keyword":"中国银行"}}]}}, "_source":["name", "city",
"fund_list.name", "fund_list.custodian"]}'
{
  ……
  "hits" : {
  "total" : {
    "value" : 1,
    "relation" : "eq"
  },
  "hits" : [
    {
      ……
      "_id" : "sawHtnUBIhDpB0g93R-D",
      "_score" : 4.0783315,
      "_source" : {
        "city" : "珠海市",
        "fund_list" : [
          {
```

```
              "custodian" : "中国工商银行",
              "name" : "浙江ETF"
            },
            {
              "custodian" : "中国银行",
              "name" : "科技ETF"
            },
            {
              "custodian" : "招商银行",
              "name" : "香港证券ETF"
            }
        ],
        "name" : "易方达基金管理有限公司"
      }
      ……
}
```

8.1.3 使用嵌套数据类型方法

如果需要保持数组中每个对象的独立属性，需要使用嵌套（nested）数据类型而不是对象数据类型。嵌套数据类型将整个对象映射为单个字段，并允许对其内容进行简单搜索。

提示：要维持嵌套数据类型的成本比较高，因此若查询与数组内部各个对象间的独立属性无关，使用对象数据类型是一个更好的选择。

1. 准备测试环境

采用的显式映射内容与使用对象数据类型相似，不同处只在于嵌套的对象必须声明为 nested 数据类型，原来要更改的内容如下：

```
"fund_list":{ "properties": {
```

更改后的内容如下：

```
"fund_list":{ "type":"nested", "properties": {
```

本测试需求三个文件，分别为 fund_company_fund_basic_nested_bulk_index.sh、fund_company_fund_basic_nested_mappings.json 和 fund_company_fund_basic_bulk.json。其中 fund_company_fund_basic_nested_bulk_index.sh 是个 bash 文件，运行执行以下任务：

（1）使用在 fund_company_fund_basic_nested_mappings.json 内定义的设定和分析器创建 fund_company_fund_basic_nested 索引。

（2）使用 fund_company_fund_basic_bulk.json 内编写好的批量处理文档接口指令，对 26 家基金公司管理的 50 个基金进行文档索引。而 fund_basic 数据嵌套到 fund_company 数据中。部分索引结果显示如下：

```
./fund_company_fund_basic_nested_bulk_index.sh
{"acknowledged":true,"shards_acknowledged":true,"index":"fund_company_fund_basic_nested"}{
    "took" : 53,
    "errors" : false,
    "items" : [
      {
        "index" : {
          "_index" : "fund_company_fund_basic_nested",
          "_type" : "_doc",
          "_id" : "uZ6KVHMBpfHDzdHUFv1i",
          "_version" : 1,
```

```
      "result" : "created",
      "_shards" : {
        "total" : 2,
        "successful" : 1,
        "failed" : 0
      },
      "_seq_no" : 0,
      "_primary_term" : 1,
      "status" : 201
    }
  },
  ……
}
```

(3) 获取索引 fund_company_fund_basic_nested 的文档总数,结果显示如下:

```
curl --request POST http://localhost:9200/fund_company_fund_basic_nested/_count?pretty=true
{
  "count" : 26,
  ……
}
```

2. 搜索与聚合

以下示例与 8.1.2 使用对象数据类型方法一节相同,示例涉及聚合操作并需要标明嵌套字段 fund_list 的路径。搜索结果类似于前面的示例,同样显示有 5 家公募基金。

```
curl --request POST http://localhost:9200/fund_company_fund_basic_nested/_search?pretty=true--header "Content-Type: application/json" --data $'{"query":{"bool":{"must":[{"term" : {"city" : "珠海市"}}]}}, "aggs":{"total_registered":{"nested":{"path":"fund_list"},"aggs":{"count":{"value_count":{"field":"fund_list.ts_code.keyword"}}}}}, "_source":["name", "city","fund_list.name"]}'
{
  ……
  "aggregations" : {
    "total_registered" : {
      "doc_count" : 5,
  ……
}
```

3. 搜索数组内对象

以下示例搜索基金管理公司的注册地址所在城市为"珠海市"并且其基金托管人为中国银行。示例使用参数 inner_hits,用于获取哪个嵌套对象导致实际的匹配,并在返回响应主体中列出,结果显示"科技 ETF"为实际匹配的基金。

```
curl --request POST http://localhost:9200/fund_company_fund_basic_nested/_search?pretty=true--header "Content-Type: application/json" --data $'{"query":{"bool":{"must":[{"term" : {"city" : "珠海市"}}, {"nested":{"path":"fund_list","query":{"term":{"fund_list.custodian.keyword":"中国银行"}},"inner_hits":{}}}]}},"_source":["name", "city","fund_list.name"]}'
{
  "took" : 2,
  "timed_out" : false,
  ……
  "inner_hits" : {
    "fund_list" : {
      "hits" : {
        "total" : {
          "value" : 1,
          "relation" : "eq"
        },
```

```
            "hits" : [
              {
                ......
                "_id" : "y6wstnUBIhDpB0g9Lx9l",
                "_nested" : {
                  "field" : "fund_list",
                  "offset" : 1
                "_source" : {
                "ts_code" : "159807.SZ",
                "name" : "科技 ETF",
                "custodian" : "中国银行",
        ......
  }
```

8.1.4 使用父子类关联数据类型方法

Elasticsearch 支持称为 join（关联）的特殊数据类型，以在相同索引的文档中创建简单父子角色或多层级父子角色。根据建议，仅当数据包含一对多关系，并且在一方的实体数量明显超过另一方时，才应该使用关联数据类型。如果查询性能是主要的关注点，请不要使用这种类型的查询，这是因为其性能比其他建模方法要慢得多。以下使用 join 数据类型在文档中定义一组父子类关系（relations）名为 fund_company_fund_basic。父角色名为 fund_company_role，而子角色名为 fund_basic_role。父角色的字段与子角色的字段均在相同的层次结构中，没有表示包含含义。因为字段名称 name 在两个角色文档中都使用，所以在子角色文档中的字段更改为 fund_name，基金管理人字段 management 相当于字段 name，所以被删除。使用父子类关联数据类型方法建模说明如图 8-3 所示。

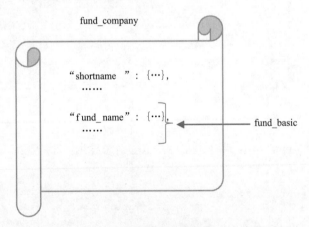

图 8-3 使用父子类关联数据类型方法建模

1.fund_company_fund_basic_join 索引的显式映射内容

采用的显式映射内容如下：

```
{
        "settings" : {"analysis": {"analyzer": {
           "my_hanlp_analyzer": {"tokenizer": "my_hanlp_tokenizer", "filter":
["unique"]},
           "comma_analyzer": {"tokenizer":"comma_tokenizer"}},
       "tokenizer": {"my_hanlp_tokenizer": {"type": "hanlp",
                  "enable_custom_config": true,
                  " enable_custom_dictionary ": false},
```

```
                    "comma_tokenizer":{"type":"pattern", "pattern":"[,]"}}}},
        "mappings" : {"dynamic": false, "properties": {
            "fund_company_fund_basic": {
                "type":"join",
                "relations": {"fund_company_role":"fund_basic_role"}
            },
            "name": {"type": "keyword"},
            "shortname": {"type": "keyword"},
            "province": {"type": "keyword"},
            "city": {"type": "keyword"},
            "address": {"type": "keyword"},
            "phone": {"type": "text", "analyzer":"classic"},
            "office": {"type": "keyword"},
            "website": {"type": "keyword"},
            "chairman": {"type": "keyword"},
            "manager": {"type": "keyword"},
            "setup_date": {"type": "date", "format": "yyyyMMdd"},
            "end_date": {"type": "date", "format": "yyyyMMdd"},
                "main_business": {"type": "text", "fields": {"keyword": {"type": "keyword"}}, "analyzer":"my_hanlp_analyzer"},
            "org_code": {"type": "keyword"},
            "credit_code": {"type": "keyword"},
            "reg_capital": {"type": "integer"},
            "employees": {"type": "integer"},
            "location": {"type": "geo_point" },
            "benchmark": {"type": "keyword"},
            "status": {"type": "keyword"},
            "fund_type": {"type": "keyword"},
            "invest_type": {"type": "keyword"},
            "type": {"type": "keyword"},
            "market": {"type": "keyword"},
            "c_fee": {"type": "float"},
            "issue_amount": {"type": "float"},
            "m_fee": {"type": "float"},
            "min_amount": {"type": "float"},
            "p_value": {"type": "float"},
            "duration_year": {"type": "float"},
            "exp_return": {"type": "float"},
            "issue_date": {"type": "date", "format": "yyyyMMdd"},
            "list_date": {"type": "date", "format": "yyyyMMdd"},
            "purc_startdate": {"type": "date", "format": "yyyyMMdd"},
            "redm_startdate": {"type": "date", "format": "yyyyMMdd"},
            "found_date": {"type": "date", "format": "yyyyMMdd"},
            "due_date": {"type": "date", "format": "yyyyMMdd"},
            "delist_date": {"type": "date", "format": "yyyyMMdd"},
                "fund_name": {"type": "text", "fields": {"keyword": {"type": "keyword"}}, "analyzer":"my_hanlp_analyzer"},
                "ts_code": {"type": "text", "fields": {"keyword": {"type": "keyword"}}, "analyzer":"my_hanlp_analyzer"},
                "trustee": {"type": "text", "fields": {"keyword": {"type": "keyword"}}, "analyzer":"my_hanlp_analyzer"},
                "custodian": {"type": "text", "fields": {"keyword": {"type": "keyword"}}, "analyzer":"my_hanlp_analyzer"}}}
    }
```

2. 准备测试环境

本测试需要 4 个文件，分别为 fund_company_fund_basic_join_bulk_index.sh、fund_company_fund_basic_join_mappings.json、fund_company_join_bulk.json 和 fund_basic_join_bulk.json。使用该 bash 文件运行执行以下任务：

（1）使用在 fund_company_fund_basic_join_mappings.json 内定义的设定和分析器创建

fund_company_fund_basic_join 索引。

（2）使用 fund_company_join_bulk.json 内编写好的批量处理文档接口指令，对 26 家基金公司进行父角色文档索引。文档索引操作时最好指定文档标识符 _id，此标识符将在子角色文档索引操作时，标明用作路由值和父角色文档标识符。同时，文档内容必须标明此文档在父子类关系 fund_company_fund_basic 中的父角色为 fund_company_role。以下只显示博时基金管理有限公司的文档索引内容。

```
{"index":{"_id":"754PVnMBpfHDzdHUa_3t"}}
{"name": "博时基金管理有限公司", "shortname": "博时基金", "province": "广东",
"city": "深圳市", "address": "广东省深圳市福田区深南大道 7088 号招商银行大厦 29 层",
"phone": "86-755-83169999,86-95105568", "office": "广东省深圳市福田区益田路 5999
号基金大厦 21 层", "website": "www.bosera.com", "chairman": "张光华", "manager":
"江向阳", "reg_capital": 25000, "setup_date": "19980713", "end_date": null,
"employees": 323, "main_business": null, "org_code": "710922202", "credit_code":
"91440300710922202N", "location":{ "lat": 22.5431, "lon": 114.0579}, "fund_company_
fund_basic":{"name":"fund_company_role"} }
```

（3）使用 fund_basic_join_bulk.json 内编写好的批量处理文档接口指令，对 50 个基金进行子角色文档索引。文档索引操作时必须标明路由值（设为父角色文档标识符）。同时文档内容必须标明此文档在父子类关系 fund_company_fund_basic 中的子角色为 fund_basic_role，并附上父文档标识符（parent）的值，以下只显示博时基金管理有限公司管理的基金恒生湾区文档内容。

```
{"index":{"routing":"754PVnMBpfHDzdHUa_3t"}}
    {"fund_company_fund_basic":{"name":"fund_basic_role", "parent":"754PVnMBp
fHDzdHUa_3t"},"ts_ccde":"159809.SZ","fund_name":"恒生湾区","custodian":"招商银行",
"fund_type":"股票型",
    "found_date":"20200430","due_date":null,"list_date":"20200521","issue_date":"20200330",
"delist_date":null,"issue_amount":3.9117,"m_fee":0.15,"c_fee":0.05,"duration_year":null,
    "p_value":1.0,"min_amount":0.1,"exp_return":null,"benchmark":
"恒生沪深港通大湾区综合指数收益率","status":"L","invest_type":"被动指数型","type":"契
约型开放式",
    "trustee":null,"purc_startdate":"20200521","redm_startdate":"20200521","market":"E"}
```

部分索引结果显示如下：

```
./fund_company_fund_basic_join_bulk_index.sh
{"acknowledged":true,"shards_acknowledged":true,"index":"fund_company_fund_basic_join"}
{
  "took" : 43,
  "errors" : false,
  "items" : [
    {
      "index" : {
        "_index" : "fund_company_fund_basic_join",
        "_type" : "_doc",
        "_id" : "754PVnMBpfHDzdHUa_3t",
      ……
},
{
  "took" : 29,
  "errors" : false,
  "items" : [
    {
      "index" : {
        "_index" : "fund_company_fund_basic_join",
        "_type" : "_doc",
        "_id" : "46xptnUBIhDpB0g9hx9O",
```

```
    ......
}
```

（4）获取索引 fund_company_fund_basic_join 的文档总数，结果显示如下：

```
curl --request POST http://localhost:9200/fund_company_fund_basic_join/_count?pretty=true
{
    "count" : 76,
    ......
}
```

8.1.5 父子类关联数据类型查询方法

如果一个索引是用关联（join）数据类型建立，查询方式和搜索结果与之前看到的不同。以下例子根据所需的搜索请求构建查询并检查搜索结果。

1. 搜索父角色文档的字段

以下示例使用词条（term）查询父角色文档中的字段 city，亦即注册地址所在城市为珠海市的基金管理公司。搜索结果有两家，易方达基金管理有限公司和广发基金管理有限公司。文档内容包括父角色文档和子角色文档。另外还有父子关系名称 fund_company_fund_basic 及父角色 fund_company_role。

提示：搜索父角色文档的字段，不显示子角色文档的内容。

```
curl --request POST http://localhost:9200/fund_company_fund_basic_join/_search?pretty=true --header "Content-Type: application/json" --data $'{"query":{"term":{"city":{"value":"珠海市"}}} }'
{
  ......
  "hits" : {
    "total" : {
      "value" : 2,
      "relation" : "eq"
    },
    "max_score" : 2.3795462,
    "hits" : [
      {
        "_index" : "fund_company_fund_basic_join",
        "_type" : "_doc",
        "_id" : "8Z4PVnMBpfHDzdHUa_3t",
        "_score" : 2.3795462,
        "_source" : {
          "name" : "易方达基金管理有限公司",
          "shortname" : "易方达基金",
          "province" : "广东",
          "city" : "珠海市",
          ......
          "fund_company_fund_basic" : {
            "name" : "fund_company_role"
          }
      ......
}
```

2. 搜索子角色文档的字段

以下示例使用匹配（match）查询子角色文档中的字段 fund_name，亦即在子角色文档中基金简称中，匹配查询字符串 800。获得最终匹配结果有两个基金，基金简称为 800ETF 和中证 800ETF 基金。文档内容包括父子关系名称 fund_company_fund_basic、子角色 fund_ basic_role 及父角色文档标识符。

> 提示：搜索子角色文档的字段，不显示父角色文档的内容。

```
curl --request POST http://localhost:9200/fund_company_fund_basic_join/_search?
pretty=true --header "Content-Type: application/json" --data
$'{"query":{"match":{"fund_name":"800"}}}'
{
  ......
  "hits" : {
    "total" : {
      "value" : 2,
      "relation" : "eq"
    },
    "max_score" : 3.4306374,
    "hits" : [
      {
        "_index" : "fund_company_fund_basic_join",
        "_type" : "_doc",
        "_id" : "wp6xWXMBpfHDzdHUY_6G",
        "_score" : 3.4306374,
        "_routing" : "9p4PVnMBpfHDzdHUa_3t",
        "_source" : {
          "fund_company_fund_basic" : {
            "name" : "fund_basic_role",
            "parent" : "9p4PVnMBpfHDzdHUa_3t"
          },
          ......
}
```

3. 使用父角色文档标识符搜索子角色文档

以下示例使用父角色文档标识符 9p4PVnMBpfHDzdHUa_3t 查询其子角色文档。结果显示子角色有两个基金，基金简称为 800ETF 和中证 800ETF 基金。

```
curl --request POST http://localhost:9200/fund_company_fund_basic_join/_search?
pretty=true --header "Content-Type: application/json" --data
$'{"query":{"parent_id":{"type":"fund_basic_role", "id": "9p4PVnMBpfHDzdHUa_3t"}}}'
{
  ......
  "hits" : {
    "total" : {
      "value" : 2,
      "relation" : "eq"
    },
    "max_score" : 3.0910425,
    "hits" : [
      {
        "_index" : "fund_company_fund_basic_join",
        "_type" : "_doc",
        "_id" : "6qxptnUBIhDpB0g9hx9O",
        "_score" : 3.0910425,
        "_routing" : "9p4PVnMBpfHDzdHUa_3t",
        "_source" : {
          "fund_company_fund_basic" : {
            "name" : "fund_basic_role",
            "parent" : "9p4PVnMBpfHDzdHUa_3t"
          },
          "ts_code" : "159802.SZ",
          "fund_name" : "800ETF",
          "custodian" : "中国工商银行",
          ......
}
```

4. 使用父子关系自动生成的附加字段同时获取父角色和子角色文档

以下示例使用自动生成的附加字段 fund_company_fund_basic # fund_company_role。该字段的名称是父子类关系名称后跟#号和父角色名称。其值是父角色文档标识符，用于维护相关的所有文档。可以用来查询所有与此文档标识符相关的文档。结果显示父角色文档（广发基金管理有限公司）和所有子角色文档（800ETF 和芯片基金），总共三个文档。

```
curl --request POST http://localhost:9200/fund_company_fund_basic_join/_search?
pretty=true --header "Content-Type: application/json" --data $'{
"query":{"match":{"fund_company_fund_basic#fund_company_role":"9p4PVnMBpfHDzdHUa_3t"}}}'
{
  ......
  "hits" : {
    "total" : {
      "value" : 3,
      "relation" : "eq"
    },
    "max_score" : 3.0910425,
    "hits" : [
      {
        "_index" : "fund_company_fund_basic_join",
        "_type" : "_doc",
        "_id" : "9p4PVnMBpfHDzdHUa_3t",
        "_score" : 3.0910425,
        "_source" : {
          "name" : "广发基金管理有限公司",
          "shortname" : "广发基金",
          "province" : "广东",
          "city" : "珠海市",
          ......
        }
```

5. 使用 has_parent 查询相关的子角色文档

使用 has_parent 查询并返回匹配的子角色文档。示例搜索基金管理公司注册地址所在城市为珠海市的基金。在请求主体中标明父子关系名称 fund_company_fund_basic 及父角色 fund_company_role。然后使用词条（term）查询父角色文档中注册地址所在城市为珠海市。搜索结果有 5 个公募基金。

提示：返回内容不包括父角色文档内容。

```
curl --request POST http://localhost:9200/fund_company_fund_basic_join/_search?
pretty=true --header "Content-Type: application/json" -data $'{"query":{"has_parent":{
"parent_type":"fund_company_role", "query":{"term":{"city":{"value":"珠
海市"}}}}}}'
{
  ......
  "hits" : {
    "total" : {
      "value" : 5,
      "relation" : "eq"
    },
    "max_score" : 1.0,
    "hits" : [
      {
        "_index" : "fund_company_fund_basic_join",
        "_type" : "_doc",
        "_id" : "5axptnUBIhDpB0g9hx9O",
        "_score" : 1.0,
        "_routing" : "8Z4PVnMBpfHDzdHUa_3t",
```

```
            "_source" : {
              "fund_company_fund_basic" : {
                "name" : "fund_basic_role",
                "parent" : "8Z4PVnMBpfHDzdHUa_3t"
              },
              "ts_code" : "159803.SZ",
              "fund_name" : "浙江 ETF",
              "custodian" : "中国工商银行",
              "fund_type" : "股票型",
              ……
            }
```

6. 使用 has_child 查询相关的父角色文档

使用 has_child 查询并返回匹配的父角色文档。示例搜索基金管理公司,其管理的基金,由中国银行托管。在请求主体中标明父子关系名称 fund_company_fund_basic 及子角色 fund_basic_role。然后使用词条(term)查询子角色文档中基金托管人为中金公司。搜索结果有 1 家基金管理公司,公司名称为建信基金。

提示:返回内容不包括子角色文档内容。

```
curl --request POST http://localhost:9200/fund_company_fund_basic_join/_search?
pretty=true --header "Content-Type: application/json" --data
$'{"query":{"has_child":{"type":"fund_basic_role", "query":{"term":{
"custodian.keyword":{"value":"中金公司"}}}}}}'
  {
    ……
    "hits" : {
      "total" : {
        "value" : 1,
        "relation" : "eq"
      },
      "max_score" : 1.0,
      "hits" : [
        {
          "_index" : "fund_company_fund_basic_join",
          "_type" : "_doc",
          "_id" : "-54PVnMBpfHDzdHUa_3t",
          "_score" : 1.0,
          "_source" : {
            "name" : "建信基金管理有限责任公司",
            "shortname" : "建信基金",
            "province" : "北京",
            ……
          }
```

此类型查询支持一些参数,例如使用参数 max_children 限制与查询匹配的子文档的最大数量、使用参数 min_children 限制与查询匹配的子文档的最小数量,以及使用参数 score_mode,从匹配的子文档中如何返回相关性得分分数(选项有 min、max、avg、sum 和 none),有兴趣的读者可以参考官网。

8.2 实际应用场景操作

尽管关联数据类型支持一些非常有用的查询,但这对性能产生重大影响。要使用关联数据类型建模还需要非常小心,因为没有存在父角色文档的子角色文档也可以进行索引操作。

另外，如果要删除父项，则删除所有子项不是自动级联（cascading）任务，需要手动清理干净。子角色文档可以重新索引／添加／删除。但是重新索引父角色文档不被允许。如果要求更强的搜索性能，请使用嵌套的数据类型。但是，嵌套数据类型方法也有明显的缺陷，如果需要进行更新嵌套字段中的值，就需要重新索引嵌套字段中的所有嵌套值，不能只在嵌套字段中更改其中一个嵌套值。相反，关联数据类型方法将比较方便，因为更新父角色文档或子角色文档的字段都是允许的。

如果所需的方案要求子级聚合（children aggregation）以获取父角色和子角色文档的合并数据的拆分级别信息，那么确实需要使用关联数据类型作为数据建模的方法。

第 9 章 聚合框架

Elasticsearch 的两个主要关键功能是搜索和数据分析。聚合可以看作是在一组文档上建立数据汇总与信息分析的工作单元。聚合框架由模块组成,以构建由搜索查询选择的数据的复杂摘要。框架非常简单,扩展性强,易于使用且功能强大。聚合框架可以细分为四个主要的家族如图 9-1 所示。在开始研习聚合操作之前,需要按以下用例准备数据。

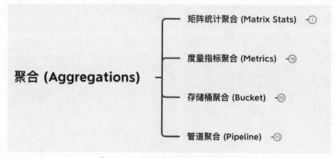

图 9-1 四个主要的聚合家族

9.1 基金净值和基金持仓样本文档

因为本章涉及数据分析,在这里引进在第二章中提到由 Tushare 提供的公募基金每日的净值和每季度的基金持仓等相关数据。并随机挑选了基金代码 159801.SZ 及 159995.SZ 两支基金用于测试。

1. 准备公募基金每日净值的测试环境

本章提供了 fund_nav_bulk.json、fund_nav_bulk_index.sh 和 fund_nav_mappings.json 3 个文件用于基金净值文档索引操作。fund_nav_bulk_index.sh 是个 bash 文件,运行执行以下任务:

(1) 使用在 fund_nav_mappings.json 内定义的设定和分析器创建 fund_nav 索引。字段与表 2-3 所示的公募基金净值数据列表一一对应,采用的显式映射内容如下:

```
{
        "mappings" : {"dynamic": false, "properties": {
        "ts_code":{"type":"text", "fields":{"keyword":{"type":"keyword"}}},
        "ann_date": {"type": "date", "format": "yyyyMMdd"},
        "end_date": {"type": "date", "format": "yyyyMMdd"},
        "unit_nav": {"type": "float"},
        "accum_nav": {"type": "float"},
```

```
            "accum_div": {"type": "float"},
            "net_asset": {"type": "float"},
            "total_netasset": {"type": "float"},
            "adj_nav": {"type": "float"}}}
}
```

(2) 使用 fund_nav_bulk.json 内编写好的批量处理文档接口指令,对 159801.SZ 及 159995.SZ 两支基金的净值数据进行文档索引,时间段大约是在 2020 年 01 月 21 日至 2020 年 07 月 18 日之间。执行 fund_nav_bulk_index.sh 后的部分索引结果显示如下:

```
./fund_nav_bulk_index.sh
{"acknowledged":true,"shards_acknowledged":true,"index":"fund_nav"}
{
  "took" : 40,
  "errors" : false,
  "items" : [
    {
      "index" : {
        "_index" : "fund_nav",
        "_type" : "_doc",
        "_id" : "rZ_6YnMBpfHDzdHUxAAm",
        "_version" : 1,
        "result" : "created",
        "_shards" : {
          "total" : 2,
          "successful" : 1,
          "failed" : 0
        },
        "_seq_no" : 0,
        "_primary_term" : 1,
        "status" : 201
      }
    },
    ......
}
```

(3) 测试索引 fund_nav 内数据的数量,结果显示总共 224 条记录。

```
curl --request POST http://localhost:9200/fund_nav/_count?pretty=true
```

2. 准备每季度基金持仓的测试环境

本章提供了 fund_portfolio_bulk.json、fund_portfolio_bulk_index.sh 和 fund_portfolio_mappings.json 3 个文件用于基金持仓文档索引操作。其中 fund_portfolio_bulk_index.sh 是个 bash 文件,运行执行以下任务:

(1) 使用在 fund_portfolio_mappings.json 内定义的设定和分析器创建 fund_portfolio 索引。字段与表 2-5 所示的公募基金持仓数据列表一一对应,采用的显式映射内容如下:

```
{
    "mappings" : {"dynamic": false, "properties": {
        "ts_code":{"type":"text", "fields":{"keyword":{"type":"keyword"}}},
        "ann_date": {"type": "date", "format": "yyyyMMdd"},
        "end_date": {"type": "date", "format": "yyyyMMdd"},
        "symbol":{"type":"text", "fields":{"keyword":{"type":"keyword"}}},
        "mkv": {"type": "float"},
        "amount": {"type": "float"},
        "stk_mkv_ratio": {"type": "float"},
        "stk_float_ratio": {"type": "float"}}}}
```

(2) 使用 fund_portfolio_bulk.json 内编写好的批量处理文档接口指令,对基金代码

159801.SZ、159995.SZ 及 515850.SH 三只基金的持仓数据进行文档索引。这三只基金的持仓数据披露日如下：

- 基金代码 159801.SZ 为 2020 年 02 月 13 日及 2020 年 04 月 21 日。
- 基金代码 159995.SZ 为 2020 年 02 月 05 日及 2020 年 04 月 22 日。
- 基金代码 515850.SH 为 2020 年 02 月 18 日、2020 年 04 月 22 日及 2020 年 07 月 21 日。

执行 fund_portfolio_bulk_index.sh 后的部分索引结果显示如下：

```
./fund_portfolio_bulk_index.sh
{"acknowledged":true,"shards_acknowledged":true,"index":"fund_portfolio"}
{
  "took" : 179,
  "errors" : false,
  "items" : [
    {
      "index" : {
        "_index" : "fund_portfolio",
        "_type" : "_doc",
        "_id" : "GzLUpXMBTmRt5jA4gn-I",
        "_version" : 1,
        "result" : "created",
        "_shards" : {
          "total" : 2,
          "successful" : 1,
          "failed" : 0
        },
        "_seq_no" : 0,
        "_primary_term" : 1,
        "status" : 201
      }
    ......
}
```

（3）测试索引 fund_portfolio 内数据的数量，结果显示总共 80 条记录。

```
curl --request POST http://localhost:9200/fund_portfolio/_count?pretty=true
{
  "count" : 80,
  "_shards" : {
    "total" : 1,
    "successful" : 1,
    "skipped" : 0,
    "failed" : 0
  }
}
```

9.2 聚合查询语法

聚合分析功能对于数据分析方案尤其重要。Elasticsearch 在同一层次结构上支持多个聚合。另外，聚合的强大功能之一是嵌入聚合的能力，又称为子聚合。父聚合和子聚合的语法相同。子聚合在上一级聚合的各个存储桶的上下文中运行。子聚合使能够将聚合不断地细化到任意的层次，以复杂的聚合逻辑来解决问题。以下使用正则表达式语法演示了聚合查询的基本语法结构。aggs 关键字是 aggregations 的缩写。单词 aggs_name_1 是给定的最上层聚合名称。单词 aggs_type 用于定义聚合的类型，聚合主体则指定聚合的条件。

```
"aggs":{
```

```
"aggs_name_1": {
    "aggs_type": { <aggs_name_1 的聚合主体>}
    [,"aggs" : {[子聚合]+}]?
}
[,"aggs_name_2":{...}]*
}
```

9.3 矩阵统计聚合

矩阵统计（matrix_stats）聚合是一系列的功能，提供一种可以同时操作多个字段并以矩阵形式生成字段结果的方法。矩阵聚合计算一组给定文档字段上的数字统计指标，如表 9-1 所示。举一个例子，在 fund_nav 索引中搜索基金代码为 159801.SZ 并使用参数 fields 指定字段，包含复权单位净值（adj_nav）和单位净值（unit_nav）的数组，进行矩阵聚合操作以获取统计报告。统计结果显示复权单位净值和单位净值有很强的关系。

表 9-1 矩阵聚合统计指标列表

统计指标名称	中文名称	描述
count	计数	测量的样本数
mean	平均值	从样品测得该字段的平均值
variance	方差	被测字段的值距离其平均值有多远。方差越大，距离平均值也就越大
skewness	偏度	描述了分布的形状。它测量字段的值分布的不对称度
kurtosis	峰度	描述了分布的形状。它测量分布的尾部重量。随着尾巴变轻，峰度减小。随着尾巴变重，峰度增加
covariance	协方差	衡量两个字段之间的联合可变性。正值表示它们的值沿相同方向移动，反之则背离
correlation	相关性	衡量两个字段之间关系的强度。有效值在 [-1, 1] 之间。-1 表示负相关、1 表示正相关，而 0 则表示没有可识别的关系

```
curl --request POST http://localhost:9200/fund_nav/_search?pretty=true
--header "Content-Type: application/json" --data $'{"query":{"term":{
"ts_code.keyword":{"value":"159801.SZ"}}}, "aggs":{"159801_SZ_STATS":{
"matrix_stats":{"fields":["adj_nav","unit_nav"]}}}, "size":0}'
{
    ……
    "aggregations" : {
        "159801_SZ_STATS" : {
            "doc_count" : 110,
            "fields" : [
                {
                    "name" : "adj_nav",
                    "count" : 110,
                    "mean" : 1.0730609146031467,
                    "variance" : 0.017408811181482859,
                    "skewness" : 1.1114695058451345,
                    "kurtosis" : 4.1056383859685015,
                    "covariance" : {
                        "adj_nav" : 0.017408811181482859,
                        "unit_nav" : 0.017408811181482859
                    },
                    "correlation" : {
                        "adj_nav" : 1.0,
                        "unit_nav" : 1.0000000000000002
                    }
                },
                {
                    "name" : "unit_nav",
```

```
    ......
}
```

9.4 度量指标聚合

度量指标聚合（metrics aggregations）提供了数学运算函数运算，可对一个或多个给定字段中的值执行运算，并有助于分析文档的分组集合。度量指标聚合系列大约有16种不同的数学运算方式如图9-2所示，以下各小节一一列出。

图9-2 度量指标聚合系列

9.4.1 最大值聚合

最大值（max）聚合提供给定的数字字段值的最大值，以下示例找出基金代码为159995.SZ在基金净值索引fund_nav中找出复权单位净值的最大值。最大值聚合名称max_adj_nav的结果显示，在114个交易日中，其最高的复权单位净值曾达到1.524。

```
curl --request POST http://localhost:9200/fund_nav/_search?pretty=true
--header "Content-Type: application/json" --data $'{"query":{"term":{
"ts_code.keyword":"159995.SZ"}},"aggs": {"max_adj_nav": {"max": {
"field":"adj_nav"}}},"size":0}'
{
  ......
  "hits" : {
    "total" : {
      "value" : 114,
      "relation" : "eq"
    },
    "max_score" : null,
```

```
        "hits" : [ ]
    },
    "aggregations" : {
        "max_adj_nav" : {
            "value" : 1.5240999460220337
        }
        ......
    }
}
```

9.4.2 最小值聚合

最小值（min）聚合提供给定的数字字段值的最小值，以下示例找出基金代码为159995.SZ 在基金净值索引 fund_nav 中找出复权单位净值的最小值。最小值聚合名称 min_adj_nav 的结果显示，在 114 个交易日中，其最低的复权单位净值 min_adj_nav 曾少于面值 1.0。

```
curl --request POST http://localhost:9200/fund_nav/_search?pretty=true
--header "Content-Type: application/json" --data $'{"query":{"term":{
"ts_code.keyword":"159995.SZ"}},"aggs": {"min_adj_nav": {"min": {
"field":"adj_nav"}}}, "size":0}'
{
    ......
    "hits" : {
        "total" : {
            "value" : 114,
            "relation" : "eq"
        },
        "max_score" : null,
        "hits" : [ ]
    },
    "aggregations" : {
        "min_adj_nav" : {
            "value" : 0.9027000069618225
        }
        ......
    }
}
```

9.4.3 总和聚合

总和（sum）聚合提供给定的数字字段值的总和，以下示例找出基金代码为159995.SZ 在基金持仓索引 fund_portfolio 中找出披露日（ann_date）为 2020 年 02 月 05 日的总投资额，在此使用了布尔查询。总投资额是持有股票市值（mkv）的总和。总和聚合名称 total_investment 的结果显示，当季度共持有 10 种不同的股票，总投资额 total_investment 为 1,550,443,008 元。

```
curl --request POST http://localhost:9200/fund_portfolio/_search?pretty=true
--header "Content-Type: application/json" --data $'{"query":{"bool":{"must":[{
"term":{"ts_code.keyword":"159995.SZ"}},{"term":{"ann_date":"20200205"}}]}},
"aggs": {"total_investment": {"sum": {"field":"mkv"}}}, "size":0}'
{
    ......
    "hits" : {
        "total" : {
            "value" : 10,
            "relation" : "eq"
        },
        "max_score" : null,
        "hits" : [ ]
    },
    "aggregations" : {
        "total_investment" : {
```

```
      "value" : 1.550443008E9
    }
  }
  ……
}
```

9.4.4 值计数聚合

值计数（value_count）聚合提供给定的字段有值（没有缺少值）的笔数。在基金净值索引 fund_nav 中合计资产净值 total_netasset 字段，有很多文档是缺少值的。以下示例找出基金代码 159995.SZ 在基金净值索引 fund_nav 中，有多少文档在该字段是有值的。值计数聚合名称 total_netasset_data 的结果显示，共有 114 个文档，但只有 3 个文档在该字段是有值的。

```
curl --request POST http://localhost:9200/fund_nav/_search?pretty=true
--header "Content-Type: application/json" --data $'{"query":{"term":{
"ts_code.keyword":"159995.SZ"}},"aggs": {"total_netasset_data": {
"value_count": {"field":"total_netasset"}}}, "size":0}'
{
  ……
  "hits" : {
    "total" : {
      "value" : 114,
      "relation" : "eq"
    },
    "max_score" : null,
    "hits" : [ ]
  },
  "aggregations" : {
    "total_netasset_data" : {
      "value" : 3
    }
  ……
}
```

9.4.5 平均值聚合

以下测试在基金索引 fund_basic 中使用平均值（avg）聚合操作，计算基金管理费的平均值。平均值聚合名称 AVG_M_FEE 的结果显示，50 个基金的平均管理费为 0.419。如果字段可能缺少值，则可使用参数 missing 指定替换值。

```
curl --request POST http://localhost:9200/fund_basic/_search?pretty=true
--header "Content-Type: application/json" --data $'{"aggs":{"AVG_M_FEE":{
"avg":{"field":"m_fee"}}}, "size":0}'
{
  ……
  "hits" : {
    "total" : {
      "value" : 50,
      "relation" : "eq"
    },
    ……
  },
  "aggregations" : {
    "AVG_M_FEE" : {
      "value" : 0.4190000033378601
    }
  }
}
```

9.4.6 加权平均值聚合

根据公式 Σ（field1×field2）/Σ（field2）计算两个给定字段数值的加权平均值（weighted_avg）。由于当前各个索引并没有类似的加权值字段，因此在示例中两个没有多大意义的字段 m_fee 和 c_fee，从基金索引 fund_basic 随机选取用于测试。加权平均值聚合名称 weighted_avg_example 的结果显示，加权平均值为 0.488 67。如果字段可能缺少值，则可使用参数 missing 指定替换值。

```
curl --request POST http://localhost:9200/fund_basic/_search?pretty=true
--header "Content-Type: application/json" --data $'{"aggs": {"weighted_avg_example":
{"weighted_avg": {"value":{"field":"m_fee"}, "weight": {"field": "c_fee"}}}},
"size":0}'
{
  ……
  "aggregations" : {
    "weighted_avg_example" : {
      "value" : 0.48867102836524945
    }
  ……
}
```

9.4.7 基数聚合

基数（cardinality）聚合的目的是获取某字段值分类后类别的总数。根据官网的说明，类别的总数是一个近似值。以下示例在索引 fund_basic 中查找那 50 个基金所属于的管理公司的数量。基数聚合名称 num_of_management_company 的结果显示，共有 26 个基金管理公司。如果字段可能缺少值，则可使用参数 missing 指定替换值。

```
curl --request POST http://localhost:9200/fund_basic/_search?pretty=true
--header "Content-Type: application/json" --data $'{"aggs": {
"num_of_management_company": {"cardinality": {"field":"management.keyword"}}},
"size":0}'
{
  ……
  "hits" : {
    "total" : {
      "value" : 50,
      "relation" : "eq"
    },
    "max_score" : null,
    "hits" : [ ]
  },
  "aggregations" : {
    "num_of_management_company" : {
      "value" : 26
    }
  ……
}
```

9.4.8 统计聚合

统计（stats）聚合提供给定的数字字段值的简单统计值，包括最小值、最大值、总和、计数和平均值。如果字段可能缺少值，则可使用参数 missing 指定替换值。以下示例执行基金代码为 159995.SZ 在基金净值索引 fund_nav 中的复权单位净值的统计聚合，聚合名称为

adj_nav_stats。

```
curl --request POST http://localhost:9200/fund_nav/_search?pretty=true
--header "Content-Type: application/json" --data $'{"query":{"term":{
"ts_code.keyword":"159995.SZ"}},"aggs": {"adj_nav_stats": {"stats": {
"field":"adj_nav"}}}, "size":0}'
{
  ……
  "hits" : {
    "total" : {
      "value" : 114,
      "relation" : "eq"
    },
    "max_score" : null,
    "hits" : [ ]
  },
  "aggregations" : {
    "adj_nav_stats" : {
      "count" : 114,
      "min" : 0.9027000069618225,
      "max" : 1.5240999460220337,
      "avg" : 1.0975113997333927,
      "sum" : 125.11629956960678
    }
    ……
}
```

9.4.9 扩展统计聚合

扩展统计（extended_stats）聚合除了提供给定的数字字段值的简单统计外，还包括平方和、方差、标准差和标准差范围（相对于平均值的 ±2 标准差）。如果字段可能缺少值，则可使用参数 missing 指定替换值。以下示例执行基金代码为 159995.SZ 在基金净值索引 fund_nav 中的复权单位净值扩展统计聚合，聚合名称为 adj_nav_extended_stats。

```
curl --request POST http://localhost:9200/fund_nav/_search?pretty=true
--header "Content-Type: application/json" --data $'{"query":{"term":{
"ts_code.keyword":"159995.SZ"}},"aggs": {"adj_nav_extended_stats": {
"extended_stats": {"field":"adj_nav"}}}, "size":0}'
{
  ……
  "hits" : {
    "total" : {
      "value" : 114,
      "relation" : "eq"
    },
    "max_score" : null,
    "hits" : [ ]
  },
  "aggregations" : {
    "adj_nav_extended_stats" : {
      "count" : 114,
      "min" : 0.9027000069618225,
      "max" : 1.5240999460220337,
      "avg" : 1.0975113997333927,
      "sum" : 125.11629956960678,
      "sum_of_squares" : 139.3309062918887,
      "variance" : 0.017669659840237523,
      "std_deviation" : 0.13292727274806146,
      "std_deviation_bounds" : {
        "upper" : 1.3633659452295157,
        "lower" : 0.8316568542372698
```

```
        }
        ……
}
```

9.4.10 中位数绝对偏差聚合

中位数绝对偏差（median_absolute_deviation）聚合是一种类似于标准差的方差度量，用于确定数据是否可能有异常值，或者可能不是正态分布。如果字段可能缺少值，则可使用参数 missing 指定替换值。以下示例利用基金代码 159995.SZ 在基金净值索引 fund_nav 中 114 个交易日中的复权单位净值，测试中位数绝对偏差值。中位数绝对偏差聚合名称 adj_nav_mad 的结果显示为 0.06940001249313354。

提示： 中位数绝对偏差的数学公式为 median（|median（X）- Xi|）。

```
curl --request POST http://localhost:9200/fund_nav/_search?pretty=true
--header "Content-Type: application/json" --data $'{"query":{"term":{
"ts_code.keyword":"159995.SZ"}}, "aggs": {"adj_nav_mad": {
"median_absolute_deviation":{"field":"adj_nav"}}}, "size":0}'
{
  ……
  "aggregations" : {
    "adj_nav_mad" : {
      "value" : 0.06940001249313354
    }
  }
  ……
}
```

9.4.11 百分位聚合

百分位（percentiles）聚合用于将给定的数字字段值生成百分位数，将文档分组。如果字段可能缺少值，则可使用参数 missing 指定替换值。此外，布尔参数 keyed 以百分位和文档分组百分率，按键值对显示结果，默认为 true。以下示例将列出基金净值索引 fund_nav 下，基金代码为 159995.SZ 的复权单位净值百分位数情况。百分位聚合名称 adj_nav_percentiles 的结果显示，在 114 个交易日中，超过 75% 的复权单位净值大于面值 1.0。

提示： 低于 25% 的数据小于 1，亦即 75% 的数据大于 1。

```
curl --request POST http://localhost:9200/fund_nav/_search?pretty=true
--header "Content-Type: application/json" --data $'{"query":{"term":{
"ts_code.keyword":"159995.SZ"}}, "aggs": {"adj_nav_percentiles": {
"percentiles":{"field":"adj_nav"}}}, "size":0}'
{
  ……
  "aggregations" : {
  "adj_nav_percentiles" : {
    "values" : {
      "1.0" : 0.9027000069618225,
      "5.0" : 0.9316200137138367,
      "25.0" : 1.000599980354309,
      "50.0" : 1.0858500003814697,
      "75.0" : 1.1490999460220337,
      "95.0" : 1.3927799701690673,
      "99.0" : 1.5114279794692993
    }
  }
  ……
}
```

若是设置布尔参数 keyed 为 false,则百分位和文档分组百分率将分开,并以键值对的值显示,结果如下:

```
curl --request POST http://localhost:9200/fund_nav/_search?pretty=true
--header "Content-Type: application/json" --data $'{"query":{"term":{
"ts_code.keyword":"159995.SZ"}}, "aggs": {"adj_nav_percentiles": {
"percentiles":{"field":"adj_nav", "keyed":false}}}, "size":0}'
{
……
  "aggregations" : {
    "adj_nav_percentiles" : {
      "values" : [
        {
          "key" : 1.0,
          "value" : 0.9027000069618225
        },
        {
          "key" : 5.0,
          "value" : 0.9316200137138367
        },
        ……
}
```

9.4.12 百分位等级聚合

百分位等级(percentile_ranks)聚合首先将给定的数字字段值生成百分位数聚合,将文档分组。然后从给定的数组,将数组内每一个数值按照百分位数分布值,转换为百分位数列出。支持参数 keyed 和 missing,与 9.4.11 节百分位聚合所描述的相同。从上一节百分位聚合的结果,若复权单位净值为 1.085 85,大约位于百分位 50%。以下示例利用百分位等级聚合,在基金净值索引 fund_nav 下,设定基金代码 159995.SZ 的复权单位净值为 1.085 85,查询其相对的百分位等级。百分位等级聚合名称 adj_nav_percentile_ranks 的结果显示,百分位约为 50% 的等级。

```
curl --request POST http://localhost:9200/fund_nav/_search?pretty=true
--header "Content-Type: application/json" --data $'{"query":{"term":{"
ts_code.keyword":"159995.SZ"}}, "aggs": {"adj_nav_percentile_ranks": {
"percentile_ranks":{"field":"adj_nav", "values":[1.08585]}}}, "size":0}'
{
  ……
  "aggregations" : {
    "adj_nav_percentile_ranks" : {
      "values" : {
        "1.08585" : 49.99999975212878
      }
    ……
}
```

9.4.13 地理重心聚合

地理重心(geo_centroid)聚合是一种根据 geo_point 数据类型的字段值,计算该字段所有坐标值中的矩心。以下示例在基金公司索引 fund_company 的 location 字段中,找出 26 家基金管理公司的地理重心。地理重心聚合名称 location_geo_centroid 的结果显示,地理重心位置为 {28.726865373229465,117.19189995541595}。

```
curl --request POST http://localhost:9200/fund_company/_search?pretty=true
```

```
--header "Content-Type: application/json" --data $'{"aggs": {
"location_geo_centroid": {"geo_centroid":{"field":"location"}}}, "size":0}'
{
  ......
  "hits" : {
    "aggregations" : {
      "location_geo_centroid" : {
        "location" : {
          "lat" : 28.726865373229465,
          "lon" : 117.19189995541595
        },
        "count" : 26
      }
    }
  ......
}
```

9.4.14 地理边界聚合

地理边界（geo_bounds）聚合是一种根据 geo_point 数据类型的字段值，计算边界框。在基金公司索引 fund_company 中，location 字段为基金公司注册地址所在城市的经度和纬度，数据类型是 geo_point。以下示例找出 26 家基金管理公司的地理边界。地理边界聚合名称 location_geo_bound 的结果显示，地理边界位于左上角和右下角的两个点的地理位置为 {{39.90419996436685,112.54889993928373} 和 {22.27099999319762, 121.54399992898107}}。

```
curl --request POST http://localhost:9200/fund_company/_search?pretty=true
--header "Content-Type: application/json" --data $'{"aggs": {"location_geo_bound":
{"geo_bounds":{"field":"location"}}}, "size":0}'
{
......
  "aggregations" : {
    "location_geo_bound" : {
      "bounds" : {
        "top_left" : {
          "lat" : 39.90419996436685,
          "lon" : 112.54889993928373
        },
        "bottom_right" : {
          "lat" : 22.27099999319762,
          "lon" : 121.54399992898107
        }
      }
    }
  ......
}
```

9.4.15 最热点聚合

最热点（top_hits）聚合旨在用作子聚合器，按先前形成个别的文档分组中，列出最热门的群组。支持参数 from、size 和 sort。默认情况下参数 size 为 3，而参数 from 为 0。亦即返回前三个匹配项。这些参数在个别的文档分组内进行操作。top_hits 聚合支持参数 _source，只返回指定的文档部分内容。以下示使用 9.5.7 节介绍的词条聚合（terms），列出索引 fund_basic 下的基金管理费最常见的前两个收费额，在每个收费额（桶）里面只显示 1 个文档。最热点聚合名称 m_fee_top_hits 的结果显示，在 50 个基金中，一般最常见的管理费额是 0.5 和 0.15。基金管理费为 0.5 有 27 个基金，而 0.15 有 12 个基金。

提示：返回结果中显示两个计数结果，doc_count_error_upper_bound 及 sum_other_doc_count。doc_count_error_upper_bound 提供了有关潜在文档错误的计数，这是因为并非所有文档都包括在该操作内。sum_other_doc_count 提供了所有不属于响应的文档总数，正如聚合结果报告 sum_other_doc_count 有 11 个文档，亦即 50-27-12=11。

```
curl --request POST http://localhost:9200/fund_basic/_search?pretty=true
--header "Content-Type: application/json" --data $'{"aggs": {"m_fees":
{"terms":{"field":"m_fee", "size":2}, "aggs":{"m_fee_top_hits":{"top_hits":{
"_source":["name","m_fee"], "size":1}}}}}, "size":0}'
{
  ……
  "aggregations" : {
    "m_fees" : {
      "doc_count_error_upper_bound" : 0,
      "sum_other_doc_count" : 11,
      "buckets" : [
        {
          "key" : 0.5,
          "doc_count" : 27,
          "m_fee_top_hits" : {
            "hits" : {
              "total" : {
                "value" : 27,
                "relation" : "eq"
              },
              "max_score" : 1.0,
              "hits" : [
                {
                  "_index" : "fund_basic",
                  "_type" : "_doc",
                  "_id" : "57A88nUBp1RIvERas6zK",
                  "_score" : 1.0,
                  "_source" : {
                    "m_fee" : 0.5,
                    "name" : "浙江 ETF"
                  }
                }
              ]
            }
          }
        },
        {
          "key" : 0.15000000596046448,
          "doc_count" : 12,
          "m_fee_top_hits" : {
            ……
}
```

9.4.16 脚本式度量指标聚合

脚本式度量指标（scripted_metric）聚合是一种执行脚本以聚合分析某字段值。聚合操作可以细分为四个阶段 init_script（初始化脚本）、map_script（映射脚本）、combine_script（合并脚本）及 reduce_script（归纳脚本）。所有脚本返回值的数据类型必须为基本类型、字符串、键值对或数组。以下简短描述这四种类型的脚本：

（1）init_script（初始化脚本）

此脚本是选项，用于初始化需要使用的变数，包括一个名为 state 的对象，用于存储分片级别的改动。

(2) map_script（映射脚本）

在文档级别执行的脚本，这意味着该文档匹配查询要求，被包括在聚合中。

(3) combine_script（合并脚本）

当所有文档都收集在分片节点内时，在分片级别执行的脚本。经过处理后，将值或对象返回到协调节点。

(4) reduce_script（归纳脚本）

当所有分片的文档都收集在协调节点时，在协调节点级别执行的脚本。名为 state 的对象储存分片节点返回结果的集合。

以下示例在基金索引 fund_basic 的字段中，计算每个基金的 m_fee 与 c_fee 的总和，找出最高值。在 init_script 脚本中初始化 state.fees 变量为列表数据类型。在 map_script 脚本中计算每个文档 m_fee 与 c_fee 的总和，然后添加到 state.fees 数组内。在 combine_script 脚本中找出个别分片级别的最高值。在 reduce_script 脚本中找出所有分片节点中的最高值。脚本式度量指标聚合名称 highest_fee 的结果显示，最高值为 1.4。

提示：脚本可能会导致搜索速度降低。如果经常在搜索操作中使用脚本来转换索引数据，建议通过第 10 章介绍的摄取节点管道处理，在索引期间处理好转换操作。这样可以加快搜索速度。

```
curl --request POST http://localhost:9200/fund_basic/_search?pretty=true
--header "Content-Type: application/json" --data $'{"aggs": {
 "highest_fee": {"scripted_metric":{"init_script": "state.fees = new ArrayList()",
 "map_script":"double total_fee=doc.m_fee.value+doc.c_fee.value; state.fees.add(total_fee)",
 "combine_script":"return Collections.max(state.fees)","reduce_script":
"return Collections.max(states)"}}}, "size":0}'
{
  ……
  "aggregations" : {
    "highest_fee" : {
      "value" : 1.400000050663948
    }
   ……
}
```

9.5 存储桶聚合

在介绍最热点（top_hits）聚合曾讨论过个别群组的形成，并可作为顶层的存储桶聚合用作分组。在本节将介绍存储桶（bucket）聚合，它提供将文档集以准则划分为组（存储桶）的机制，该准则确定当前的文档是否落入其中一组。存储桶聚合系列大约有 25 种类型如图 9-3 所示，以下各小节一一列出说明。

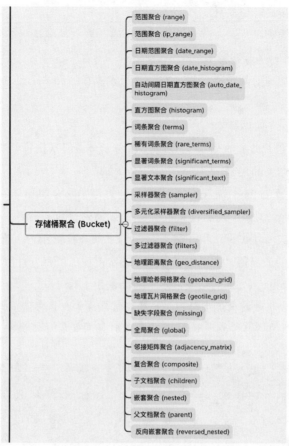

图 9-3　存储桶聚合系列

9.5.1　范围聚合

范围（range）聚合的目的是将对每个给定的字段 field 范围内，对文档进行计数。由参数 from 和 to 指定范围值，from 是给定的起始值（包含），而 to 是给定的结束值（不包含）。用数据类型为数组的参数 ranges 指定多个范围。此聚合支持参数 key，可对每个范围自定义给定名称。此外还支持布尔参数 keyed，默认为 false，就是对范围名称以键值对的值显示聚合结果。以下示例在基金索引 fund_basic 下，对管理费进行范围统计。假设基金管理费 m_free 分为三个范围，例如小于 0.5、0.5 至 1.0 及大于 1.0。范围聚合名称 m_fee_ranges 的结果显示，第一个范围的基金总数为 20，第二个范围的基金总数 28，而最后一个范围的基金总数为 2。

提示：布尔参数 keyed 为 false 时（默认值），聚合结果 buckets 的数据类型为数组。范围名称在个别存储桶内。

```
curl --request POST http://localhost:9200/fund_basic/_search?pretty=true
--header "Content-Type: application/json" --data $'{"aggs":{"m_fee_ranges":{"range":
{"field":"m_fee","ranges":[{"key":"first_range", "to":0.5}, {"key":"second_range",
"from":0.5 , "to":1.0}, {"key":"third_range", "from":1.0}]}}}, "size":0}'
{
```

```
    ......
    "aggregations" : {
      "m_fee_ranges" : {
        "buckets" : [
          {
            "key" : "first_range",
            "to" : 0.5,
            "doc_count" : 20
          },
          {
            "key" : "second_range",
            "from" : 0.5,
            "to" : 1.0,
            "doc_count" : 28
          },
          {
            "key" : "third_range",
            "from" : 1.0,
            "doc_count" : 2
          }
        ]
    ......
}
```

若是设置 keyed 为 true，聚合结果 buckets 的数据类型为 Map，范围名称以键值对的键（个别存储桶的名称）显示聚合结果，结果如下：

```
curl --request POST http://localhost:9200/fund_basic/_search?pretty=true
  --header "Content-Type: application/json" --data $'{"aggs":{"m_fee_ranges":{"range":
  {"field":"m_fee","ranges":[{"key":"first_range", "to":0.5}, {"key":"second_range",
  "from":0.5 , "to":1.0}, {"key":"third_range", "from":1.0}], "keyed":true}}},
"size":0}'
{
    ......
    "aggregations" : {
      "m_fee_ranges" : {
        "buckets" : {
          "first_range" : {
            "to" : 0.5,
            "doc_count" : 20
          },
          "second_range" : {
            "from" : 0.5,
            "to" : 1.0,
            "doc_count" : 28
          },
          "third_range" : {
            "from" : 1.0,
            "doc_count" : 2
          }
    ......
}
```

9.5.2 IP 范围聚合

与 9.5.1 节范围聚合类似，IP 范围（ip_range）聚合的目的是用于 IP 数据类型字段以基于 IP 范围对文档进行计数。支持布尔参数 keyed，与 9.5.1 节所描述的相同。由于当前建立的索引并没有具备该数据类型，以下仅显示一个虚构的例子。IP 范围在 192.168.1.1 和 192.168.1.255 之间。

```
{"aggs" : {"scan_ips" : {"ip_range" : {"field" : "ipaddr","ranges": [{ "from" : "192.168.1.1" },{ "to" : "192.168.1.255" }]}}}}
```

9.5.3 日期范围聚合

与 9.5.1 节范围聚合类似，区别在于日期范围（date_range）聚合使用日期格式。字段范围值由参数 from 和 to 指定，from 是给定的起始日期（包含），而 to 是给定的结束日期（不包含）。字段范围值可以使用 1.6 节接口用法约定说明中的"格式化日期的数学表示方式"。支持参数 ranges、key 和 keyed，与 9.5.1 节范围聚合所描述的相同。另外还支持参数 format 和参数 missing。参数 format 提供输入的自定义日期，请参考 1.6 节接口用法约定说明中的"日期字符串"。

以下示例在基金净值索引 fund_nav 下，查询基金代码为 159995.SZ，对公告日期字段（ann_date）以日期范围聚合操作，自定义日期格式为"yyyyMMdd"。设置两个时间段从 20200101 开始到 20200201 和从 20200701 开始。第一个时间段给定名称为 first_period，而第二个时间段名称为 second_period。日期范围聚合名称 ann_date_2_periods 的结果显示，在第一个时间段交易日总数为 2，而在第二个时间段为交易日总数 14。

提示：聚合结果内的 from 和 to 值，被换为 UTC 以来的毫秒数值，from_as_string 为其日期格式的字符串。

```
curl --request POST http://localhost:9200/fund_nav/_search?pretty=true
--header "Content-Type: application/json" --data $'{"query":{"term":{
"ts_code.keyword":"159995.SZ"}}, "aggs":{"ann_date_2_periods":{"date_range": {"field":
"ann_date","format":"yyyyMMdd", "ranges":[{"key":"first_period",
"from":"20200101",
"to":"20200201"},{"key":"second_period","from":"20200701"}]}}}, "size":0}'
{
  ......
  "aggregations" : {
    "ann_date_2_periods" : {
      "buckets" : [
        {
          "key" : "first_period",
          "from" : 1.5778368E12,
          "from_as_string" : "20200101",
          "to" : 1.5805152E12,
          "to_as_string" : "20200201",
          "doc_count" : 2
          "key" : "second_period",
          "from" : 1.5935616E12,
          "from_as_string" : "20200701",
          "doc_count" : 14
        }
      ......
  }
}
```

若是设置 keyed 为 true，聚合结果 buckets 的数据类型为 JSON 对象，范围名称以键值对的键（个别存储桶的名称）显示聚合结果，结果如下：

```
curl --request POST http://localhost:9200/fund_nav/_search?pretty=true
--header "Content-Type: application/json" --data $'{"query":{"term":{"ts_code.
keyword":"159995.SZ"}}, "aggs":{"ann_date_2_periods":{"date_range": {"field":"ann_
date","format":"yyyyMMdd", "ranges":[{"key":"first_period", "from":"20200101",
"to":"20200201"},{"key":"second_period","from":"20200701"}], "keyed":true}}},
"size":0}'
```

```
{
    ......
    "aggregations" : {
        "ann_date_2_periods" : {
            "buckets" : {
                "first_period" : {
                    "from" : 1.5778368E12,
                    "from_as_string" : "20200101",
                    "to" : 1.5805152E12,
                    "to_as_string" : "20200201",
                    "doc_count" : 2
                },
                "second_period" : {
                    "from" : 1.5935616E12,
                    "from_as_string" : "20200701",
                    "doc_count" : 14
                }
            }
        }
    }
}
```

9.5.4 直方图聚合

直方图（histogram）聚合的目的是将字段值（value）分为固定间隔（interval）的范围，并聚合在一起计算文档数。起始范围是 0，或是给定少于固定间隔的偏移量（参数 offset）。范围距离由间隔值指定，使用的公式为：

```
bucket_key = Math.floor((value-offset) / interval) * interval + offset。
```

提供参数 min_doc_count，可指定间隔最低文档数。若文档数少于该数，则略过该间隔。支持参数 order 设定排序方向。支持参数 extended_bounds，可指定范围的下限（min）和上限（max）。另外还支持参数 missing 和 keyed。

以下示例在基金净值索引 fund_nav 下，查询基金代码为 159995.SZ，获得复权单位净值统计后的直方图。这个测试设定间隔为 0.1 和文档数降序排序。参数 min_doc_count 设置为 20。直方图聚合名称 adj_nav_histogram 的结果显示复权单位净值大多分布在 1.0、0.9 和 1.1 之间。

提示：存储桶聚合可以根据 _key（存储桶的名称），_count（文档数）或其子集合的值进行排序，以下示例使用文档数进行演示。

```
curl --request POST http://localhost:9200/fund_nav/_search?pretty=true --header
"Content-Type: application/json" --data $'{"query":{"term":{"ts_code.keyword":
{"value":"159995.SZ"}}}, "aggs":{"adj_nav_histogram": {"histogram": {"field":"adj_
nav", "interval":0.1, "order":{"_count":"desc"}, "min_doc_count":20}}}, "size":0}'
{
    ......
    "aggregations" : {
        "adj_nav_histogram" : {
            "buckets" : [
                {
                    "key" : 1.0,
                    "doc_count" : 42
                },
                {
                    "key" : 0.9,
                    "doc_count" : 27
```

```
      },
      {
        "key" : 1.1,
        "doc_count" : 27
      }
    ]
  }
}
```

9.5.5　日期直方图聚合

日期直方图（date_histogram）聚合将日期值字段分组为具有固定时间单位间隔的范围（存储桶），并对文档进行计数。范围距离由间隔值指定，有两个参数可以指定其时间单位：

- 参数 calendar_interval 可指定分钟（m 或 1m）、小时（h 或 1h）、日（d 或 1d）、周（w 或 1w）、月（M 或 1M）、季度（q 或 1q）、和年（y 或 1y）。
- 参数 fixed_interval 可指定毫秒（Xms）、秒（Xs）、分钟（Xm）、小时（Xh）和日（Xd）。其中 X 是一个正整数。

支持参数 keyed、format 和 missing，与 9.5.3 节日期范围聚合所描述的相同。支持参数 min_doc_count，与 9.5.4 节直方图聚合所描述的相同。另外还支持参数 offset，可对日期起始值以 ± 时间单位来进行调整正偏移或负偏移，时间单位为 1.6 节接口用法约定说明中的"持续时间的单位"。

以下示例在基金净值索引 fund_nav 下，查询基金代码为 159995.SZ，对公告日期字段以日期直方图聚合操作，设置时间单位 interval 为每周（1w）。日期直方图聚合名称 ann_date_2_periods 的结果显示总共 25 周。

提示： 20200127 这周因文档数为 0，而参数 min_doc_count 设置为 1，所以不包含在聚合结果中。

```
http://localhost:9200/fund_nav/_search?pretty=true --header "Content-Type:
application/json" --data $'{"query":{"term":{"ts_code.keyword":"159995.SZ"}},
"aggs":{"ann_date_histogram":{"date_histogram": {"field":"ann_
date","format":"yyyyMMdd",
"calendar_interval":"1w", "min_doc_count":1}}}, "size":0}'
{
  ......
  "aggregations" : {
    "ann_date_histogram" : {
    "buckets" : [
      {
        "key_as_string" : "20200120",
        "key" : 1579478400000,
        "doc_count" : 2
      },
      {
        "key_as_string" : "20200203",
        "key" : 1580688000000,
        "doc_count" : 2
      },
      ......
    }
```

9.5.6 自动间隔日期直方图聚合

自动间隔日期直方图（auto_date_histogram）聚合类似于日期直方图聚合。区别在自动间隔日期直方图使用给定数量的间隔（存储桶）而不是固定时间单位间隔。并且自动选择最佳实现间隔目标。返回的间隔数将小于或等于给定的数量。如果未指定间隔数，则默认值为 10。提供参数 minimum_interval 可指定最小舍入间隔，单位为 year、month、day、hour、minute 和 second。另外还支持参数 format 和 missing，与 9.5.3 节日期范围聚合所描述的相同。

以下示例在基金净值索引 fund_nav 下，查询基金代码为 159995.SZ，对公告日期字段以自动间隔日期直方图聚合操作，使用默认间隔数，结果显示最佳实现间隔（interval）为 7，从 20200201 开始至 20200701 结束。每个月的第一天为开始日。在返回的响应主体中，字段 key 为 Epoch 以来的毫秒数、key_as_string 为日期格式以及 doc_count 为该时间间隔的文档计数。自动间隔日期直方图聚合名称 auto_ann_date_histogram 的结果显示系统设置最佳间隔 interval 为 1M（一个月）。

提示：自动间隔设置的数目亦即存储桶 buckets 的数目。

```
curl --request POST http://localhost:9200/fund_nav/_search?pretty=true
--header "Content-Type: application/json" --data $'{"query":{"term":{
"ts_code.keyword":"159995.SZ"}}, "aggs":{"auto_ann_date_histogram":{
"auto_date_histogram": {"field":"ann_date","format":"yyyyMMdd"}}}, "size":0}'
{
  ……
  "aggregations" : {
    "auto_ann_date_histogram" : {
      "buckets" : [
        {
          "key_as_string" : "20200101",
          "key" : 1577836800000,
          "doc_count" : 2
        },
        {
          "key_as_string" : "20200201",
          "key" : 1580515200000,
          "doc_count" : 17
        },
        ……
      ],
      "interval" : "1M"
    }
  ……
}
```

9.5.7 词条聚合

词条（terms）聚合的目的是根据字段值进行分组，然后提供每个字段值的文档计数。根据官网信息，每个组的文档计数是大约数。支持参数 min_doc_count，与 9.5.4 节直方图聚合所描述的相同。参数 size 可设定应从整体分组中返回多少个分组，默认情况下只返回文档计数最多的分组。而参数 shard_size 可指定协调节点应从每个分片中，请求多少个分组。shard_size 默认值为 (size×1.5 + 10)。还有参数 include 及 exclude 可设定包含或排除的词条或词条数组，另外还支持参数 missing 和 order。

以下示例在基金公司索引 fund_company 下，查询以城市 city 字段分组获得在该市注册的基金公司数量，但只报告前 4 名注册公司最多的城市。词条聚合名称 city_group 的结果显示分别前 4 名为上海市、深圳市、北京市和珠海市。

```
curl --request POST http://localhost:9200/fund_company/_search?pretty=true
    --header "Content-Type: application/json" --data $'{"aggs": { "city_
group":{"terms"
    :{"field":"city","size":4}}}, "size":0}'
    {
      ......
      "aggregations" : {
        "city_group" : {
          "doc_count_error_upper_bound" : 0,
          "sum_other_doc_count" : 4,
          "buckets" : [
            {
              "key" : "上海市",
              "doc_count" : 10
            },
            {
              "key" : "深圳市",
              "doc_count" : 8
            },
            {
              "key" : "北京市",
              "doc_count" : 2
            },
            {
              "key" : "珠海市",
              "doc_count" : 2
            }
          ]
          ......
    }
```

词条聚合提供 collect_mode 参数，可控制聚合的集合模式，提高性能。选项有 breadth_first（广度优先）和 depth_first（深度优先），当分组的基数为未知数或是请求的大小大于分组的基数时，广度优先为默认模式。这是由于将子级聚合的计算程序延迟到父级聚合得到结果之后，可能会更有效率。

9.5.8 稀有词条聚合

稀有词条（rare_terms）聚合类似于词条聚合，区别是以找出文档计数最稀有的字段值分组，概念上类似于以文档计数 _count 升序排序的词条聚合。同样每个组的文档计数是大约数。提供参数 max_doc_count 可指定成为一分组的最多文档数，还有参数 include 及参数 exclude 可设定包含或排除的词条或词条数组。另外还支持参数 missing。

以下示例在基金公司索引 fund_company 下，查询以城市值字段分组获得在该市注册的基金公司数量，但只报告前 4 名注册公司最少的城市。稀有词条聚合名称 city_group 的结果显示分别为南宁市、天津市、太原市和宁波市。

```
curl --request POST http://localhost:9200/fund_company/_search?pretty=true
    --header "Content-Type: application/json" --data $'{"aggs": { "city_group":{
"rare_terms":{"field":"city"}}}, "size":0}'
    {
```

```
    ……
    "aggregations" : {
      "city_group" : {
        "buckets" : [
          {
            "key" : "南宁市",
            "doc_count" : 1
          },
          {
            "key" : "天津市",
            "doc_count" : 1
          },
          {
            "key" : "太原市",
            "doc_count" : 1
          },
          {
            "key" : "宁波市",
            "doc_count" : 1
          }
        ]
      ……
    }
```

9.5.9 显著词条聚合

显著词条（significant_terms）聚合类似于词条聚合，但并非仅仅找出最常出现的字段值分组，区别是能够找出异乎寻常频率出现的字段值。显著的含义在于相关的一组特定的文档中找出最相关的字词。例如在前景（特定集合范围）和背景（索引范围）集合之间测得不寻常的重大变化。此聚合不支持浮点数值（float）字段，并且文档计数是近似值。支持参数size、shard_size 和 min_doc_count，与 9.5.7 词条聚合所描述的相同。为求更有效率，此聚合固定使用广度优先（breadth_first）模式处理子聚合。

以下示例找出在基金索引 fund_basic 中，基金托管人和基金管理人之间，有什么有趣的关联需要关注？显著词条聚合名称 custodian 的结果显示在 50 家基金里面，招商银行总共承担 12 家基金的托管人。而博时基金总共管理 6 个基金，有 4 个基金的托管人为招商银行。而其他基金管理人或托管人没有什么关联需要关注。

```
curl --request POST http://localhost:9200/fund_basic/_search?pretty=true
--header "Content-Type: application/json" --data $'{ "aggs": {"custodian":
{"terms":{"field": "custodian.keyword"}, "aggs":{"significant_management":{
"significant_terms":{"field":"management.keyword"}}}}}, "size":0}'
{
  ……
  "aggregations" : {
    "custodian" : {
      "doc_count_error_upper_bound" : 0,
      "sum_other_doc_count" : 5,
      "buckets" : [
        {
          "key" : "招商银行",
          "doc_count" : 12,
          "significant_management" : {
            "doc_count" : 12,
            "bg_count" : 50,
            "buckets" : [
              {
```

```
            "key" : "博时基金",
            "doc_count" : 4,
            "score" : 0.59259259259259259926,
            "bg_count" : 6
         }
      ......
}
```

9.5.10 显著文本聚合

显著文本（significant_text）聚合类似于显著词条聚合，区别是以找出文本类型字段上异乎寻常频率出现的词汇。聚合操作时会即时对文本内容进行重新分析，还会过滤文本的噪声部分，避免扭曲统计信息。此聚合存在一些限制，例如不支持在它下层的子聚合，也不支持嵌套对象及文档计数是近似值。支持参数 size、shard_size 和 min_doc_count，与 9.5.7 词条聚合所描述的相同。

以下示例在基金索引 fund_basic 中，找出基金简称包含 ETF 字符串的基金，进行显著文本聚合。操作时参数 min_doc_count 设为 10。显著文本聚合名称 significant_fund_text 的结果显示，除了 50 家基金简称里面，有 35 家包含 ETF 字符串外，没有发现不寻常的现象。

```
curl --request POST http://localhost:9200/fund_basic/_search?pretty=true
--header "Content-Type: application/json" --data $'{"query":{"match":{"name":
"ETF"}}, "aggs":{"significant_fund_text":{"significant_text":{"field":"name",
"min_doc_count":10}}}, "size":0}'
{
  ......
  "aggregations" : {
  "significant_fund_text" : {
     "doc_count" : 35,
     "bg_count" : 50,
     "buckets" : [
        {
           "key" : "ETF",
           "doc_count" : 35,
           "score" : 0.428857142857142866,
           "bg_count" : 35
        }
      ......
}
```

9.5.11 采样器聚合

采样器（sampler）聚合类可通过指定参数 shard_size 值（默认值为 100）限制子聚合的采样数（采样只包括最高评分文档的样本）。将分析重点放在高相关性样本上，从而降低聚合的运行成本，尤其对显著文本聚合非常有效。以下示例在基金公司索引 fund_company 中，首先使用采样器聚合收集文档，并以省份分组（terms 聚合），找出省份内注册的基金公司数量是否有不寻常发现（significant_terms 聚合）。聚合操作中设定 shard_size 值为 13（总数的一半）以降低运行成本。采样器聚合名称 sampling 的结果显示，在 13 个样本中，其词条子聚合名称 significant_province 的结果为广东省占 5 家、上海市占 4 家，而其他 4 个城市各占 1 家。在词条子聚合下，显著词条子聚合名称 significant_city 的结果显示，在广东省 5 家中深圳市占 3 家，而注册在深圳市原来有 8 家。注册在上海市原来有 10 家。

```
    curl --request POST http://localhost:9200/fund_company/_search?pretty=true
  --header "Content-Type: application/json" --data $'{"aggs":{"sampling":
{"sampler":
    {"shard_size":13},"aggs": {"significant_province": {"terms":{"field": "province"},
    "aggs":{"significant_city":{"significant_terms":{"field":"city"}}}}}}}, "size":0}'
{
  ……
  "aggregations" : {
    "sampling" : {
      "doc_count" : 13,
      "significant_province" : {
        "doc_count_error_upper_bound" : 0,
        "sum_other_doc_count" : 0,
        "buckets" : [
          {
            "key" : "广东",
            "doc_count" : 5,
            "significant_city" : {
              "doc_count" : 5,
              "bg_count" : 26,
              "buckets" : [
                {
                  "key" : "深圳市",
                  "doc_count" : 3,
                  "score" : 0.5699999999999998,
                  "bg_count" : 8
                }
              ]
            }
          },
          {
            "key" : "上海",
            "doc_count" : 4,
            "significant_city" : {
              "doc_count" : 4,
              "bg_count" : 26,
              "buckets" : [
                {
                  "key" : "上海市",
                  "doc_count" : 4,
                  "score" : 1.5999999999999999,
                  "bg_count" : 10
                }
              ]
            }
          },
          ……
        ]
      }
    }
  }
}
```

9.5.12 多元化采样器聚合

多元化采样器（diversified_sampler）聚合类似于采样器（sampler）聚合。除了可通过指定参数 shard_size 值限制子聚合的采样数外（采样只包括最高评分文档的样本），还可指定参数 max_docs_per_value 值去限制相同得分值的文档匹配数量。这样可减少样品池的偏差分布，并将分析重点放在高相关性样本上，从而降低聚合的运行成本。以下示例类似在采样器聚合的示例，但使用了多元化采样器聚合。示例使用相同的参数值，但再增加一个参数 max_docs_per_value 来限制相同值的文档匹配数量为 3。故此上海市的采样样本少了一个（从 4 家减少到 3 家），但是深圳市的采样样本没有变化，仍然是 3 家。这测试显示了减少样品池的偏差分布的现象。

```
curl --request POST http://localhost:9200/fund_company/_search?pretty=true
--header "Content-Type: application/json" --data $'{"aggs":{"sampling":
{"diversified_sampler": {"field": "city","shard_size":13,"max_docs_per_value":3},
"aggs": {"significant_province": {"terms":{"field": "province"},
"aggs":{"significant_city":{"significant_terms":{"field":"city"}}}}}}}, "size":0}'
{
  ......
    {
      "key" : "上海",
      "doc_count" : 3,
      "significant_city" : {
        "doc_count" : 3,
        "bg_count" : 26,
        "buckets" : [
          {
            "key" : "上海市",
            "doc_count" : 3,
            "score" : 1.5999999999999999,
            "bg_count" : 10
          }
        ]
      }
    },
}
```

9.5.13 过滤器聚合

过滤器（filter）聚合的目的是筛选适合指定规范的文档并提供总数，通常这是确定文档范围的预筛选步骤。以下示例在基金公司索引 fund_company 中，筛选注册地为珠海市的基金公司。过滤器聚合名称"珠海市 count"的结果显示在结果显示有 2 家。

```
curl --request POST http://localhost:9200/fund_company/_search?pretty=true
--header "Content-Type: application/json" --data $'{"aggs":{"珠海市 count":{"filter":{
"term":{"city":"珠海市"}}}},"size":0}'
{
  ......
  "aggregations" : {
    "珠海市 count" : {
      "doc_count" : 2
    }
    ......
}
```

9.5.14 多过滤器聚合

多过滤器（filters）聚合类似于过滤器（filter）聚合，区别是可给定多个过滤器。目的是根据个别过滤器筛选，提供各个符合过滤器规范的文档总数。以下示例在基金公司索引 fund_company 下，筛选注册地为珠海市的基金公司和深圳市的基金公司。多过滤器聚合名称 filters_demo 的结果显示珠海市有 2 家，而深圳市有 8 家。

```
curl --request POST http://localhost:9200/fund_company/_search?pretty=true
 --header "Content-Type: application/json" --data $'{"aggs":{"filters_
demo":{"filters"
:{"filters":[{"term":{"city":"珠海市"}},{"term":{"city":"深圳市"}}]}}},"size":0}'
{
  ......
```

```
"aggregations" : {
  "filters_demo" : {
    "buckets" : [
      {
        "doc_count" : 2
      },
      {
        "doc_count" : 8
      }
    ]
    ......
  }
}
```

9.5.15 地理距离聚合

地理距离（geo_distance）聚合是一种基于 geo_point 数据类型的字段值，类似范围（range）聚合，将对每个给定的地理距离范围内，对文档进行计数。提供参数 origin（原点）、ranges（范围）和 unit（距离单位）标示地理距离范围。原点可以使用经纬度对象格式，例如 {"lat":28.96,"lon":117.54}、字串格式"28.96,117.54"和基于 GeoJson 的数组格式。距离单位默认为 m（米），也可以设置为 1.6 节接口用法约定说明的"距离单位"。与范围聚合相同，范围可以通过参数 from 和 to 指定。此外可通过参数 key，对每个范围自定义给定名称。

以下示例在基金公司索引 fund_company 的 location 字段中，找出距离地理重心（"28.96,117.54"）450 公里内注册的基金公司。地理距离聚合名称 register_density 的结果显示只有一家。

```
curl --request POST http://localhost:9200/fund_company/_search?pretty=true
--header "Content-Type: application/json" --data $'{"aggs":{"register_density":{
"geo_distance": { "field": "location", "origin": {"lat":28.96, "lon":117.54},
"ranges": [ { "to": 450 , "key": "距离450公里远"}], "unit": "km"}}}, "size": 0} '
{
......
  "aggregations" : {
    "register_density" : {
      "buckets" : [
        {
          "key" : "距离450公里远",
          "from" : 0.0,
          "to" : 450.0,
          "doc_count" : 1
        }
      ]
    ......
  }
}
```

9.5.16 地理哈希网格聚合

Geohash 是一种地理编码系统，可将地理位置编码为一小段字母和数字字符串。地理哈希网格（geohash_grid）聚合是一种基于 geo_point 数据类型的字段值，将文档分配到以 Geohash 网格为单元的分组。聚合返回的结果以稀疏（sparse）形式只显示有匹配文档的网格。网格的精度可使用参数 precision 给定，精度介于 1 到 12 之间，精度明确定义的尺寸如表 9-2 所示，默认值为 5。参数 size 可设定从整体分组中返回多少个分组，默认为 10 000。而参数 shard_size 可指定协调节点从每个分片中，请求多少个分组，默认为 max(10,(size×分片数量))。

另外参数 bounds 可指定范围限制在所提供的地理边界内。

以下示例在基金公司索引 fund_company 的 location 字段中，测试地理哈希网格聚合，找出网格精度为 2 时，基金公司注册分布情况。地理哈希网格聚合名称 grid_distribution 的结果显示网格 wt 有 11 家、网格 ws 有 8 家，而其他网格则有 1 或 2 家。

提示：网格编号 wt、ws 和 wx 等是 Geohash 地理位置编码的网格。

```
curl --request POST http://localhost:9200/fund_company/_search?pretty=true
--header "Content-Type: application/json" --data $'{"aggs":{"grid_distribution"
:{"geohash_grid":{"field":"location", "precision":2}}},"size":0}'
{
  ……
  "aggregations" : {
    "grid_distribution " : {
      "buckets" : [
        {
          "key" : "wt",
          "doc_count" : 11
        },
        {
          "key" : "ws",
          "doc_count" : 8
        },
        {
          "key" : "wx",
          "doc_count" : 2
        },
        {
          "key" : "ww",
          "doc_count" : 2
        },
        {
          "key" : "we",
          "doc_count" : 2
        },
        {
          "key" : "wk",
          "doc_count" : 1
        }
      ……
}
```

表 9-2　网格精度列表

精度值	网格面积＝宽度 × 长度	单位
1	5 009.4×4 992.6	km
2	1 252.3×624.1	km
3	156.5×156	km
4	39.1×19.5	km
5	4.9×4.9	km
6	1.2×0.609 4	km
7	152.9×152.4	m
8	38.2×19	m
9	4.8×4.8	m
10	1.2×0.595	m
11	14.9×14.9	cm
12	3.7×1.9	cm

9.5.17 地理瓦片网格聚合

地理瓦片网格（geotile_grid）聚合是一种基于 geo_point 数据类型的字段值，将文档分配到以网格为单元的分组。可使用参数 precision 给定精度，介于 0 和 29 之间的缩放级别，默认值为 7。参数 size 和 shard_size 的含义和用法与地理哈希网格聚合相同。另外参数 bounds 可指定范围限制在所提供的地理边界内。聚合返回的结果以稀疏（sparse）形式只显示有匹配文档的网格。而网格采用的表示方法为瓦片坐标系统，用 {zoom}/{x}/{y} 格式标记，其中 zoom 等于用户指定的精度。{x}/{y} 是 Geotile 地理位置编码的网格。

以下示例在基金公司索引 fund_company 的 location 字段中，测试地理瓦片网格聚合，找出网格精度为 2 时，基金公司注册分布情况。地理瓦片网格聚合名称 grid_distribution 的结果显示网格 8/214/104 有 10 家、网格 8/209/111 有 8 家，而其他网格则有 1 或 2 家。

```
curl --request POST http://localhost:9200/fund_company/_search?pretty=true
--header "Content-Type: application/json" --data $'{"aggs":{"grid_distribution":{
"geotile_grid":{"field":"location", "precision":8}}},"size":0}'
{
  ......
  "aggregations" : {
    "grid_distribution" : {
      "buckets" : [
        {
          "key" : "8/214/104",
          "doc_count" : 10
        },
        {
          "key" : "8/209/111",
          "doc_count" : 8
        },
        {
          "key" : "8/210/97",
          "doc_count" : 2
        },
        {
          "key" : "8/208/111",
          "doc_count" : 2
        },
        {
          "key" : "8/214/105",
          "doc_count" : 1
        },
      ......
}
```

9.5.18 缺失字段聚合

缺失字段（missing）聚合与 9.4.4 节的值计数（value_count）聚合是互补的。目的是计算缺失给定字段的文档。缺失字段包括显式的空值（null）、空数组（[]）和不存在的字段。以下示例在基金净值索引 fund_nav 下，找出基金代码为 159995.SZ 有多少文档在字段 total_netasset 是没有值的。缺失字段聚合名称 total_missing_netasset 的结果显示，在 114 个文档中共有 111 个文档在该字段是没有值的。

```
curl --request POST http://localhost:9200/fund_nav/_search?pretty=true
--header "Content-Type: application/json" --data $'{"query":{"term":{
```

```
"ts_code.keyword":"159995.SZ"}},"aggs": {"total_missing_netasset": {
"missing": {"field":"total_netasset"}}}, "size":0}'
{
  ......
  "hits" : {
    "total" : {
      "value" : 114,
      "relation" : "eq"
    },
    "max_score" : null,
    "hits" : [ ]
  },
  "aggregations" : {
    "total_missing_netasset" : {
      "doc_count" : 111
    }
  ......
}
```

9.5.19 全局聚合

全局（global）聚合的目的是略过查询句子的搜索条件，来对索引中的所有文档执行聚合。因此，可以将全局聚合的结果与经由匹配查询的聚合结果进行比较。以下测试在基金索引 fund_basic 中，计算博时基金管理下的基金平均管理费（聚合名称博时基金_avg_m_fee）。与此同时计算全局的基金平均管理费做比较（聚合名称 global_avg_aggs）。聚合结果显示，50 个基金的管理费平均为 0.419，而博时管理下的基金管理费平均为 0.258。

提示：平均聚合名称 global_avg_aggs 是全局聚合名称 global_avg_m_fee 子聚合，因此不受搜索条件的影响。

```
curl --request POST http://localhost:9200/fund_basic/_search?pretty=true
--header "Content-Type: application/json" --data $'{"query":{"term":{
"management.keyword":"博时基金"}},"aggs":{"global_avg_m_fee":{"global":{},
"aggs":{"global_avg_aggs":{"avg":{"field":"m_fee"}}}}, "博时基金_avg_m_fee":{
"avg":{"field":"m_fee"}}},"size":0}'
{
  ......
  "hits" : {
    "total" : {
      "value" : 6,
      "relation" : "eq"
    },
    ......
  },
  "aggregations" : {
    "博时基金_avg_m_fee" : {
      "value" : 0.25833333532015484
    },
    "global_avg_m_fee" : {
      "doc_count" : 50,
      "global_avg_aggs" : {
        "value" : 0.4190000033378601
      }
    }
  ......
}
```

9.5.20 邻接矩阵聚合

邻接矩阵（adjacency_matrix）聚合可以定义任何数量的过滤器，并以过滤器之间的邻接矩阵形式提供存储桶结果，这意味着可以将任何过滤器在聚合操作中一起应用。以下示例搜索基金持仓 fund_portfolio 索引，使用两个不同的过滤器 filter1 及 filter2，filter1 过滤掉持仓占股票市值比例少于 10.0 的基金，而 filter2 过滤掉持仓占流通股本比例少于 2.0 的基金。邻接矩阵聚合名称 stk_mkv_vs_stk_float 的结果显示，有 10 个基金符合 filter1，而另外有 8 个基金符合 filter2。符合 filter1 及 filter2 的基金只有一个，其基金代码为 159995.SZ，其仓占股票市值比为 10.67，而持仓占流通股本比为 2.66。

```
curl --request POST http://localhost:9200/fund_portfolio/_search?pretty=true
--header "Content-Type: application/json" --data $'{"aggs": { "stk_mkv_vs_stk_float":
{"adjacency_matrix": {"filters": {"filter1": {"range":{"stk_mkv_
ratio":{"gte":10.0}}}
,"filter2": {"range":{"stk_float_ratio":{"gte":2.0}}}}}}},"size":0}'
{
  ......
  "aggregations" : {
    "stk_mkv_vs_stk_float" : {
      "buckets" : [
        {
          "key" : "filter1",
          "doc_count" : 10
        },
        {
          "key" : "filter1&filter2",
          "doc_count" : 1
        },
        {
          "key" : "filter2",
          "doc_count" : 8
        }
      ]
      ......
  }
```

9.5.21 复合聚合

复合（composite）聚合将输入的文档，根据参数 sources 给定的一个或者多个聚合值，组合构建存储桶，作为复合聚合结果的内容。支持 sources 的聚合目前有 terms、histogram 和 date_histogram 聚合。存储桶支持分页，细节可以参考下面例子。支持参数 order 设定排序方向。默认情况下，缺失字段的文档将忽略。但可通过设定参数 missing_bucket 为 true，包括在响应中。

以下示例在基金持仓索引 fund_portfolio 下，首先对占股票市值比字段使用范围查询，返回大于 10.5 的基金共 7 个。如果返回记录数很大，则需要设置参数 size。由于稍后需要演示使用 after 参数获得下一批结果，复合聚合返回结果参数 size 设为 3 个存储桶。存储桶内容包括基金代码和股票代码。

（1）发出第一个复合聚合请求，聚合名称 mkv_ratio_analysis 的结果显示总共有 3 个存储桶。在此批次中返回 6 个。在返回的响应主体中字段 after_key 会用作获取下一批结果的识别戳。

提示：搜索结果 hits.total 为 7 个文档，作为复合聚合的输入。相同复合聚合结果会在同一存储桶内，如同文档在存储桶的分布 2,1,2。

```
curl --request POST http://localhost:9200/fund_portfolio/_search?pretty=true
```

```
        --header "Content-Type: application/json" --data $'{"query":{"range":{"stk_mkv_ratio":{
   "gte":10.5}}}, "aggs":{"mkv_ratio_analysis":{"composite":{"sources":[{"ts_code":{
   "terms":{"field":"ts_code.keyword"}}},{"stock_symbol":{"terms":{"field":
   "symbol.keyword"}}}], "size":3}}}, "size":0}'
{
  ......
    "hits" : {
      "total" : {
        "value" : 7,
      },
      "relation" : "eq"
    },
    ......
    "aggregations" : {
      "mkv_ratio_analysis" : {
        "after_key" : {
          "ts_code" : "159995.SZ",
          "stock_symbol" : "603986.SH"
        },
        "buckets" : [
          {
            "key" : {
              "ts_code" : "159801.SZ",
              "stock_symbol" : "603986.SH"
            },
            "doc_count" : 2
          },
          {
            "key" : {
              "ts_code" : "159995.SZ",
              "stock_symbol" : "603160.SH"
            },
            "doc_count" : 1
          },
          {
            "key" : {
              "ts_code" : "159995.SZ",
              "stock_symbol" : "603986.SH"
            },
            "doc_count" : 2
          }
        ]
      ......
    }
```

（2）发出第二个复合聚合请求，内容类似于第一个请求。使用参数 after 并设值为第一个复合聚合结果 after_key 的内容。聚合 mkv_ratio_analysis 结果显示总共有一个存储桶及 2 个文档。

```
        curl --request POST http://localhost:9200/fund_portfolio/_search?pretty=true
        --header "Content-Type: application/json" --data $'{"query":{"range":{
        "stk_mkv_ratio":{"gte":10.5}}}, "aggs":{"mkv_ratio_analysis":{"composite":{
        "sources":[{"ts_code":{"terms":{"field":"ts_code.keyword"}}},{"stock_symbol":{
        "terms":{"field":"symbol.keyword"}}}],"after":{"ts_code":"159995.SZ",
        "stock_symbol":"603986.SH"},"size":3}}}, "size":0}'
{
    ......
    "aggregations" : {
      "mkv_ratio_analysis" : {
        "after_key" : {
```

```
          "ts_code" : "515850.SH",
          "stock_symbol" : "600030.SH"
        },
        "buckets" : [
          {
            "key" : {
              "ts_code" : "515850.SH",
              "stock_symbol" : "600030.SH"
            },
            "doc_count" : 2
          }
        ]
......
}
```

9.5.22 子文档聚合

子文档（children）聚合通过关联（join）数据类型字段，将子角色文档聚合在一起。示例在索引 fund_company_fund_basic_join 下进行，各个聚合名称和其使用的相关聚合如下：

- 聚合名称 in_city 使用 terms 聚合，并调用父角色文档的字段 city 分组。
- 聚合名称 total_funds_in_city 使用 children 子文档聚合，配合子角色名称 fund_basic_role，将父角色文档和子角色文档相关联，计算每个父角色文档的 city 里面有多少子角色文档（亦即基金）。

聚合名称 in_city 结果显示，注册地址所在城市，在上海市有 10 家基金管理公司（聚合名称 in_city 的 bucket 的 doc_count）共 15 个基金（聚合名称 total_funds_in_city 结果的 bucket 的 doc_count），在深圳市有 8 家基金管理公司共 19 个基金。在搜索结果中，hits 为 76，因为索引 fund_company_fund_basic_join 共有 76 个文档。其他注册地址所在城市分组请参考结果。

提示：索引 fund_company_fund_basic_join 采用的显式映射内容、父子类关系及其角色，请参考 8.1.4 节使用父子类关联数据类型方法。

```
curl --request POST http://localhost:9200/fund_company_fund_basic_join/_search?
pretty=true --header "Content-Type: application/json" --data $'{"aggs": {
"in_city":{"terms":{"field":"city"}, "aggs":{"total_funds_in_city": {"children":
{"type": "fund_basic_role"}}}}}, "size":0}'
{
  ......
  "hits" : {
    "total" : {
      "value" : 76,
      "relation" : "eq"
    },
    ......
  },
  "aggregations" : {
    "in_city" : {
      "doc_count_error_upper_bound" : 0,
      "sum_other_doc_count" : 0,
      "buckets" : [
        {
          "key" : "上海市",
          "doc_count" : 10,
          "total_funds_in_city" : {
            "doc_count" : 15
```

```
          }
        },
        {
          "key" : "深圳市",
          "doc_count" : 8,
          "num_of_funds_in_city" : {
             "doc_count" : 19
          }
      ……
}
```

9.5.23 嵌套聚合

嵌套（nested）聚合对声明为 nested 数据类型的字段进行聚合，由 path 参数给定嵌套对象字段的路径。以下示例使用与 9.5.22 节子文档聚合相同的范例，计数每个城市有多少基金管理公司。示例在索引 fund_company_fund_basic_nested 下进行，各个聚合名称和其使用的相关聚合如下：

- 聚合名称 in_city 使用 terms 聚合，并调用根 / 主文档的字段 city 分组。
- 聚合名称 total_funds_in_city 使用 nested 聚合和嵌套字段 fund_list，在各个 city 里面对 fund_list 字段（基金）进行分组。

聚合名称 in_city 结果显示大致与 9.5.22 节子文档聚合结果接近相同。在搜索结果中，hits 为 26 是因为索引 fund_company_fund_basic_nested 只有 26 个文档。

提示：索引 fund_company_fund_basic_nested 采用的显式映射内容、嵌套关系 fund_list 及其路径 fund_list.ts_code，请参考 8.1.3 节使用嵌套数据类型方法。

```
curl --request POST http://localhost:9200/fund_company_fund_basic_nested/_search?pretty=true--header "Content-Type: application/json" --data
$'{ "aggs":{"in_city":{"terms":{"field":"city"}, "aggs":{"total_funds_in_city":{"nested":{"path":"fund_list"}}}}}, "size":0}'
{
  ……
  "hits" : {
    "total" : {
      "value" : 26,
      "relation" : "eq"
    },
    "max_score" : null,
    "hits" : [ ]
  },
  "aggregations" : {
    "in_city" : {
      "doc_count_error_upper_bound" : 0,
      "sum_other_doc_count" : 0,
      "buckets" : [
        {
          "key" : "上海市",
          "doc_count" : 10,
          "total_funds_in_city" : {
             "doc_count" : 15
          }
        },
        {
          "key" : "深圳市",
          "doc_count" : 8,
          "total_funds" : {
             "doc_count" : 19
```

```
                    }
              },
       ......
}
```

9.5.24 父文档聚合

父文档（parent）聚合通过关联（join）数据类型字段，将父角色文档聚合在一起。以下示例在索引 fund_company_fund_basic_join 下进行，计算每个管理费额分组里面，各省有多少个基金管理公司和基金。各个聚合名称和其使用的相关聚合如下：

- 聚合名称 fund_in_m_fee 使用 terms 聚合，并调用子角色文档的字段 m_fee 字段分组。
- 聚合名称 company_in_m_fee 使用 parent 聚合和子角色名称 fund_basic_role，将子角色文档和父角色文档相关联，计算每个子角色管理费额里面有多少父角色文档（亦即基金管理公司）。
- 聚合名称 company_in_province_in_m_fee 使用 terms 聚合，并调用 province 字段，按照各省列出基金管理公司数量。

聚合名称 fund_in_m_fee 结果显示，基金管理费额 0.5 分组里，总共有 27 个基金（fund_in_m_fee 的 bucket 的 doc_count）；共 16 家基金管理公司（聚合名称 company_in_m_fee 的 bucket 的 doc_count）；公司注册地址所在省份（聚合名称 company_in_province_in_m_fee 的 bucket 的 doc_count）分别为广东省有 8 家，上海市有 5 家，而北京市、天津市和山西省各有 1 家。在搜索结果中，hits 为 76，因为索引 fund_company_fund_basic_join 共有 76 个文档。其他管理费额分组请参考聚合结果。

提示：有父子角色及其关系，请参考 8.1.4 节使用父子类关联数据类型方法。

```
curl --request POST http://localhost:9200/fund_company_fund_basic_
join/_search?pretty=true--header "Content-Type: application/json" --data
$'{"aggs":{"fund_in_m_fee":{"terms":{"field":"m_fee"} , "aggs":{"company_in_m_fee":{
"parent":{"type": "fund_basic_role"}, "aggs":{"company_in_province_in_m_fee":
{"terms":{"field":"province"}}}}}}},"size":0}'
{
  ......
  "hits" : {
    "total" : {
      "value" : 76,
      "relation" : "eq"
    },
    ......
    "aggregations" : {
      "fund_in_m_fee" : {
        "doc_count_error_upper_bound" : 0,
        "sum_other_doc_count" : 0,
        "buckets" : [
          {
            "key" : 0.5,
            "doc_count" : 27,
            "company_in_m_fee" : {
              "doc_count" : 16,
              "company_in_province_in_m_fee" : {
                "doc_count_error_upper_bound" : 0,
                "sum_other_doc_count" : 0,
                "buckets" : [
                  {
```

```
                    "key" : "广东",
                    "doc_count" : 8
                },
                {
                    "key" : "上海",
                    "doc_count" : 5
                },
                {
                    "key" : "北京",
                    "doc_count" : 1
                },
                {
                    "key" : "天津",
                    "doc_count" : 1
                },
                {
                    "key" : "山西",
                    "doc_count" : 1
                }
            ]
        }
    },
    ......
}
```

9.5.25 反向嵌套聚合

反向嵌套（reserve_nested）聚合，目的是对嵌套对象中声明为 nested 数据类型的字段进行文档的反向分组，反向的意思，就是对根/主文档进行聚合分组，请参考 8.1.3 节使用嵌套数据类型方法。以下示例在索引 fund_company_fund_basic_nested 下进行，使用与 9.5.24 节父文档聚合相同的范例，计算每个管理费额分组里面，各省有多少个基金管理公司和基金。各个聚合名称和其使用的相关聚合如下：

- 聚合名称 funds 使用 nested 聚合，对声明为 nested 数据类型的字段（亦即对基金）进行分组。
- 聚合名称 fund_in_m_fee 使用 terms 聚合，并调用字段 fund_list.m_fee 进行分组。
- 聚合名称 company_in_m_fee 使用 reserve_nested 聚合，对 nested 数据类型的 fund_list 进行文档的反向（亦即对基金管理公司）分组。
- 聚合名称 company_in_province_in_m_fee 使用 terms 聚合，并调用 province 字段，按照各省列出基金管理公司数量。

由于这个聚合操作相当复杂，中间结果也同时显示。从开始到聚合名称 company_in_m_fee 的聚合结果显示，在搜索结果中，hits 为 26，是因为索引 fund_company_fund_basic_nested 只有 26 个文档。在聚合结果中，聚合名称 total_funds 的 doc_count 为 50，因为嵌套字段 fund_list 总共有 50 基金。在基金管理费额 0.5 分组里，总共有 27 个基金（fund_in_m_fee 的 bucket 的 doc_count）；共 16 家基金管理公司（聚合名称 company_in_m_fee 的 bucket 的 doc_count）。

提示：聚合名称 company_in_m_fee 的反向嵌套聚合，是由该管理费额下的基金，反向找出其基金管理公司。

```
curl --request POST http://localhost:9200/fund_company_fund_basic_
nested/_search?pretty=true--header "Content-Type: application/json" --data
$'{"aggs":{"total_funds":{"nested":{"path":"fund_list"}, "aggs":{"fund_in_m_
fee": {"terms":{"field":"fund_list.m_fee"},"aggs":{"company_in_m_fee":{
"reverse_nested":{}}}}}}}, "size":0}'
{
  ......
  "hits" : {
    "total" : {
      "value" : 26,
      "relation" : "eq"
    },
    ......
  "aggregations" : {
    "total_funds" : {
      "doc_count" : 50,
      "fund_in_m_fee" : {
        "doc_count_error_upper_bound" : 0,
        "sum_other_doc_count" : 0,
        "buckets" : [
          {
            "key" : 0.5,
            "doc_count" : 27,
            "company_in_m_fee" : {
              "doc_count" : 16
            }
          },
          {
            "key" : 0.15000000596046448,
            "doc_count" : 12,
            "company_in_m_fee" : {
              "doc_count" : 8
            }
          },
          ......
  }
}
```

从开始到聚合名称 company_in_province_in_m_fee 的聚合操作结果，与 9.5.24 节父文档聚合结果接近相同。

提示：索引 fund_company_fund_basic_nested 采用的显式映射内容、嵌套关系 fund_list 及其路径 fund_list.ts_code，请参考 8.1.3 节使用嵌套数据类型方法。

```
curl --request POST http://localhost:9200/fund_company_fund_basic_nested/_
search?pretty=true --header "Content-Type: application/json" --data $'{"aggs":{
"funds":{"nested":{"path":"fund_list"}, "aggs":{"fund_in_m_
fee":"terms":{"field":
"fund_list.m_fee"}, "aggs":{"company_in_m_fee":{"reverse_nested":{}, "aggs":{
"company_in_province_in_m_fee":{"terms":{"field":"province"}}}}}}}}, "size":0}'
{
  ......
  "hits" : {
    "total" : {
      "value" : 26,
      "relation" : "eq"
    },
    ......
  "aggregations" : {
    "total_funds" : {
      "doc_count" : 50,
      "fund_in_m_fee" : {
        "doc_count_error_upper_bound" : 0,
        "sum_other_doc_count" : 0,
```

```
      "buckets" : [
        {
          "key" : 0.5,
          "doc_count" : 27,
          "company_in_m_fee" : {
            "doc_count" : 16,
            "company_in_province_in_m_fee" : {
              "doc_count_error_upper_bound" : 0,
              "sum_other_doc_count" : 0,
              "buckets" : [
                {
                  "key" : "广东",
                  "doc_count" : 8
                },
                {
                  "key" : "上海",
                  "doc_count" : 5
                },
                {
                  "key" : "北京",
                  "doc_count" : 1
                },
                {
                  "key" : "天津",
                  "doc_count" : 1
                },
                {
                  "key" : "山西",
                  "doc_count" : 1
                }
              ]
            }
          }
        },
        ......
}
```

9.6 管道聚合

管道（pipeline）聚合的目的是使上一阶段的聚合结果，作为输入传递给下一阶段的聚合操作。为了传递结果，提供了参数 buckets_path，可指定管道输入的来源，从而首尾可以链接在一起。要定义管道输入的来源，需要遵循以下语法：

"buckets_path":<聚合名称> [<聚合分隔符><聚合名称>] * [<指标分隔符><指标>]

在此聚合分隔符为符号大于（>），指标分隔符为英语符号句号（.），而指标是上一阶段聚合结果的度量名称。而符号 * 表示括号 [] 内的语法可出现 0 次或多次。

管道聚合可以分为两个类型：parent 和 sibling。如果该管道聚合为其父级聚合输出聚合结果，则称为 parent。如果该管道聚合为同级别聚合输出聚合结果，则称为 sibling。父级聚合意味着与该聚合处于不同级别。同级类型意味着两个聚合处于同一级别。管道聚合不能有子聚合，但可依靠 buckets_path 链接两个不同级别的聚合在一起。管道聚合系列大约有 15 种不同的类型如图 9-4 所示，以下各小节一一列出说明。

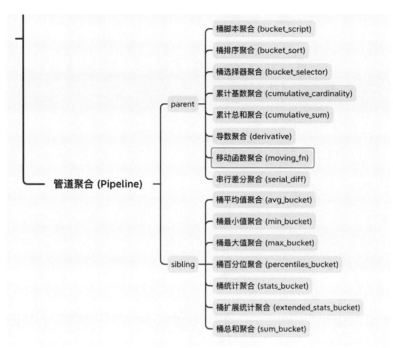

图 9-4 管道聚合系列

9.6.1 桶平均值聚合

桶平均值（avg_bucket）聚合的目的是根据上一阶段的聚合结果，计算得出所有存储桶的平均值。支持参数 format，可应用于此聚合输出值的格式，请参考 1.6 节接口用法约定说明中的"日期字符串"。另外还支持参数 gap_policy 处理数据间隙，该参数有两个选项，默认为 skip，意味跳过没有数据的存储桶。另一个选项 insert_zeros 是用 0 值替换缺失值。以下示例找出基金代码为 159801.SZ，在基金净值索引 fund_nav 中找出复权单位净值的月平均值和整体平均值。复权单位净值使用的日期为（end_date）截止日期，这是当天的数值。各个聚合名称和其使用的相关聚合如下：

- 聚合名称 monthly_avg_adj_nav 使用 avg 聚合，并调用 adj_nav 字段。
- 聚合名称 overall_avg_adj_nav 使用 avg_bucket 聚合，并使用 buckets_path 调用 monthly_avg_adj_nav 聚合的结果。

日期直方图聚合名称 monthly_avg_report 结果显示，整体的平均值为 1.0838375410833272。

提示：桶平均值聚合名称 overall_avg_adj_nav 使用的 buckets_path 为聚合名称 monthly_avg_adj_nav 的结果。因为涉及两个聚合，所以 buckets_path 使用的方式为 monthly_avg_report>monthly_avg_adj_nav。由于桶平均值聚合属于 sibling 类型，所以聚合名称 monthly_avg_report 和聚合名称 overall_avg_adj_nav 在同一级别，其聚合结果也在同一级别。

```
curl --request POST http://localhost:9200/fund_nav/_search?pretty=true
--header "Content-Type: application/json" --data $'{"query":{"term": {
"ts_code.keyword": {"value":"159801.SZ"}}},"aggs":{"monthly_avg_report":{
"date_histogram":{"field":"end_date","calendar_interval":"1M"}, "aggs":{
```

```
"monthly_avg_adj_nav":{"avg":{"field":"adj_nav"}}}}, "overall_avg_adj_nav":{
"avg_bucket":{"buckets_path":"monthly_avg_report>monthly_avg_adj_
nav"}}},"size":0}'
    {
      ……
      "aggregations" : {
        "monthly_avg_report" : {
          "buckets" : [
            ……
            {
              "key_as_string" : "20200701",
              "key" : 1593561600000,
              "doc_count" : 13,
              "monthly_avg_adj_nav" : {
                "value" : 1.3259461659651537
              }
            }
          ]
        },
        "overall_avg_adj_nav" : {
          "value" : 1.0838375410833272
        }
      ……
    }
```

9.6.2 桶最小值聚合

桶最小值（min_bucket）聚合的目的是根据上一阶段的聚合结果，找出所有存储桶的最小值。支持参数 format 和 gap_policy，与 9.6.1 桶平均值聚合所描述的相同。以下示例找出基金代码为 159801.SZ，在基金净值索引 fund_nav 中找出复权单位净值的月最小值和整体最小值。各个聚合名称和其使用的相关聚合如下：

- 聚合名称 monthly_min_adj_nav 使用 min 聚合，并调用 adj_nav 字段。
- 聚合名称 overall_min_adj_nav 使用 min_bucket 聚合，并使用 buckets_path 调用 monthly_min_adj_nav 聚合的结果。

日期直方图聚合名称 monthly_min_report 结果显示，整体最小值发生在分组为 2020 年 03 月 01 日，值为 0.8801000118255615。

提示：桶最小值聚合名称 overall_ min _adj_nav 使用的 buckets_path 为 monthly_ min _adj_nav 聚合名称的结果。因为涉及两个聚合，所以 buckets_path 使用的形式为 monthly_ min _report>monthly_ min _adj_nav。由于桶最小值聚合属于 sibling 类型，所以聚合名称 monthly_ min _report 和聚合名称 overall_min_adj_nav 在同一级别，其聚合结果也在同一级别。

```
    curl --request POST http://localhost:9200/fund_nav/_search?pretty=true
--header "Content-Type: application/json" --data $'{"query":{"term":{
"ts_code.keyword": {"value":"159801.SZ"}}},"aggs":{"monthly_min_report":{
"date_histogram":{"field":"end_date","calendar_interval":"1M"}, "aggs":{
"monthly_min_adj_nav":{"min":{"field":"adj_nav"}}}}, "overall_min_adj_nav":{
"min_bucket":{"buckets_path":"monthly_min_report>monthly_min_adj_
nav"}}},"size":0}'
    {
      ……
      "aggregations" : {
        "monthly_min_report" : {
          "buckets" : [
            ……
```

```
          {
            "key_as_string" : "20200701",
            "key" : 1593561600000,
            "doc_count" : 13,
            "monthly_min_adj_nav" : {
              "value" : 1.1633000373840332
            }
          }
        ]
      },
      "overall_min_adj_nav" : {
        "value" : 0.8801000118255615,
        "keys" : [
          "20200301"
        ]
      }
    ......
}
```

9.6.3 桶最大值聚合

桶最大值（max_bucket）聚合的目的是根据上一阶段的聚合结果，找出所有存储桶的最大值。支持参数 format 和 gap_policy，与 9.6.1 桶平均值聚合所描述的相同。以下示例找出基金代码为 159801.SZ，在基金净值索引 fund_nav 中找出复权单位净值的月最大值和整体最大值。各个聚合名称和其使用的相关聚合如下：

- 聚合名称 monthly_max_adj_nav 使用 max 聚合，并调用 adj_nav 字段。
- 聚合名称 overall_max_adj_nav 使用 max_bucket 聚合，并使用 buckets_path 调用 monthly_max_adj_nav 聚合的结果。

日期直方图聚合名称 monthly_max_report 结果显示，整体最大值发生在分组为 2020 年 07 月 01 日，值为 1.485700011253357。

```
curl --request POST http://localhost:9200/fund_nav/_search?pretty=true
--header "Content-Type: application/json" --data $'{"query":{"term": {
"ts_code.keyword": {"value":"159801.SZ"}}},"aggs":{"monthly_max_report":{
"date_histogram":{"field":"end_date","calendar_interval":"1M"}, "aggs":{
"monthly_max_adj_nav":{"max":{"field":"adj_nav"}}}}, "overall_max_adj_nav":{
"max_bucket":{"buckets_path":"monthly_max_report>monthly_max_adj_
nav"}}},"size":0}'
{
  ......
  "aggregations" : {
    "monthly_max_report" : {
      "buckets" : [
      ......
        {
          "key_as_string" : "20200701",
          "key" : 1593561600000,
          "doc_count" : 13,
          "monthly_max_adj_nav" : {
            "value" : 1.485700011253357
          }
        }
      ]
    },
    "overall_max_adj_nav" : {
      "value" : 1.485700011253357,
      "keys" : [
```

```
            "20200701"
        ]
    ......
}
```

9.6.4 桶百分位聚合

桶百分位（percentiles_bucket）聚合的目的是根据上一阶段的聚合结果生成百分位数。支持参数 format 和 gap_policy，与 9.6.1 桶平均值聚合所描述的相同。和百分位聚合一样提供参数 keyed。另外还支持参数 percents 可列出要计算的百分位数。以下示例找出基金代码为 159801.SZ，在基金净值索引 fund_nav 中找出复权单位净值的月平均值和月平均值的桶百分位分布情况。

- 聚合名称 monthly_avg_adj_nav 使用 avg 聚合，并调用 adj_nav 字段。
- 聚合名称 monthly_avg_pecentiles 使用 percentiles_bucket 聚合，并使用 buckets_path 调用 monthly_avg_adj_nav 聚合的结果。

日期直方图聚合名称结果显示，超过 75%的复权单位净值的月平均值大于面值 1.0。

```
curl --request POST http://localhost:9200/fund_nav/_search?pretty=true --header
"Content-Type: application/json" --data $'{"query":{"term": {"ts_code.keyword":
{"value":"159801.SZ"}}},"aggs":{"monthly_report":{"date_histogram": {"field":
"end_date","calendar_interval":"1M"}, "aggs":{"monthly_avg_adj_nav": {
"avg":{"field":"adj_nav"}}}}, "monthly_avg_pecentiles":{"percentiles_bucket":{
"buckets_path": "monthly_report>monthly_avg_adj_nav"}}},"size":0}'
{
  ......
  "aggregations" : {
  "monthly_report" : {
      "buckets" : [
          ......
          {
              "key_as_string" : "20200701",
              "key" : 1593561600000,
              "doc_count" : 13,
              "monthly_avg_adj_nav" : {
                  "value" : 1.3259461659651537
              }
          }
      ]
  },
  "monthly_avg_pecentiles" : {
      "values" : {
          "1.0" : 0.9431761900583903,
          "5.0" : 0.9431761900583903,
          "25.0" : 1.0013999938964844,
          "50.0" : 1.0611889097425673,
          "75.0" : 1.1740000064556415,
          "95.0" : 1.3259461659651537,
          "99.0" : 1.3259461659651537
      }
  ......
}
```

9.6.5 桶统计聚合

桶统计（stats_bucket）聚合的目的是根据上一阶段的聚合结果作出统计。统计指标有最小值、最大值、总和、计数和平均值。支持参数 format 和 gap_policy，与 9.6.1 桶平均值聚合所描述的相同。以下示例找出基金代码为 159801.SZ，在基金净值索引 fund_nav 中找出复权单位净值的月平均值和月平均值的桶统计。各个聚合名称和其使用的相关聚合如下：

- 聚合名称 monthly_avg_adj_nav 使用 min 聚合，并调用 adj_nav 字段。
- 聚合名称 overall_stats_monthly_avg 使用 stats_bucket 聚合，并使用 buckets_path 调用 monthly_avg_adj_nav 聚合的结果。

日期直方图聚合名称 monthly_report 结果显示，总共共 7 个复权单位净值的月平均值。最小为 0.9431761900583903，最大为 1.3259461659651537 和整体为 1.0838375410833272。月平均的总和值忽略不提。

```
curl --request POST http://localhost:9200/fund_nav/_search?pretty=true --header "Content-Type: application/json" --data $'{"query":{"term": {"ts_code.keyword": {"value":"159801.SZ"}}},"aggs":{"monthly_report":{"date_histogram": {"field":"end_date","calendar_interval":"1M"}, "aggs":{"monthly_avg_adj_nav": {"avg":{"field":"adj_nav"}}}}, "overall_stats_monthly_avg":{"stats_bucket":{"buckets_path":"monthly_report>monthly_avg_adj_nav"}}},"size":0}'
{
  ……
  "aggregations" : {
    "monthly_report" : {
      "buckets" : [
        ……
        {
          "key_as_string" : "20200701",
          "key" : 1593561600000,
          "doc_count" : 13,
          "monthly_avg_adj_nav" : {
            "value" : 1.3259461659651537
          }
        }
      ]
    },
    "overall_stats_monthly_avg " : {
      "count" : 7,
      "min" : 0.9431761900583903,
      "max" : 1.3259461659651537,
      "avg" : 1.0838375410833272,
      "sum" : 7.58686278758329
    }
  }
}
```

9.6.6 桶扩展统计聚合

桶扩展统计（extended_stats_bucket）聚合类似桶统计聚合，不同处在于除了提供基本的统计指标外，还包括平方和、方差、标准差和标准差范围（相对于平均值的 ±2 标准差）。支持参数 format 和 gap_policy，与 9.6.1 桶平均值聚合所描述的相同。另外还支持参数 sigma 供设置标准差范围。以下示例找出基金代码为 159801.SZ，在基金净值索引 fund_nav 中找出复权单位净值的月平均值和月平均值的桶扩展统计。各个聚合名称和其使用的相关聚合如下：

- 聚合名称 monthly_avg_adj_nav 使用 avg 聚合，并调用 adj_nav 字段。
- 聚合名称 overall_extended_stats_monthly_avg 使用 extended_stats_bucket 聚合，并使用

buckets_path 调用 monthly_avg_adj_nav 聚合的结果。

日期直方图聚合名称 monthly_report 结果显示如下，不再详细描述。

```
curl --request POST http://localhost:9200/fund_nav/_search?pretty=true --header
"Content-Type: application/json" --data $'{"query":{"term": {"ts_code.keyword":
{"value":"159801.SZ"}}},"aggs":{"monthly_report":{"date_histogram": {"field":
"end_date","calendar_interval":"1M"}, "aggs":{"monthly_avg_adj_nav": {"avg":{
"field":"adj_nav"}}}}, "overall_extended_stats_monthly_avg" : {"extended_stats_
bucket":{
"buckets_path": "monthly_report>monthly_avg_adj_nav"}}},"size":0}'
{
   ……
   "aggregations" : {
     "monthly_report" : {
       "buckets" : [
         ……
         {
           "key_as_string" : "20200701",
           "key" : 1593561600000,
           "doc_count" : 13,
           "monthly_avg_adj_nav" : {
             "value" : 1.3259461659651537
           }
         }
       ]
     },
     "overall_extended_stats_monthly_avg" : {
       "count" : 7,
       "min" : 0.9431761900583903,
       "max" : 1.3259461659651537,
       "avg" : 1.0838375410833272,
       "sum" : 7.58686278758329,
       "sum_of_squares" : 8.323618005239414,
       "variance" : 0.014384471001220487,
       "std_deviation" : 0.11993527838472084,
       "std_deviation_bounds" : {
         "upper" : 1.3237080978527689,
         "lower" : 0.8439669843138855
       }
   ……
}
```

9.6.7 桶总和聚合

桶总和（sum_bucket）聚合的目的是根据上一阶段的聚合结果来计算总和。支持参数 format 和 gap_policy，与 9.6.1 桶平均值聚合所描述的相同。以下示例找出基金代码为 159801.SZ，在基金净值索引 fund_nav 中找出复权单位净值的月平均值和文档总数。各个聚合名称和其使用的相关聚合如下：

- 聚合名称 monthly_avg_adj_nav 使用 avg 聚合，并使用 adj_nav 字段。
- 聚合名称 total_doc 使用 sum_bucket 聚合，并使用 buckets_path 调用 monthly_report 聚合结果的文档总数 _count。

日期直方图聚合名称 monthly_report 的结果显示，文档总数为 110。

```
curl --request POST http://localhost:9200/fund_nav/_search?pretty=true
--header "Content-Type: application/json" --data $'{"query":{"term": {
"ts_code.keyword":{"value":"159801.SZ"}}},"aggs":{"monthly_report":{
"date_histogram": {"field":"end_date","calendar_interval":"1M"}, "aggs":{
```

```
"monthly_avg_adj_nav":{"avg":{"field":"adj_nav"}}}},"total_doc":{
"sum_bucket":{"buckets_path": "monthly_report._count"}}},"size":0}'
{
   ……
   "aggregations" : {
     "monthly_report" : {
       "buckets" : [
       ……
         {
           "key_as_string" : "20200701",
           "key" : 1593561600000,
           "doc_count" : 13,
           "monthly_avg_adj_nav" : {
             "value" : 1.3259461659651537
           }
         }
       ]
     },
     "total_doc" : {
       "value" : 110.0
     }
   }
}
```

9.6.8 桶脚本聚合

桶脚本（bucket_script）聚合的目的是为上一个阶段中的聚合结果，对每个存储桶执行用户定义的脚本。提供参数 gap_policy，与 9.6.1 桶平均值聚合所描述的相同。另外还支持参数 script 可在脚本编写指令内容。

以下示例找出基金代码为 159801.SZ，在基金净值索引 fund_nav 中找出复权单位净值每月的最大值和最小值，然后找出该月的数据范围（range = max - min）。各个聚合名称和其使用的相关聚合如下：

- 聚合名称 monthly_max_adj_nav 使用 max 聚合，并调用 adj_nav 字段。
- 聚合名称 monthly_min_adj_nav 使用 min 聚合，并调用 adj_nav 字段。
- 聚合名称 monthly_range_adj_nav 使用 bucket_script 聚合，并调用 monthly_max_adj_nav 聚合和 monthly_min_adj_nav 聚合的结果进行减法运算。

与 9.6.8 节桶脚本聚合示例相同，在 date_histogram 聚合中使用参数 min_doc_count 设置最小文档数为 1。日期直方图聚合名称 monthly_range_report 的结果显示如下。

提示：由于桶脚本聚合属于 parent 类型，所以聚合名称 monthly_range_report 和聚合名称 monthly_range_adj_nav 不在同一级别。其聚合结果也不在同一级别。

```
curl --request POST http://localhost:9200/fund_nav/_search?pretty=true
--header "Content-Type: application/json" --data $'{"query":{"term": {
"ts_code.keyword":{"value":"159801.SZ"}}},"aggs":{"monthly_range_report":{
"date_histogram":{"field":"end_date","calendar_interval":"1M", "min_doc_
count":"1"},
 "aggs":{"monthly_max_adj_nav": {"max":{"field":"adj_nav"}},"monthly_min_adj_nav":{
 "min":{"field":"adj_nav"}},"monthly_range_adj_nav":{"bucket_script":{"buckets_
path":{
 "min_value":"monthly_min_adj_nav ", "max_value":" monthly_max_adj_nav
"},"script":
 "params.max_value - params.min_value"}}}}},"size":0}'
 {
```

```
......
    "aggregations" : {
    "monthly_range_report" : {
      "buckets" : [
        {
          "key_as_string" : "20200101",
          "key" : 1577836800000,
          "doc_count" : 2,
          "monthly_max_adj_nav" : {
            "value" : 1.0027999877929688
          },
          "monthly_min_adj_nav" : {
            "value" : 1.0
          },
          "monthly_range_adj_nav" : {
            "value" : 0.00279998779296875
          }
      ......
}
```

9.6.9 桶选择器聚合

桶选择器（bucket_selector）聚合的目的是为上一个阶段中的聚合结果，选择合适条件的存储桶。支持参数 format 和 gap_policy，与 9.6.1 桶平均值聚合所描述的相同。另外还支持参数 script，可在脚本编写选择合适条件的存储桶。

以下示例找出基金代码为 159801.SZ，在基金净值索引 fund_nav 中找出复权单位净值每月的最大值和最小值，然后只选择保留复权单位净值数据范围大于 0.3（max – min > 0.3）的月份。各个聚合名称和其使用的相关聚合如下：

- 聚合名称 monthly_max_adj_nav 使用 max 聚合，并调用 adj_nav 字段。
- 聚合名称 monthly_min_adj_nav 使用 min 聚合，并调用 adj_nav 字段。
- 聚合名称 monthly_select_range 使用 bucket_selector 聚合，并调用 monthly_max_adj_nav 聚合和 monthly_min_adj_nav 聚合的结果首先进行减法运算，然后将相差与 0.3 比较，成为选择标准。

与 9.6.8 节桶脚本聚合示例相同，在 date_histogram 聚合中使用参数 min_doc_count 设置最小文档数为 1。日期直方图聚合名称 monthly_select_range_report 的结果显示，2020 年 02 月和 2020 年 07 月，在这两个月数据范围均大于 0.3。

提示：由于桶选择器聚合属于 parent 类型，所以聚合名称 monthly_select_range_report 和聚合名称 monthly_select_range 不在同一级别，其聚合结果也不在同一级别。聚合名称 monthly_select_range 的结果保留在 buckets 的存储桶内。

```
curl --request POST http://localhost:9200/fund_nav/_search?pretty=true --header "Content-Type: application/json" --data $'{"query":{"term": {"ts_code.keyword": {"value":"159801.SZ"}}},"aggs":{"monthly_select_range_report":{"date_histogram":
{"field":"end_date","calendar_interval":"1M", "min_doc_count":"1"}, "aggs":{
"monthly_max_adj_nav":{"max":{"field":"adj_nav"}}, "monthly_min_adj_nav":{"min":{
"field":"adj_nav"}},"monthly_select_range":{"bucket_selector":{"buckets_path":{
"min_value":"monthly_min_adj_nav ", "max_value":" monthly_max_adj_nav "},
"script":"(params.max_value - params.min_value)>0.3"}}}}},"size":0}'
    {
```

```
    ......
    "aggregations" : {
    "monthly_select_range_report" : {
      "buckets" : [
        {
          "key_as_string" : "20200201",
          "key" : 1580515200000,
          "doc_count" : 13,
          "monthly_max_adj_nav" : {
            "value" : 1.3597999811172485
          },
          "monthly_min_adj_nav" : {
            "value" : 1.0296000242233276
          }
        },
      ......
}
```

9.6.10 桶排序聚合

桶排序（bucket_sort）聚合的目的是为上一个阶段中的聚合结果，对存储桶内给定的聚合名称内部的值进行排序。支持参数 gap_policy，与 9.6.1 桶平均值聚合所描述的相同。另外还支持参数 from、size 和 sort。参数 from 和 sort 用法和含义类似于搜索请求。参数 size 在默认情况下，将返回父集合所有匹配项。

以下示例找出基金代码为 159801.SZ，在基金净值索引 fund_nav 中找出复权单位净值每月的最大值，然后按照每月的最大值降序排序。聚合名称 monthly_ranking_adj_nav 的排序结果在 buckets 的存储桶内。

各个聚合名称和其使用的相关聚合如下：

- 聚合名称 monthly_max_adj_nav 使用 max 聚合，并调用 adj_nav 字段。
- 聚合名称 monthly_ranking_adj_nav 使用 bucket_sort 聚合，并调用 monthly_max_adj_nav 聚合的结果降序排列。

与 9.6.8 节桶脚本聚合示例相同，在 date_histogram 聚合中使用参数 min_doc_count 设置最小文档数为 1。日期直方图聚合名称 monthly_ranking_report 的结果显示，所有复权单位净值的最大值出现在 2020 年 07 月，排名第一。

```
curl --request POST http://localhost:9200/fund_nav/_search?pretty=true
--header "Content-Type: application/json" --data $'{"query":{"term": {
"ts_code.keyword":{"value":"159801.SZ"}}},"aggs":{"monthly_ranking_report":{
"date_histogram": {"field":"end_date","calendar_interval":"1M", "min_doc_count":"1"},
"aggs":{"monthly_max_adj_nav":{"max":{"field":"adj_nav"}},"monthly_ranking_adj_nav":{
"bucket_sort":{"sort":[{"monthly_max_adj_nav":{"order":"desc"}}]}}}}},"size":0}'
    {
    ......
    "aggregations" : {
        "monthly_ranking_report" : {
          "buckets" : [
            {
              "key_as_string" : "20200701",
              "key" : 1593561600000,
              "doc_count" : 13,
              "monthly_max_adj_nav" : {
                "value" : 1.485700011253357
              }
            },
    ......
```

}

9.6.11 累计基数聚合

累计基数（cumulative_cardinality）聚合的目的是为上一个阶段中的基数聚合结果，计算每个存储桶的累计基数值。参数 min_doc_count 与 9.5.4 直方图聚合所描述的相同，而且还必须设置为默认值 0。支持参数 format，与 9.6.1 桶平均值聚合所描述的相同。

以下示例在基金持仓索引 fund_portfolio 下，计算出在所有基金中，每月涉及投资组合内变更的基金数目 num_of_invested_funds。然后再找出每月累计曾投资的基金数目 cumulative_invested_funds（也就是累计基数聚合）。聚合名称 num_of_invested_funds 和聚合名称 cumulative_invested_funds 的结果在 buckets 的存储桶内。

各个聚合名称和其使用的相关聚合如下：

- 聚合名称 num_of_invested_funds 使用 cardinality 聚合，并调用 symbol.keyword 字段。
- 聚合名称 cumulative_invested_funds 使用 cumulative_cardinality 聚合，并使用 buckets_path 调用 num_of_invested_funds 聚合的结果。

日期直方图聚合名称 total_invested_funds 的结果汇总到表 9-3。

```
curl --request POST http://localhost:9200/fund_portfolio/_search?pretty=true
--header "Content-Type: application/json" --data $'{"aggs":{"total_invested_
funds":{
"date_histogram": {"field":"ann_date","calendar_interval":"1M"},"aggs":{
"num_of_invested_funds":{"cardinality":{"field":"symbol.keyword"}},
"cumulative_invested_funds":{"cumulative_cardinality":{"buckets_path":
"num_of_invested_funds"}}}}},"size":0}'
{
  ......
  "aggregations" : {
    "total_invested_funds" : {
      "buckets" : [
        {
          "key_as_string" : "20200201",
          "key" : 1580515200000,
          "doc_count" : 32,
          "num_of_invested_funds" : {
            "value" : 20
          },
          "cumulative_invested_funds" : {
            "value" : 20
          }
        },
        ......
}
```

表 9-3 累计基数聚合示例结果汇总

日期	每月投资组合内变更的笔数	每月投资组合内涉及变更的基金数目	累计曾投资的基金数目
20200201	32	20	20
20200301	0	0	20
20200401	35	25	25
20200501	0	0	25
20200601	0	0	25
20200701	15	15	31

9.6.12 累计总和聚合

累计总和（cumulative_sum）聚合的目的是根据上一阶段的聚合结果来计算存储桶的累加总和值。支持参数 format，与 9.6.1 桶平均值聚合所描述的相同。以下示例在基金持仓索引 fund_portfolio 下，计算出在所有基金中，每月涉及投资组合内变更的基金数目 monthly_num_of_changes。然后再找出每月变更的基金数目累计总和 cumulative_num_of_changes。聚合名称 monthly_num_of_changes 和聚合名称 cumulative_num_of_changes 的结果在 buckets 的存储桶内。

各个聚合名称和其使用的相关聚合如下：

- 聚合名称 monthly_num_of_changes 使用 value_count 聚合，并调用 symbol.keyword 字段。
- 聚合名称 cumulative_num_of_changes 使用 cumulative_sum 聚合，并使用 buckets_path 调用 monthly_num_of_changes 聚合的结果。

日期直方图聚合名称 num_of_changes 的结果如下，显示从 2020 年 02 月至 2020 年 07 月，总共 32 个变更。

```
curl --request POST http://localhost:9200/fund_portfolio/_search?pretty=true
--header "Content-Type: application/json" --data $'{"aggs":{"num_of_changes":{
"date_histogram": {"field":"ann_date","calendar_interval":"1M"},"aggs":{
"monthly_num_of_changes":{"value_count":{"field":"symbol.keyword"}},
"cumulative_num_of_changes":{"cumulative_sum":{"buckets_path":
"monthly_num_of_changes"}}}}},"size":0}'
{
  ......
  "aggregations" : {
    "num_of_changes " : {
      "buckets" : [
        {
          "key_as_string" : "20200201",
          "key" : 1580515200000,
          "doc_count" : 32,
          "monthly_num_of_changes" : {
            "value" : 32
          },
          "cumulative_num_of_changes" : {
            "value" : 32.0
          }
        ......
}
```

9.6.13 导数聚合

导数（derivative）聚合的目的是根据上一阶段的聚合结果来计算相邻存储桶之间的导数值（一阶导数或二阶导数）。支持参数 format 和 gap_policy，与 9.6.1 桶平均值聚合所描述的相同。以下示例在基金净值索引 fund_nav 中，找出基金代码为 159801.SZ 的复权单位净值的一阶导数和二阶导数。聚合名称 1st_derivative 和聚合名称 2nd_derivative 的结果在 buckets 的存储桶内。

各个聚合名称和其使用的相关聚合如下：

- 聚合名称 1st_derivative 使用 derivative 聚合，并使用 buckets_path 调用 daily_adj_nav 聚合的结果。

- 聚合名称 2nd_derivative 使用 derivative 聚合，并使用 buckets_path 调用 1st_derivative 聚合的结果。

日期直方图聚合名称 daily_derivative_report 的结果如下，由于数据过多，仅显示前三天的数据。

与 9.6.8 节桶脚本聚合示例相同，在 date_histogram 聚合中使用参数 min_doc_count 设置最小文档数为 1。因为需要使用每日 adj_nav 的值，所以使用总和聚合 daily_adj_nav（每日只有一个 adj_nav 记录）。

```
curl --request POST http://localhost:9200/fund_nav/_search?pretty=true
--header "Content-Type: application/json" --data $'{"query":{"term": {
"ts_code.keyword": {"value":"159801.SZ"}}},"aggs":{"daily_derivative_report":{
"date_histogram":{"field":"end_date","calendar_interval":"1d", "min_doc_
count":"1"},
"aggs":{"daily_adj_nav":{"sum":{"field":"adj_nav"}}, "1st_
derivative":{"derivative":{
"buckets_path":"daily_adj_nav"}},"2nd_derivative":{"derivative":{"buckets_path":
"1st_derivative"}}}}},"size":0}'
{
  ......
  "aggregations" : {
    "daily_derivative_report" : {
      "buckets" : [
        {
          "key_as_string" : "20200120",
          "key" : 1579478400000,
          "doc_count" : 1,
          "daily_adj_nav" : {
            "value" : 1.0
          }
        },
        {
          "key_as_string" : "20200123",
          "key" : 1579737600000,
          "doc_count" : 1,
          "daily_adj_nav" : {
            "value" : 1.0027999877929688
          },
          "1st_derivative" : {
            "value" : 0.00279998779296875
          }
        },
        {
          "key_as_string" : "20200207",
          "key" : 1581033600000,
          "doc_count" : 1,
          "daily_adj_nav" : {
            "value" : 1.0405000448226929
          },
          "1st_derivative" : {
            "value" : 0.03770005702972412
          },
          "2nd_derivative" : {
            "value" : 0.03490006923675537
          }
        },
        ......
    }
}
```

9.6.14 移动函数聚合

执行移动函数（moving_fn）聚合之前，必须首先执行直方图或日期直方图聚合。然后根据上一阶段的聚合结果，按照给定滑动窗口参数 window 的大小，执行参数 script 指定的脚本。另外还支持参数 shift，可以平移滑动窗口，默认值为 0。滑动窗口的右边界位于当前数据之前的位置。如果设定 shift = 1 向右平移一位，则当前数据将包括在计算内。如果设定 shift = window/2，则当前数据在窗口中间的位置。Elasticsearch 预先构建了许多可在移动函数聚合内使用的函数，所有移动函数聚合可以作为父级类型管道聚合，以下示例说明各个移动函数。

1. 移动最大值、最小值和范围总和

根据每个窗口中的数据集计算相对应的移动最大值、移动最小值和移动范围总和。若是数据值为 null 和 NaN 值将被忽略，但是对空窗口及所有值均为 null 或 NaN 的情况下，则返回 NaN 作为最后结果。以下示例在基金净值索引 fund_nav 中，找出基金代码为 159801.SZ 的复权单位净值的 5 个交易日相关的移动值。各个聚合名称和其使用的相关函数如下：

- 聚合名称 moving_max 使用函数 MovingFunctions.max()，并调用 daily_adj_nav 聚合的结果。
- 聚合名称 moving_min 使用函数 MovingFunctions.min()，并调用 daily_adj_nav 聚合的结果。
- 聚合名称 moving_range 使用 bucket_script 聚合，并调用 moving_max 聚合和 moving_min 聚合的结果。
- 聚合名称 moving_sum_range 使用函数 MovingFunctions.sum()，并调 moving_range 聚合的结果。移动范围等于移动最大值和移动最小值的差。

与 9.6.8 节桶脚本聚合示例相同，在 date_histogram 聚合中使用参数 min_doc_count 设置最小文档数为 1。日期直方图聚合名称 daily_report 的结果如下，由于数据过多，仅显示最后一天的数据。

```
curl --request POST http://localhost:9200/fund_nav/_search?pretty=true
--header "Content-Type: application/json" --data $'{"query":{"term": {
"ts_code.keyword":{"value":"159801.SZ"}}},"aggs":{"daily_report":{
"date_histogram":{"field":"end_date","calendar_interval":"1d",
"min_doc_count":"1"}, "aggs":{"daily_adj_nav": { "sum" : {"field": "adj_nav"}},
"moving_max":{"moving_fn":{"buckets_path": "daily_adj_nav", "window":5, "shift":1,
"script":"MovingFunctions.max(values)"}},"moving_min":{"moving_fn":{
"buckets_path":"daily_adj_nav", "window":5, "shift":1,"script":
"MovingFunctions.min(values)"}},"moving_range":{"bucket_script": {"buckets_path": {
"max":"moving_max", "min":"moving_min" }, "script":"params.max-params.min"}},
"moving_sum_range":{"moving_fn":{"buckets_path":"moving_range", "window":5,
"shift":1,"script":"MovingFunctions.sum(values)"}}}}},"size":0}'
{
  ......
  "aggregations" : {
    "daily_report" : {
      "buckets" : [
      ......
      {
        "key_as_string" : "20200717",
        "key" : 1594944000000,
        "doc_count" : 1,
        "daily_adj_nav" : {
          "value" : 1.2588000297546387
```

```
        },
        "moving_max" : {
          "value" : 1.485700011253357
        },
        "moving_min" : {
          "value" : 1.2588000297546387
        },
        "moving_range" : {
          "value" : 0.22689998149871826
        },
        "moving_sum_range" : {
          "value" : 0.8352000713348389
        }
      ......
}
```

2. 移动平均值

根据每个窗口中的数据集，以对应的模型（简单移动平均或者线性加权移动平均）计算的移动平均值。与未加权模型的区别是，线性加权模型将线性权重按照窗口中的出现顺序分配给系列中的点，较早出现的点对平均值的影响较小。null 和 NaN 值的处理方式与移动最小值、最大值及移动范围总和相同。以下示例在基金净值索引 fund_nav 中，找出基金代码为 159801.SZ 的复权单位净值的 5 个交易日相关的移动值。与 9.6.8 节桶脚本聚合示例相同，在 date_histogram 聚合中使用参数 min_doc_count 设置最小文档数为 1。各个聚合名称和其使用的相关函数如下：

- 聚合名称 moving_avg 使用函数 MovingFunctions.unweightedAvg()，并调用 daily_adj_nav 聚合的结果。
- 聚合名称 moving_lw_avg 使用函数 MovingFunctions.linearWeightedAvg()，并调用 daily_adj_nav 聚合的结果。

日期直方图聚合名称 daily_report 的结果如下，由于数据过多，仅显示最后一天的数据。

```
curl --request POST http://localhost:9200/fund_nav/_search?pretty=true --header
"Content-Type: application/json" --data $'{"query":{"term":{"ts_code.keyword":
{"value":"159801.SZ"}}},"aggs":{"daily_report":{"date_histogram":{"field":
"end_date","calendar_interval":"1d","min_doc_count":"1"},"aggs":{"daily_adj_nav":{
"sum":{"field":"adj_nav"}},"moving_avg":{"moving_fn":{"buckets_path":
"daily_adj_nav", "window":5, "shift":1, "script":"MovingFunctions.
unweightedAvg(values)"}}
,"moving_lw_avg":{"moving_fn":{"buckets_path": "daily_adj_nav", "window":5,
"shift":1,
"script":"MovingFunctions.linearWeightedAvg (values)"}}}},"size":0}'
{
  "aggregations" : {
    "daily_report" : {
      "buckets" : [
      ......
        {
          "key_as_string" : "20200717",
          "key" : 1594944000000,
          "doc_count" : 1,
          "daily_adj_nav" : {
            "value" : 1.2588000297546387
          },
          "moving_avg" : {
            "value" : 1.3730200052261352
          },
```

```
            "moving_lw_avg" : {
              "value" : 1.2467500045895576
            }
......
}
```

3. 标准差

根据每个窗口中的数据集计算相对应的标准差。数据值为 null 和 NaN 值将被忽略，但是对空窗口及所有值均为 null 或 NaN 的情况下，则返回 0 作为最后结果。参数 avg 可以指定平均值计算方法，例如简单或者线性加权移动平均值。以下示例在基金净值索引 fund_nav 中，找出基金代码为 159801.SZ 的复权单位净值的 5 个交易日移动标准差。聚合名称 moving_stdDev 使用函数 MovingFunctions.stdDev()，并调用 daily_adj_nav 聚合的结果。与 9.6.8 节桶脚本聚合示例相同，在 date_histogram 聚合中使用参数 min_doc_count 设置最小文档数为 1。日期直方图聚合名称 daily_report 的结果如下，由于数据过多，仅显示最后一天的数据。

```
curl --request POST http://localhost:9200/fund_nav/_search?pretty=true
--header "Content-Type: application/json" --data $'{"query":{"term": {
"ts_code.keyword":{"value":"159801.SZ"}}},"aggs":{"daily_report":{
"date_histogram":{"field":"end_date","calendar_interval":"1d","min_doc_count":"1"},
"aggs":{"daily_adj_nav": {"sum" : {"field": "adj_nav"}}, "moving_stdDev":{
"moving_fn":{"buckets_path": "daily_adj_nav", "window":5, "shift":1,"script":
"MovingFunctions.stdDev(values, MovingFunctions.linearWeightedAvg(values))"}}}}}
,"size":0}'
{
  "aggregations" : {
    "daily_report" : {
      "buckets" : [
         ......
         {
           "key_as_string" : "20200717",
           "key" : 1594944000000,
           "doc_count" : 1,
           "daily_adj_nav" : {
             "value" : 1.2588000297546387
           },
           "moving_stdDev" : {
             "value" : 0.15765529272600945
           }
         ......
         }
```

4. 指数加权移动平均值

根据每个窗口中的数据子集，计算每个窗口中数据子集的指数加权移动平均值。与线性加权移动平均函数相似，指数加权移动平均模型按照窗口中的出现顺序，按给定的参数以指数衰减速度，降低其重要性分配给系列中的点，亦即较早出现的点对平均值的影响较小。指数加权移动平均模型可以细分为单指数 (ewma)、二次指数 (holt) 和三次指数 (holtWinters)。单指数需要给定参数 alpha。二次指数需要给定参数 alpha 及 beta。三次指数需要给定参数 alpha、beta、gamma、period 及 multiplicative。null 和 NaN 值的处理方式与移动平均值相同。

以下示例在基金净值索引 fund_nav 中，找出基金代码为 159801.SZ 的复权单位净值的 5 个交易日各个指数加权移动平均聚合。各个指数函数的参数设置为 window=5、alpha=0.3、 beta=0.2、gamma=0.1、period=2 及 multiplicative=false。参数 window 必须至少是参数 period 的两倍。

各个聚合名称和其使用的相关函数如下：

- 聚合名称 ewma 使用函数 MovingFunctions.ewma()，并调用 daily_adj_nav 聚合的结果。
- 聚合名称 holt 使用函数 MovingFunctions.holt()，并调用 daily_adj_nav 聚合的结果。
- 聚合名称 holtWinters 使用函数 MovingFunctions.holtWinters()，并调用 daily_adj_nav 聚合的结果。

与 9.6.8 节桶脚本聚合示例相同，在 date_histogram 聚合中使用参数 min_doc_count 设置最小文档数为 1。日期直方图聚合名称 daily_report 的结果如下，由于数据过多，仅显示最后一天的数据。

```
curl --request POST http://localhost:9200/fund_nav/_search?pretty=true
--header "Content-Type: application/json" --data $'{"query":{"term": {
"ts_code.keyword":{"value":"159801.SZ"}}},"aggs":{"daily_report":{
"date_histogram":{"field":"end_date","calendar_interval":"1d","min_doc_
count":"1"},
"aggs":{"daily_adj_nav":{"sum":{"field":"adj_nav"}},"ewma":{"moving_fn":{
"buckets_path": "daily_adj_nav", "window":5,"shift":1,"script":
"MovingFunctions.ewma(values,0.3)"}}, "holt":{"moving_fn":{"buckets_path":
 "daily_adj_nav", "window":5, "shift":1,"script":"MovingFunctions.
holt(values,0.3,0.2)"}},
 "holtWinters":{"moving_fn":{"buckets_path": "daily_adj_nav", "window":5,
 "shift":1,"script":"if (values.length>=5) {MovingFunctions.
holtWinters(values,0.3,0
.2,0.1,2,false)}"}} }},"size":0}'
{
  ......
    "aggregations" : {
      "daily_report" : {
        "buckets" : [
          ......
          {
            "key_as_string" : "20200717",
            "key" : 1594944000000,
            "doc_count" : 1,
            "daily_adj_nav" : {
              "value" : 1.2588000297546387
            },
            "ewma" : {
              "value" : 1.3556092373132702
            },
            "holt" : {
              "value" : 1.3397522743511194
            },
            "holtWinters" : {
              "value" : 1.3808981605464914
            }
          ......
}
```

9.6.15 串行差分聚合

串行差分（serial_diff）聚合的目的是根据上一阶段的聚合结果来计算相邻存储桶之间，时间滞后的一系列值差。假设参数 lag 的值为 n，串行差分会计算从当前存储桶中的值减去前 n 个存储桶中的值。支持参数 format 和 gap_policy，与 9.6.1 桶平均值聚合所描述的相同。以下示例与导数聚合中的示例相似，在基金净值索引 fund_nav 中，找出基金代码 159801.SZ 的复权单位净值的串行差分聚合，设定滞后值 lag 为 1。串行差分聚合名称 daily_serial_diff_report 的结果与导数聚合的示例相同。

提示：如果 lag=1，串行差分聚合相当于一阶导数聚合。

```
curl --request POST http://localhost:9200/fund_nav/_search?pretty=true
--header "Content-Type: application/json" --data $'{"query":{"term": {
"ts_code.keyword":{"value":"159801.SZ"}}},"aggs":{"daily_serial_diff_report":{
"date_histogram":{"field":"end_date","calendar_interval":"1d", "min_doc_
count":"1"},
 "aggs":{"daily_adj_nav":{"sum":{"field":"adj_nav"}},"serial_diff_adj_nav":{
"serial_diff":{"buckets_path":"daily_adj_nav","lag":1}}}}},"size":0}'
{
  ……
  "aggregations" : {
    "daily_serial_diff_report" : {
      "buckets" : [
        {
          "key_as_string" : "20200120",
          "key" : 1579478400000,
          "doc_count" : 1,
          "daily_adj_nav" : {
            "value" : 1.0
          }
        },
        {
          "key_as_string" : "20200123",
          "key" : 1579737600000,
          "doc_count" : 1,
          "daily_adj_nav" : {
            "value" : 1.0027999877929688
          },
          "serial_diff_adj_nav" : {
            "value" : 0.00279998779296875
          }
        }
  ……
}
```

9.7　后置过滤器

通常聚合引用的搜索结果将在聚合结果之后显示，如果想在聚合之后进一步过滤搜索结果，例如显示范围比较小的搜索结果，那么可以在查询后运行后置过滤器（post_filter）。后置过滤器的操作对聚合结果没有影响。以下使用与 9.4.4 节相同的示例，在基金净值索引 fund_nav 中找出基金代码为 159995.SZ，在 total_netasset 字段有值的文档有多少个。结果显示共有 114 个文档，但只有 3 个文档在该字段是有值的。如果需要列出搜索结果，则根据参数 size 大小将文档列出。可利用后置过滤器过滤掉在该字段无值的文档，最后搜索结果只保留 3 个文档被陈列出来。

```
curl --request POST http://localhost:9200/fund_nav/_search?pretty=true
--header "Content-Type: application/json" --data $'{"query":{"term":{
"ts_code.keyword":"159995.SZ"}},"aggs": {"num_of_total_netasset": {"value_count":
{"field":"total_netasset"}}}, "post_filter":{"exists":{"field":"total_netasset"}},
"_source":["total_netasset"]}'
```

```
{
  ......
  "hits" : {
    "total" : {
      "value" : 3,
      "relation" : "eq"
    },
    ......
    {
      "_index" : "fund_nav",
      "_type" : "_doc",
      "_id" : "jJ_6YnMBpfHDzdHUxAEm",
      "_score" : 0.6755256,
      "_source" : {
        "total_netasset" : 5388463006
      }
    ......
      "aggregations" : {
      "num_of_total_netasset" : {
          "value" : 3
      }
  ......
}
```

第 10 章 摄取节点管道处理接口

在建立索引之前，可以对文档进行预处理。在 1.4 节了解 Elasticsearch 的架构中，曾经介绍四种类型的 Elasticsearch 节点，其中之一是摄取节点。若在摄取节点指定一系列管道处理器，接收节点将拦截批量请求和索引请求，转发到摄取节点，然后按照处理器的顺序执行预处理，最后将文档传递回去。默认情况下，所有节点都启用摄取节点管道处理器 (Ingest pipeline processor) 模式，也可以在节点的配置文件中禁用摄取节点的功能。

10.1 摄取节点接口

基本的摄取节点（_ingest）接口包括摄取管道的创建、更新、模拟和删除。由参数 processors 指定的处理器数组，按流水线式顺序执行。可以使用索引设置 index.default_pipeline 去定义默认的摄取管道，也可以使用另一个索引设置 index.final_pipeline 去定义最终摄取节点管道，前题是摄取管道必须先定义好。以下各节详细说明接口用法。

10.1.1 创建或更新接口

要创建或是更新摄取节点管道，需要使用 PUT 请求，并给定管道识别符。在请求主体中使用参数 processors 来描述处理器数组内各个处理器将执行哪些功能。如果管道是先前创建的，则是一个更新请求，将会覆盖原始内容。另外可使用参数 description 存储相关功能的描述。创建或更新接口请求语法如下：

```
PUT /_ingest/pipeline/管道识别符
{
  "description" : "...",
  "processors" : [ ... ]
}
```

以下示例创建一个摄取节点管道命名为 ohlc_pipeline，计算 OHLC 平均值，成为一个新字段。OHLC 其定义如下：

ohlc = open + high + low + close /4

创建识别符为 ohlc_pipeline 的摄取节点管道示例命令行如下：

```
curl --request PUT http://localhost:9200/_ingest/pipeline/ohlc_pipeline?pretty=true --header "Content-Type: application/json" --data $'{"description":"计算和索引OHLC字段", "processors":[{"script":"ctx.ohlc=(ctx.open+ctx.high+ctx.low+ctx.close)/4"}]}'
```

```
{ "acknowledged" : true }
```

10.1.2 读取接口

对摄取节点管道接口使用 GET 请求，并给定管道识别符来检索管道的定义。

```
GET /_ingest/pipeline/管道识别符
```

以下示例读取摄取节点管道识别符为 ohlc_pipeline 的定义：

```
curl --request GET http://localhost:9200/_ingest/pipeline/ohlc_pipeline?pretty=true
{
  "ohlc_pipeline" : {
    "description" : "脚本处理器",
    "processors" : [
      {
        "script" : "ctx.ohlc=(ctx.open+ctx.high+ctx.low+ctx.close)/4"
      }
    ]
  }
}
```

10.1.3 模拟接口

对摄取节点管道模拟（_ingest/pipeline/ 管道识别符 /_simulate）接口使用 POST 请求，并给定管道识别符及通过指定带有测试文档数组的参数 docs 来测试。也可以在请求主体中使用参数 pipeline 并同时定义管道以进行测试。模拟接口请求语法如下：

```
POST /_ingest/pipeline/管道识别符/_simulate
GET  /_ingest/pipeline/管道识别符/_simulate
POST /_ingest/pipeline/_simulate
GET  /_ingest/pipeline/_simulate
```

以下示例在请求主体中使用参数 docs 数组来指定测试文档。数组内每个文档必须由参数 _source 引用。测试文档为基金代码 159801.SZ 在 2020 年 07 月 31 日的基金交易行情。

1. 使用 ohlc_pipeline 摄取节点管道进行模拟

结果显示新字段 ohlc 计算结果为 1.31175。

```
curl --request POST http://localhost:9200/_ingest/pipeline/ohlc_pipeline/_simulate?pretty=true--header "Content-Type: application/json" --data
$'{"docs":[{"_source":{ "ts_code": "159801.SZ", "trade_date": 20200731, "pre_close": 1.296, "open": 1.295, "high": 1.333, "low": 1.29, "close": 1.329, "change": 0.033, "pct_chg": 2.5463, "vol": 1743608.88, "amount": 229883.049 }}]}'
{
  "docs" : [
    {
      "doc" : {
        "_index" : "_index",
        "_type" : "_doc",
        "_id" : "_id",
        "_source" : {
          "amount" : 229883.049,
          "ohlc" : 1.31175,
          "change" : 0.033,
          "trade_date" : 20200731,
          "pre_close" : 1.296,
          "high" : 1.333,
          "vol" : 1743608.88,
          "ts_code" : "159801.SZ",
          "low" : 1.29,
```

```
          "close" : 1.329,
          "pct_chg" : 2.5463,
        },
        "_ingest" : {
          "timestamp" : "2020-08-03T20:44:43.795555Z"
        }
      ……
}
```

2. 使用参数 pipeline 即时定义摄取节点管道进行模拟

在请求主体中使用与 ohlc_pipeline 的摄取节点管道相同的内容，即时定义以进行测试。结果显示除了时间戳记字段外，其他与项目 1 的相同。

```
curl --request POST http://localhost:9200/_ingest/pipeline/_simulate?pretty=true
--header "Content-Type: application/json" --data $'{"pipeline":{"description":"
脚本处理器","processors":[{"script":"ctx.ohlc=(ctx.open+ctx.high+ctx.low+ctx.
close)/4"}]},"docs":[{"_source":{ "ts_code": "159801.SZ", "trade_date": 20200731,
"pre_close": 1.296, "open": 1.295, "high": 1.333, "low": 1.29, "close": 1.329,
"change": 0.033, "pct_chg": 2.5463, "vol": 1743608.88, "amount": 229883.049 }}]}'
```

10.1.4 删除接口

对摄取节点管道接口使用 DELETE 请求删除管道，管道识别符可以使用通配符星号来删除多个管道。

```
DELETE _ingest/pipeline/管道识别符
```

以下示例对摄取节点管道 ohlc_pipeline 进行删除，如果管道标识符不存在，则返回状态码 404（找不到）。

```
curl --request DELETE http://localhost:9200/_ingest/pipeline/ohlc_
pipeline?pretty=true
  { "acknowledged" : true }
```

10.2 摄取管道处理器

摄取管道支持约 30 多种处理器 processors，在文档进行预处理过程中，每个处理器都按照定义的方式在管道中执行。每个处理器都需要进行参数配置，表 10-1 中先描述的一些常用参数，然后逐项介绍常用的处理器。

表 10-1 摄取管道处理器常用参数

参数名称	描述
field	大部分处理器都需要使用此参数指定操作字段名称
if	大部分处理器都支持此参数，可以定义处理器的执行条件
ignore_failure	此参数为布尔数据类型，默认为 false。当设定为 true 时，可以静默地忽略目前处理器的失败情况，并继续执行下一个处理器
ignore_missing	此参数为布尔数据类型，默认为 false。当设定为 true 时，即使参数 field 所指向的字段在索引文档中为空值或没有值时，也可以安静地忽略目前处理器的失败情况，在不修改文档的情况下静默地退出
on_failure	用于定义后备处理器列表，当原处理器在管道中发生错误并暂停执行时，可以执行这些后备处理器
Tag	用于指定一个字符串来标识处理器
target_field	用于指定目的字段的名称默认值取决于各个处理器

10.2.1 附加处理器

附加（append）处理器将参数 value 指定的值 Y 附加到参数 field 指定的字段 X。若字段 X 是数组数据类型，则将 Y 附加到数组 X。若字段 X 是纯量数据类型，则将 X 转换为数组，然后将 Y 附加到数组 X。若文档没有包含字段 X，则创建 X 为空数组，然后将 Y 附加到数组 X。若 Y 是数组数据类型，则将 Y 的每一项附加到数组 X。此外，附加处理器还支持可选的参数有 if、on_failure、ignore_failure 和 tag。

以下示例使用附加处理器，附加一个字段 indicators，值为字符串 up。测试文档为基金代码 159801.SZ 在 2020 年 07 月 31 日的基金交易行情。结果显示文档内出现新字段 indicators，为数组数据类型，内有一个元素，其值是字符串 up。

```
curl --request POST http://localhost:9200/_ingest/pipeline/_simulate?pretty=true
--header "Content-Type: application/json" --data $'{"pipeline":{"description":
"附加处理器示例", "processors":[{"append":{ "field":"indicators","value":"up"}}]},
"docs":[{"_source":{ "ts_code": "159801.SZ", "trade_date": 20200731, "pre_close":
1.296, "open": 1.295, "high": 1.333, "low": 1.29, "close": 1.329, "change": 0.033,
"pct_chg": 2.5463, "vol": 1743608.88, "amount": 229883.049 }}]}'
{
  "docs" : [
    {
      "doc" : {
        "_index" : "_index",
        "_type" : "_doc",
        "_id" : "_id",
        "_source" : {
          "amount" : 229883.049,
          "change" : 0.033,
          "indicators" : [
            "up"
          ],
          "trade_date" : 20200731,
          "pre_close" : 1.296,
          "high" : 1.333,
          "vol" : 1743608.88,
          "ts_code" : "159801.SZ",
          "low" : 1.29,
          "close" : 1.329,
          "pct_chg" : 2.5463,
          "open" : 1.295
        },
      ......
}
```

10.2.2 删除处理器

删除（remove）处理器将参数 field 指定的字段 X（字段 X 可以是内含文档字段的数组或一个文档字段），从文档中删除那些字段。此外，删除处理器还支持可选的参数有 ignore_missing、if、on_failure、ignore_failure 和 tag。以下示例使用删除处理器，删除数组内的元素 pre_close，测试文档为基金代码 159801.SZ 在 2020 年 07 月 31 日的基金交易行情。结果显示文档已经没有字段 pre_close。

```
curl --request POST http://localhost:9200/_ingest/pipeline/_simulate?pretty=true
--header "Content-Type: application/json" --data $'{"pipeline":{"description":
"删除处理器", "processors":[{"remove":{"field": [ "pre_close"]}}]},"docs":[{"_source":{
"ts_code": "159801.SZ", "trade_date": 20200731, "pre_close": 1.296, "open": 1.295,
"high": 1.333, "low": 1.29, "close": 1.329, "change": 0.033, "pct_chg": 2.5463,
"vol": 1743608.88, "amount": 229883.049 }}]}'
{
  "docs" : [
    {
      "doc" : {
        "_index" : "_index",
        "_type" : "_doc",
        "_id" : "_id",
        "_source" : {
          "amount" : 229883.049,
          "change" : 0.033,
          "trade_date" : 20200731,
          "high" : 1.333,
          "vol" : 1743608.88,
          "ts_code" : "159801.SZ",
          "low" : 1.29,
          "close" : 1.329,
          "pct_chg" : 2.5463,
          "open" : 1.295
        },
        ......
    }
```

10.2.3 重命名处理器

重命名（rename）处理器将参数 field 指定的字段 X，重命名为参数 targe_field 指定的值 Y。如果文档中不存在字段 X 或文档已经存在字段 Y，则引发错误。重命名处理器还支持可选的参数有 ignore_missing、if、on_failure、ignore_failure 和 tag。以下示例使用重命名处理器，重新命名字段 pct_chg 为 percentage_change。测试文档为基金代码 159801.SZ 在 2020 年 07 月 31 日的基金交易行情。结果显示文档的字段 pct_chg 已经被字段 percentage_change 取代。

```
curl --request POST http://localhost:9200/_ingest/pipeline/_simulate?pretty=true
--header "Content-Type: application/json" --data $'{"pipeline":{"description":
"重命名处理器", "processors":[{"rename":{"field": "pct_chg","target_field":
"percentage_change"}}]}, "docs":[{"_source":{ "ts_code": "159801.SZ", "trade_
date": 20200731,
"pre_close": 1.296, "open": 1.295, "high": 1.333, "low": 1.29, "close": 1.329,
"change": 0.033, "pct_chg": 2.5463, "vol": 1743608.88, "amount":
229883.049 }}]}'
{
  "docs" : [
    {
      "doc" : {
        "_index" : "_index",
        "_type" : "_doc",
        "_id" : "_id",
        "_source" : {
          "amount" : 229883.049,
          "change" : 0.033,
          "percentage_change" : 2.5463,
          "trade_date" : 20200731,
          "pre_close" : 1.296,
          "high" : 1.333,
          "vol" : 1743608.88,
          "ts_code" : "159801.SZ",
```

```
            "low" : 1.29,
            "close" : 1.329,
            "open" : 1.295
        },
    ......
}
```

10.2.4 小写处理器

小写（lowercase）处理器将参数 field 指定的字段 X 的值，转换为小写，默认情况下覆盖原来字段内容。若参数 target_field 存在并指定字段 Y，则创建字段 Y 并设定为 X 的小写值。若字段 Y 存在，则字段 Y 的值将重置为 X 的小写值。小写处理器还支持可选的参数有 ignore_missing、if、on_failure、ignore_failure 和 tag。以下示例使用小写处理器，将字段 ts_code 的字符串值转换为小写。测试文档为基金代码 159801.SZ 在 2020 年 07 月 31 日的基金交易行情。结果显示文档的字段 ts_code 已经被转换为小写。

```
curl --request POST http://localhost:9200/_ingest/pipeline/_simulate?pretty=true
--header "Content-Type: application/json" --data $'{"pipeline":{"description":
"小写处理器", "processors":[{"lowercase":{"field":"ts_code"}}]},"docs":[{"_source":{
"ts_code": "159801.SZ", "trade_date": 20200731, "pre_close": 1.296, "open": 1.295,
"high": 1.333, "low": 1.29, "close": 1.329, "change": 0.033, "pct_chg": 2.5463,
"vol": 1743608.88, "amount": 229883.049 }}]}'
{
  "docs" : [
    {
      "doc" : {
        "_index" : "_index",
        "_type" : "_doc",
        "_id" : "_id",
        "_source" : {
          "amount" : 229883.049,
          "change" : 0.033,
          "trade_date" : 20200731,
          "pre_close" : 1.296,
          "high" : 1.333,
          "vol" : 1743608.88,
          "ts_code" : "159801.sz",
          "low" : 1.29,
          "close" : 1.329,
          "pct_chg" : 2.5463,
          "open" : 1.295
        },
    ......
}
```

10.2.5 大写处理器

大写（uppercase）处理器将参数 field 指定的字段 X 的值，转换为大写，默认情况下覆盖原来字段内容。若参数 target_field 存在并指定字段 Y，则创建字段 Y 并设定为 X 的大写值。若字段 Y 存在，则字段 Y 的值将重置为 X 的大写值。写处理器还支持可选的参数有 ignore_missing、if、on_failure、ignore_failure 和 tag。以下示例使用大写处理器，将字段 ts_code 的字符串值转换为大写。测试文档为基金代码 159801.SZ 在 2020 年 07 月 31 日的基金交易行情，并先转换字段 ts_code 为小写。结果显示文档的字段 ts_code 已经被转换为大写。

```
curl --request POST http://localhost:9200/_ingest/pipeline/_simulate?pretty=true
--header "Content-Type: application/json" --data $'{"pipeline":{"description":
"大写处理器", "processors":[{"uppercase":{"field":"ts_code"}}]},"docs":[{"_source":{
"ts_code": "159801.sz", "trade_date": 20200731, "pre_close": 1.296, "open": 1.295,
"high": 1.333, "low": 1.29, "close": 1.329, "change": 0.033, "pct_chg": 2.5463,
"vol": 1743608.88, "amount": 229883.049 }}]}'
{
  "docs" : [
    {
      "doc" : {
        "_index" : "_index",
        "_type" : "_doc",
        "_id" : "_id",
        "_source" : {
          "amount" : 229883.049,
          "change" : 0.033,
          "trade_date" : 20200731,
          "pre_close" : 1.296,
          "high" : 1.333,
          "vol" : 1743608.88,
          "ts_code" : "159801.SZ",
          "low" : 1.29,
          "close" : 1.329,
          "pct_chg" : 2.5463,
          "open" : 1.295
        },
      ......
    }
```

10.2.6 拆分处理器

拆分(split)处理器将参数 field 指定的字段 X 的值,按照参数 separator 指定的正则表达式做分隔符拆分,默认情况下覆盖原来字段内容。若参数 target_field 存在并指定字段 Y,则创建字段 Y 并设定为 X 拆分后的数组。若字段 Y 存在,则字段 Y 的值将重置为 X 拆分后的数组。拆分处理器还支持可选的参数有 ignore_missing、if、on_failure、ignore_failure 和 tag。以下示例使用拆分处理器,将字段 ts_code 的字符串值按照参数 separator 指定的英语句号(\\\\.)为分隔符拆分。然后存储在 target_field 指定的新字段 sep_ts_code。测试文档为基金代码 159801.SZ 在 2020 年 07 月 31 日的基金交易行情。结果显示文档的字段 ts_code 已经被拆分为 19801 和 SZ 两个字符串,存储在数组 sep_ts_code 中。

```
curl  --request POST http://localhost:9200/_ingest/pipeline/_simulate?pretty=true
--header
  "Content-Type: application/json" --data $'{"pipeline":{"description":"拆分处理器
", "processors":[{"split":{"field":"ts_code","separator":"\\\\.","target_field":"sep_ts_
code"}}]},
  "docs":[{"_source":{ "ts_code": "159801.SZ", "trade_date": 20200731, "pre_close":
1.296,
  "open": 1.295, "high": 1.333, "low": 1.29, "close": 1.329, "change": 0.033,
  "pct_chg": 2.5463, "vol": 1743608.88, "amount": 229883.049 }}]}'
{
    "docs" : [
      {
        "doc" : {
          "_index" : "_index",
          "_type" : "_doc",
          "_id" : "_id",
          "_source" : {
```

```
      "amount" : 229883.049,
      "sep_ts_code" : [
        "159801",
        "SZ"
      ],
      "change" : 0.033,
      "trade_date" : 20200731,
      "pre_close" : 1.296,
      "high" : 1.333,
      "vol" : 1743608.88,
      "ts_code" : "159801.SZ",
      ......
}
```

10.2.7 连接处理器

连接（join）处理器将参数 field 指定数据类型为数组的字段 X，按照参数 separator 指定的分隔符将数组内的元素连接在一起，默认情况下覆盖原来字段内容。若参数 target_field 存在并指定字段 Y，则创建字段 Y 并设定为连接后的字符串。若字段 Y 存在，则字段 Y 的值将重置为连接后的字符串。连接处理器还支持可选的参数有 if、on_failure、ignore_failure 和 tag。以下示例使用连接处理器，将字段数组 ts_code_array 的字符串值按照参数 separator 指定的英语句号（.）为分隔符。然后存储在 target_field 指定的字段 join_ts_code。测试文档为基金代码 159801.SZ 在 2020 年 07 月 31 日的基金交易行情。文档中新添加一个字段 ts_code_array，数据类型是数组，内有两个字符串 159801 及 SZ。结果显示文档新添加字段 join_ts_code，内容为字符串 159801.SZ。

```
    curl --request POST http://localhost:9200/_ingest/pipeline/_simulate?pretty=true
--header
    "Content-Type: application/json" --data $'{"pipeline":{"description":"连接处理器",
"processors":[{"join":{"field":"ts_code_array","separator":".","target_field":"join_ts_
code"}}]},
    "docs":[{"_source":{"ts_code":"159801.SZ","ts_code_array":["159801","SZ"],"trade_
date":20200731,
    "pre_close": 1.296, "open": 1.295, "high": 1.333, "low": 1.29, "close": 1.329,
"change": 0.033,
    "pct_chg": 2.5463, "vol": 1743608.88, "amount": 229883.049 }}]}'
{
  "docs" : [
    {
      "doc" : {
        "_index" : "_index",
        "_type" : "_doc",
        "_id" : "_id",
        "_source" : {
          "amount" : 229883.049,
          "join_ts_code" : "159801.SZ",
          ......
          "ts_code" : "159801.SZ",
          "low" : 1.29,
          "ts_code_array" : [
            "159801",
            "SZ"
          ],
          ......
      }
```

10.2.8 修剪处理器

修剪（trim）处理器将参数 field 指定的字段 X 的值，删除 X 的开头和结尾的空格字符，默认情况下覆盖原来字段内容。若参数 target_field 存在并指定字段 Y，则创建字段 Y 并设定为 X 修剪后的值。若字段 Y 存在，则字段 Y 的值将重置为 X 修剪后的值。修剪处理器还支持可选的参数有 ignore_missing、if、on_failure、ignore_failure 和 tag。以下示例使用修剪处理器，将字段 ts_code 的字符串值修剪。测试文档为基金代码 159801.SZ 在 2020 年 07 月 31 日的基金交易行情，并先在 ts_code 字段的开头加入空格字符以及结尾加入换行字符。结果显示文档的字段 ts_code 已经被修剪处理。

```
curl   --request  POST http://localhost:9200/_ingest/pipeline/_simulate?pretty=true
--header
  "Content-Type: application/json" --data $'{"pipeline":{"description":"修剪处理器", "pr
ocessors":[{"trim":{"field":"ts_code"}}]},"docs":[{"_source":{ "ts_code":
"159801.SZ", "trade_date": 20200731, "pre_close": 1.296, "open": 1.295, "high": 1.333,
"low": 1.29, "close": 1.329, "change": 0.033, "pct_chg": 2.5463, "vol": 1743608.88,
"amount": 229883.049 }}]}'
{
    "docs" : [
      {
        "doc" : {
          "_index" : "_index",
          "_type" : "_doc",
          "_id" : "_id",
          "_source" : {
            "amount" : 229883.049,
            "change" : 0.033,
            "trade_date" : 20200731,
            "pre_close" : 1.296,
            "high" : 1.333,
            "vol" : 1743608.88,
            "ts_code" : "159801.SZ",
            "low" : 1.29,
            "close" : 1.329,
            "pct_chg" : 2.5463,
            "open" : 1.295
          },
      ......
  }
```

10.2.9 设置处理器

设置（set）处理器将参数 field 指定的字段 X，重新设定为参数 value 指定的值 Y。如果文档中不存在字段 X，则创建字段 X 并设定为值 Y。如果参数 override 设定为 false 并且索引文档中的字段 X 具有非空值，则将跳过该处理器的操作，参数 override 默认值为 true。此外，设置处理器还支持可选的参数有 if、on_failure、ignore_failure 和 tag。value 指定的值 Y 可以是数字、字符串或是模板代码。模板代码以括号开头及结尾。括号内为文档的字段或是元字段。一个模板代表字段或是元字段的值。两个模板代表各个模板串联在一起的值。以下示例使用设置处理器演示 ohlc 字段的计算过程。测试文档为基金代码 159801.SZ 在 2020 年 07 月 31 日的基金交易行情。结果显示新增字段 ohlc_formula 为 OHLC 定义下代入相对应字段 open、high、low 和 close 的值。

提示：支持模板的设置，模板表示以"{{"开始，并以"}}"结尾，括号内为字段名称。可以包含零或多个模板。

```
    curl --request POST http://localhost:9200/_ingest/pipeline/_simulate?pretty=true
--header 
    "Content-Type: application/json" --data $'{"pipeline":{"description":"设置处理器
","processors":
    [{"set":{"field":"ohlc_formula","value":"({{open}}+{{high}}+{{low}}+{{close}})/4"}}
]},"docs":
    [{"_source":{ "ts_code": "159801.SZ", "trade_date": 20200731, "pre_close": 1.296,
"open": 1.295, "high": 1.333, "low": 1.29, "close": 1.329, "change": 0.033,
"pct_chg": 2.5463, "vol": 1743608.88, "amount": 229883.049 }}]}'
    {
      "docs" : [
        ......
        "_source" : {
          "amount" : 229883.049,
          "change" : 0.033,
          "ohlc_formula" : "(1.295+1.333+1.29+1.329)/4",
          "trade_date" : 20200731,
          "pre_close" : 1.296,
          "high" : 1.333,
          "vol" : 1743608.88,
          "ts_code" : "159801.SZ",
          "low" : 1.29,
          "close" : 1.329,
          "pct_chg" : 2.5463,
          "open" : 1.295
        },
        ......
    }
```

10.2.10 日期处理器

日期（date）处理器将参数 field 指定的字段 X，按照参数 formats 指定的日期表达式解析读入其值 Y。若参数 target_field 存在并指定字段 Z，则创建字段 Z 并设定为 Y 值。若参数 target_field 没有给定，则创建字段 @timestamp 并设定为 Y 值。其他两个支持的可选参数是时区 timezone 和区域 locale 设置。时区参数默认为 UTC，而语言环境参数为英语。日期处理器还支持可选的参数有 ignore_missing、if、on_failure、ignore_failure 和 tag。

以下示例按照参数 formats 的日期表达式读入参数 field 指定的字段 X 的值 Y，然后创建字段 trade_date_formatted 并设定为值 Y，同时设定时区 timezone 为 Asia/Shanghai 和区域 locale 为 zh-CN。测试文档为基金代码 159801.SZ 在 2020 年 07 月 31 日的基金交易行情。结果显示新增字段 trade_date_formatted 的值为 2020-07-31T00:00:00.000+08:00。其中 +08:00（GMT+8）代表格林尼治时间加上八个小时。

```
    curl --request POST http://localhost:9200/_ingest/pipeline/_simulate?pretty=true --header
    "Content-Type: application/json" --data $'{"pipeline":{"description":"日期处理器",
"processors":
    [{"date":{"field":"trade_date","formats":["yyyyMMdd"],"target_field":"trade_date_formatted",
"timezone":"Asia/Shanghai", "locale":"zh-CN"}}]},"docs":[{"_source":{ "ts_code": "159801.SZ",
"trade_date": 20200731, "pre_close": 1.296, "open": 1.295, "high": 1.333, "low": 1.29,
"close": 1.329, "change": 0.033, "pct_chg": 2.5463, "vol": 1743608.88, "amount":
229883.049}}]}'
    {
      "docs" : [
        {
```

```
          "doc" : {
            "_index" : "_index",
            "_type" : "_doc",
            "_id" : "_id",
            "_source" : {
              "amount" : 229883.049,
              "change" : 0.033,
              "trade_date_formatted" : "2020-07-31T00:00:00.000+08:00",
              "trade_date" : 20200731,
              "pre_close" : 1.296,
              "high" : 1.333,
              "vol" : 1743608.88,
              "ts_code" : "159801.SZ",
        ......
        }
```

10.2.11 脚本处理器

脚本（script）处理器可以提供用户使用参数 source 设定脚本，或是使用参数 id 指定预先存储的脚本的标识符。可以使用参数 params 指定脚本的运作参数。默认脚本编程语言 painless。如果使用其他编程语言（例如 Java），则可以使用参数 lang 来指定编程语言。脚本处理器还支持可选的参数有 if、on_failure、ignore_failure 和 tag。应用示例与 10.1.1 节创建或更新接口的示例相同，此处不再赘述。

10.2.12 丢弃处理器

丢弃（drop）处理器可以给用户设定条件，如果文档中的字段内容满足设定条件，文档会被丢弃，文档的后续操作会静默地跳过，不会返回错误。丢弃处理器还支持可选的参数有 if、on_failure、ignore_failure 和 tag。以下示例使用在 10.2.11 节脚本处理器的示例，并在脚本处理器之前加入丢弃处理器，设定条件为当文档中的字段 open、high、low 或是 close 没有值时，静默地跳过后续的。测试文档为基金代码 159801.SZ 在 2020 年 07 月 31 日的基金交易行情，并清除 open 字段。结果显示脚本处理器被静默地跳过，表示没有操作。

```
curl --request POST http://localhost:9200/_ingest/pipeline/_simulate?pretty=true
--header
  "Content-Type: application/json" --data $'{"pipeline":{"description":"丢弃处理器",
"processors":[{"drop":{"if":"ctx.open==null || ctx.close==null || ctx.high==null ||
  ctx.low==null"}}, {"script":{"source":"ctx.ohlc=(ctx.open+ctx.high+ctx.low+ctx.
close)/4"}}]},"docs":
  [{"_source":{"ts_code": "159801.SZ", "trade_date": 20200731, "pre_close": 1.296,
   "high": 1.333, "low": 1.29, "close": 1.329, "change": 0.033, "pct_chg": 2.5463,
   "vol": 1743608.88, "amount": 229883.049 }}]}'
{
  "docs" : [
    null
  ]
}
```

10.2.13 管道委托处理器

管道委托（pipeline）处理器可以给用户设定条件，将某些处理委托给参数 name 指定的管道 X 进行处理。如果管道 X 尚未定义，则处理器将引发错误。管道委托处理器还支持可选的参数有 if、on_failure、ignore_failure 和 tag。以下示例设定条件为当文档中的字段 open、high、low 及 close 都有值时，委托给在 10.1.1 节创建或更新接口定义的脚本处理器 ohlc_pipeline。测试文档为基金代码 159801.SZ 在 2020 年 07 月 31 日的基金交易行情，结果显示新字段 ohlc，计算结果为 1.31175。

提示：如果 ohlc_pipeline 定义已经在 10.2.13 节管道委托处理器中删除，需要重新创建。

```
curl --request POST http://localhost:9200/_ingest/pipeline/_simulate?pretty=true
--header
    "Content-Type: application/json" --data $'{"pipeline":{"description":"管道委托处理器",
"processors":[{"pipeline":{"if":"ctx.open!=null && ctx.close!=null && ctx.high!=null &&
    ctx.low!=null", "name":"ohlc_pipeline"}}]}, "docs":[{"_source":{ "ts_code":
"159801.SZ",
    "trade_date": 20200731, "pre_close": 1.296, "open": 1.295, "high": 1.333, "low": 1.29,
    "close": 1.329, "change": 0.033, "pct_chg": 2.5463, "vol": 1743608.88, "amount":
229883.049 }}]}'
{
    "docs" : [
      {
        "doc" : {
          "_index" : "_index",
          "_type" : "_doc",
          "_id" : "_id",
          "_source" : {
            "amount" : 229883.049,
            "ohlc" : 1.31175,
            "change" : 0.033,
            "trade_date" : 20200731,
            "pre_close" : 1.296,
            "high" : 1.333,
            "vol" : 1743608.88,
            "ts_code" : "159801.SZ",
            "low" : 1.29,
            "close" : 1.329,
            "pct_chg" : 2.5463,
            "open" : 1.295
          },
      ......
    }
```

10.2.14 故障处理器

在故障（fail）处理器内用户使用参数 if 指定失败情况下的条件，及使用参数 message 指定返回的失败消息。若是失败条件满足，失败消息将返回。故障处理器还支持可选的参数有 if、on_failure、ignore_failure 和 tag。以下示例使用在 10.2.11 节脚本处理器的示例，并在脚本处理器之前加入故障处理器，设定条件为当文档中的字段 open、high、low 或是 close 没有值时，将返回失败异常类型 fail_processor_exception 及消息"open、high、low 或是 close 没有值"。测试文档为基金代码 159801.SZ 在 2020 年 07 月 31 日的基金交易行情，并清除 open 字段。结果显示失败消息返回，没有操作脚本处理器。

```
curl --request POST http://localhost:9200/_ingest/pipeline/_simulate?pretty=true
--header
    "Content-Type: application/json" --data $'{"pipeline":{"description":"故障处理器
", "processors":
    [{"fail":{"if":"ctx.open==null || ctx.close==null || ctx.high==null || ctx.
low==null", "message":
```

```
    "open、high、low 或是 close 没有值 "}}, {"script":"ctx.ohlc=(ctx.open+ctx.high+ctx.
low+ctx.close)/4"}]],
    "docs":[{"_source":{ "ts_code": "159801.SZ", "trade_date": 20200731, "pre_
close": 1.296,
    "high": 1.333, "low": 1.29, "close": 1.329, "change": 0.033, "pct_chg": 2.5463, 
"vol": 1743608.88,
    "amount": 229883.049 }}]}'
{
  "docs" : [
    {
      "error" : {
        "root_cause" : [
          {
            "type" : "ingest_processor_exception",
            "reason" : "org.elasticsearch.ingest.common.FailProcessorException: 
open、high、low 或是 close 没有值 "
          }
        ],
        "type" : "fail_processor_exception",
        "reason" : " open、high、low 或是 close 没有值 "
      }
      ......
}
```

10.2.15 字节处理器

字节（bytes）处理器将参数 field 指定的字段 X，按照字段 X 的字符串值，若为 b、kb、mb、gb、tb 或 pb（不区分大小写）的格式，则会将 X 的字符串值转换为以字节为单位的数值。默认情况下覆盖原来字段内容。若参数 target_field 存在并指定字段 Y，则创建字段 Y 并设定为以字节为单位的数值。若字段 Y 存在，则字段 Y 的值将重置为以字节为单位的数值。字节处理器还支持可选的参数有 ignore_missing、if、on_failure、ignore_failure 和 tag。以下示例设定文档中的字段 byte_string 为字符串 123kb，并设定参数 target_field 为字段 byte_numeric。结果显示字段 byte_numeric 的数值为 125 952（123x1024）。

```
  curl    --request POST http://localhost:9200/_ingest/pipeline/_simulate?pretty=true 
--header
  "Content-Type: application/json" --data $'{"pipeline":{"description":"字节处理器
", "processors":[{"bytes":{"field":"byte_string","target_field":"byte_numeric"}}]},
  "docs":[{"_source":{ "byte_string": "123kb" }}]}'
{
  "docs" : [
    {
      "doc" : {
        "_index" : "_index",
        "_type" : "_doc",
        "_id" : "_id",
        "_source" : {
          "byte_string" : "123kb",
          "byte_numeric" : 125952
        },
        ......
    }
```

10.2.16 转换处理器

转换（convert）处理器将参数 field 指定的字段 X 的值，按照参数 type 指定的数据类型进行转换，默认情况下覆盖原来字段内容。若参数 target_field 存在并指定字段 Y，则创建字段 Y 并设定其值。若字段 Y 存在，则字段 Y 的值将重置。转换处理器还支持可选的参数有 ignore_missing、if、on_failure、ignore_failure 和 tag。以下示例设定测试文档中的字段 byte_string 为字符串 123，并设定参数 target_field 为字段 numeric 及参数 type 为 integer。结果显示转换处理器转换字符串 "123" 为 integer 数据类型并储存至字段 numeric。

```
curl    --request POST http://localhost:9200/_ingest/pipeline/_simulate?pretty=true
--header
    "Content-Type: application/json" --data $'{"pipeline":{"description":"转换处理器", "processors":[{"convert":{"field":"byte_string","type":"integer","target_field":"numeric" }}]},
    "docs":[{"_source":{ "byte_string": "123" }}]}'
{
  "docs" : [
    {
      "doc" : {
        "_index" : "_index",
        "_type" : "_doc",
        "_id" : "_id",
        "_source" : {
          "byte_string" : "123",
          "numeric" : 123
        },
      ......
    }
```

10.2.17 循环处理器

循环（foreach）处理器可以将参数 field 指定数据类型为数组的字段 X，按照参数 processor 指定的处理器，对文档中的字段 X 的数组内所有元素遍历处理。若是处理数组中任何一个元素失败，并且未指定 on_failure 参数，则循环处理器将中止执行，并且数组保留不变。循环处理器还支持可选的参数有 ignore_missing、if、on_failure、ignore_failure 和 tag。以下示例设定参数 processor 使用修剪（trim）处理器，对文档中数据类型为数组的字段 strings 内所有元素遍历处理。结果显示字段 strings 内的字符串均已经被修剪干净。

```
curl    --request POST http://localhost:9200/_ingest/pipeline/_simulate?pretty=true
--header
    "Content-Type: application/json" --data $'{"pipeline":{"description":"循环处理器", "processors":[{"foreach":{"field":"strings","processor":{"trim":{"field":"_ingest._value"}}}}]},
    "docs":[{"_source":{ "strings": [" abc ", " def\\n"]}}]}'
{
  "docs" : [
    {
      "doc" : {
        "_index" : "_index",
        "_type" : "_doc",
        "_id" : "_id",
        "_source" : {
          "strings" : [
            "abc",
            "def"
          ]
        },
        "_ingest" : {
          "_value" : null,
          "timestamp" : "2020-08-05T19:42:56.966936Z"
        }
      ......
    }
```

10.2.18 geoip 处理器

Geoip（geoip）处理器可以将参数 field 指定的字段 X，数据值为 IP 地址字符串，按照参数 database_file 指定的地理数据库（默认为 MaxMind 的 GeoLite2-City.mmdb）查询有关 IP 地址的地理位置信息。geoip 处理器可以解析 IPv4 和 IPv6 地址。若参数 target_field 存在并指定字段 Y（默认为 geoip），则创建字段 Y 并设定其值。若字段 Y 存在，则字段 Y 的值将重置。另外，还支持可选的参数有 properties，供选择要存储的 geoIP 内容的一部分属性，例如 continent_name、country_iso_code、region_iso_code、region_name、city_name 和 location。也支持可选的参数 ignore_missing。

tushare.pro 的服务 IP 地址为 103.235.227.230，以下示例设定测试文档的字段 ip_string 其值为 103.235.227.230。设置 geoip 处理器的参数 field 为字段 ip_string。参数 target_field 设置为字段 ip_info。结果显示字段 ip_info 地理位置信息为 Asia、CN-BJ、Beijing 等等。

```
curl --request POST http://localhost:9200/_ingest/pipeline/_simulate?pretty=true
--header "Content-Type: application/json" --data $'{"pipeline":{"description"
:"geoip 处 理 器 ", "processors":[{"geoip":{"field":"ip_string","target_field":"ip_
info"}}]},"docs":[{"_source":{ "ip_string": "103.235.227.230"}}]}'
    {
        "docs" : [
            {
                "doc" : {
                    ......
                    "_source" : {
                      "ip_string" : "103.235.227.230",
                      "ip_info" : {
                        "continent_name" : "Asia",
                        "region_iso_code" : "CN-BJ",
                        "region_name" : "Beijing",
                        "location" : {
                          "lon" : 116.3883,
                          "lat" : 39.9289
                        },
                        "country_iso_code" : "CN"
                      }
                    },
                    ......
            }
    }
```

10.2.19 Grok 处理器

Grok（grok）处理器可以将参数 field 指定的文本字段 X，按照数组数据类型的参数 patterns 所指定的 Logstash Grok 表达式，从 X 提取键值对，每个键值对的键将是文档的一个新字段，每个键值对的值是相对应新字段的值，并添加到索引文档的顶层。Logstash Grok 表达式类似于正则表达式，读者可以参考 Oniguruma Regular Expressions 官网。也可以使用参数 pattern_definitions 自定义指定的模式。若参数 patterns 里有个多模式，用户需要知道哪个模式匹配上，可以设置参数 trace_match 为 true（默认为 false）。则字段 _ingest.grok_match_index 将插入到匹配的文档的元数据中，并设定其数组内相对应的位置编号（位置编号从 0 开始）。grok 处理器还支持可选的参数有 ignore_missing、if、on_failure、ignore_failure 和 tag。如果提供的 grok 表达式与字段值不匹配，处理器将引发异常。

1. Logstash grok 表达式解析示例

根据 Logstash grok 表达式中的消息格式,对以下的日志消息进行解析:

```
[2020-06-27T10:34:38,093][DEBUG][o.e.a.s.TransportSearchAction] [WTW.local] All shards failed for phase: [query]
```

要解析上述日志消息的格式可以利用数个规则,例如时间戳(TIMESTAMP)、日志级别(LOGLEVEL)、空格字符(SPACE)、数据(DATA)和贪婪匹配的数据(GREEDYDATA)。使用的表达式如下:

```
[%{TIMESTAMP_ISO8601:timestamp}][%{LOGLEVEL:log-level}][%{JAVACLASS:class}]%{SPACE}[%{DATA:node}]%{SPACE}%{GREEDYDATA:content}%{SPACE}[%{GREEDYDATA:action}]
```

2. Grok 处理器的测试示例

以下示例设定测试文档的字段 message 和 patterns 所指定的模式为上述的方案。结果显示相对应的字段从字段 message 中成功提取,成为索引文档顶层的字段 node、log-level 等等。

```
    curl  --request POST http://localhost:9200/_ingest/pipeline/_simulate?pretty=true
--header
   "Content-Type: application/json" --data $'{"pipeline":{"description":"grok 处
理器", "processors":[{"grok":{"field":"message","patterns":["\\\\[%{TIMESTAMP_
ISO8601:timestamp}\\\\]
  \\\\[%{LOGLEVEL:log-level}\\\\]\\\\[%{JAVACLASS:class}\\\\]%{SPACE}\\\\[%{DATA:no
de}\\\\]
   %{SPACE}%{GREEDYDATA:content}%{SPACE}\\\\[%{GREEDYDATA:action}\\\\]"], "ignore_
missing": true}},{"remove":{"field":["message"]}}]}, "docs":[{"_source":{"message":"[2020-
06-27T10:34:38,093]
   [DEBUG][o.e.a.s.TransportSearchAction] [WTW.local] All shards failed for phase:
[query]"}}]}'
{
  "docs" : [
    {
      "doc" : {
        "_index" : "_index",
        "_type" : "_doc",
        "_id" : "_id",
        "_source" : {
          "node" : "WTW.local",
          "log-level" : "DEBUG",
          "action" : "query",
          "class" : "o.e.a.s.TransportSearchAction",
          "content" : "All shards failed for phase: ",
          "timestamp" : "2020-06-27T10:34:38,093"
        },
        ......
    }
```

10.2.20 分解处理器

类似于 grok 处理器的工作原理,分解(dissect)处理器可以将参数 field 指定的文本字段 X,按照参数 pattern 所指定的用户定义模式进行匹配,从字段 X 提取键值对,每个键值对的键将是文档的一个新字段,每个键值对的值是相对应的新字段的值,并添加到索引文档的顶层。然而,由于分解处理器的语法更简单,在一般情况下,效率比 grok 处理器更快。

参数 pattern 指定的文本,是用户以 %{key_name} 的模式指定的键及其他为非键的字符串。在分解处理过程中,从左到右比对非键的字符串与原始文本作匹配。如果匹配成功,键值对的键（key_name）会添加到索引文档中，其值从原始文本中提取。如果 key_name 的对应键不能从文本中找到，处理器将引发异常。如果引用的字段有多个匹配项，则对应的键不止一个。所有相关的值都用分隔符连接在一起。默认的分隔符是空字符串，可用参数 append_separator 指定分隔符。分解处理器还支持可选的参数有 ignore_missing、if、on_failure、ignore_failure 和 tag。

分解处理器支持几个修饰模式的字符，可以更改模式的执行方式，表 10-2 中描述这些修饰字符。

表 10-2　分解处理器模式修饰字符

修饰字符	示例	描述
->	%{key_name->}	忽略任何右侧重复字符
+	%{+key_name}%{+key_name}	按顺序将值附加到同一键内
+ 及 /n	%{+key_name/1}%{+key_name/2}	按修饰字符指定的位置 n 将值附加到同一键内
?	%{?key_name}	跳过匹配的键
* 及 &	%{*key_name}:%{&key_name}	保留原来键名及原来值
%{}	%{}	没有指定键名，表示跳过匹配的键

以下使用与 10.2.18 节 Grok 处理器相同的日志消息进行解析：

```
[%{timestamp}][%{log-level}][%{class}] [%{node}] %{content}[%{action}]
```

以下示例设定日志消息的格式与 10.2.18 节 Grok 处理器相同的测试文档字段，结果显示相对应的字段从字段 message 中成功提取，成为索引文档顶层的字段 node、log-level 等等。

```
curl --request POST http://localhost:9200/_ingest/pipeline/_simulate?pretty=true --header "Content-Type: application/json" --data $'{"pipeline":{"description":"grok 处理器 ", "processors":[{"dissect":{"field":"message","pattern":"[%{timestamp}][%{log-level}][%{class}] [%{node}] %{content}[%{action}]", "ignore_missing": true}},{"remove":{"field":["message"]}}]}, "docs":[{"_source":{"message":"[2020-06-27T10:34:38,093][DEBUG][o.e.a.s.TransportSearchAction] [WTW.local] All shards failed for phase: [query]"}}]}'
{
  "docs" : [
    {
      "doc" : {
        "_index" : "_index",
        "_type" : "_doc",
        "_id" : "_id",
        "_source" : {
          "node" : "WTW.local",
          "log-level" : "DEBUG",
          "action" : "query",
          "class" : "o.e.a.s.TransportSearchAction",
          "content" : "All shards failed for phase: ",
          "timestamp" : "2020-06-27T10:34:38,093"
        },
        ......
}
```

10.2.21　Gsub 处理器

Gsub（gsub）处理器可以将参数 field 指定的文本字段 X，按照参数 pattern 所指定的正则表达式进行匹配，匹配成功的部分文本，将被参数 replacement 所指定的字符串 Z 替换。索引文档中的字段 X 必须是字符串。如果不是，处理器将引发异常。默认情况下在原来字段覆盖替换。若参数 target_field 存在并指定字段 Y，则创建字段 Y 并设定其值。若字段 Y 存在，则字段 Y 的值将重置。gsub 处理器还支持可选的参数有 ignore_missing、if、on_failure、ignore_failure 和 tag。

以下示例设定测试文档的字段为 message、参数 patterns 所指定的模式为所有标点符号和参数 replacement 所指定的字符串为空字符串。结果显示字段 message 中所有标点符号已移除。

```
curl   --request POST http://localhost:9200/_ingest/pipeline/_simulate?pretty=true
--header
  "Content-Type: application/json" --data $'{"pipeline":{"description
":"gsub 处理器", "processors":[{"gsub":{"field":"message","pattern":"\\\\
p{Punct}","replacement":""}}]}, "docs":[{"_source":{"message":" All shards failed
for phase: [query]"}}]}'
{
   "docs" : [
     {
        "doc" : {
          "_index" : "_index",
          "_type" : "_doc",
          "_id" : "_id",
          "_source" : {
             "message" : " All shards failed for phase query"
          },
     ......
   }
```

10.2.22 HTML Strip 处理器

HTML Strip（html_strip）处理器可以将参数 field 指定的文本字段 X 内容中所有 HTML 标签替换成换行符号。索引文档中的字段 X 必须是字符串。如果不是字符串，处理器将引发异常。默认情况下在原来字段覆盖替换。若参数 target_field 存在并指定字段 Y，则创建字段 Y 并设定其值。若字段 Y 存在，则字段 Y 的值将重置。HTML 处理器还支持可选的参数有 ignore_missing、if、on_failure、ignore_failure 和 tag。

以下示例设定测试文档的字段为 text_field、内容具有标准的 HTML 标签。结果显示字段 text_field 中所有 HTML 标签已替换成换行符号 \n。

提示：HMTL tag 的 \<html\>、\<head\>、\<title\> 等，原始本来带有换行符"\n"。

```
curl   --request POST http://localhost:9200/_ingest/pipeline/_simulate?pretty=true
--header
  "Content-Type: application/json" --data $'{"pipeline":{"description":"HT
ML Strip 处理器", "processors":[{"html_strip":{"field":"text_field"}}]}, "docs":[{"_
source":{"text_field": "<html><head><title>Test</title></head><body><h1>A Test</h1></
body></html>"}}]}'
{
   "docs" : [
     {
        "doc" : {
          "_index" : "_index",
          "_type" : "_doc",
          "_id" : "_id",
          "_source" : {
             "text_field" : "\n\n\nTest\n\n\n\nA Test\n\n\n"
```

```
        }
    ......
}
```

10.2.23 URL 解码处理器

URL 解码（urldecode）处理器可以将参数 field 指定的文本字段 X 内容执行 URL 解码操作，默认情况下覆盖原来字段内容。若参数 target_field 存在并指定字段 Y，则创建字段 Y 并设定其值为解码操作后的值。若字段 Y 存在，则字段 Y 的值将重置。URL 解码处理器还支持可选的参数有 ignore_missing、if、on_failure、ignore_failure 和 tag。

以下示例设定测试文档的字段 url_encoded_string 内容为 url 编码的 Tushare 金融大数据开放社区登录网址，并设定参数 target_field 为 url_decoded_string。结果显示字段 jurl_encoded_string 的内容已解码，并且创建字段 url_decoded_string 并存储解码后的内容。原字段 url_encoded_string 则保留。

提示：URL 编码的字符"%2F"，解码后的值为"/"。

```
curl   --request POST http://localhost:9200/_ingest/pipeline/_simulate?pretty=true
--header
   "Content-Type: application/json" --data $'{"pipeline":{"description":"URL解码处理器",
   "processors": [{"urldecode":{"field":"url_encoded_string", "target_field":"url_
decoded_string"}}]}, "docs":[{"_source":{"url_encoded_string":"https://tushare.pro/
login?next=%2Fnews"}}]}'
{
   "docs" : [
     {
       "doc" : {
         "_index" : "_index",
         "_type" : "_doc",
         "_id" : "_id",
         "_source" : {
           "url_decoded_string" : "https://tushare.pro/login?next=/news",
           "url_encoded_string" : "https://tushare.pro/login?next=%2Fnews"
         },
       ......
   }
```

10.2.24 JSON 处理器

JSON（json）处理器可以将参数 field 指定的文本字段 X 内容替换成结构化的 JSON 对象。索引文档中的字段 X 必须是 JSON 字符串。如果不是 JSON 字符串，处理器将引发异常。默认情况下覆盖原来字段内容。若参数 target_field 存在并指定字段 Y，则创建字段 Y 并设定其值。若字段 Y 存在，则字段 Y 的值将重置。另外还支持布尔数据类型的参数 add_to_root，用于将转换后的对象存储在文档的根级别。当参数 add_to_root 设定为 true 时，参数 target_field 不得设置。JSON 处理器还支持可选的参数有 if、on_failure、ignore_failure 和 tag。

以下示例设定测试文档的字段为 json_string，内容具有当前 JSON 处理器部分字符串，并设定参数 add_to_root 为 true。结果显示字段 json_string 中的内容已由 JSON 字符串变换成 JSON 对象，并且创建于文档根级部，原来字段 json_string 则保留。

提示：字段 json_string 的内容是一个字符串，因此使用"\\"符号把双引号引起来。

```
    curl   --request POST http://localhost:9200/_ingest/pipeline/_simulate?pretty=true
--header
    "Content-Type: application/json" --data $'{"pipeline":{"description":"JSON 处理器
", "processors":[{"json":{"field":"json_string", "add_to_root":true}}]}, "docs":[{"_
source":{"json_string":"{\\\"pipeline\\\":{\\\"description\\\":\\\"JSON 处理器 \\\", \\\"proc
essors\\\":[{\\\"json\\\":{\\\"field\\\":\\\"json_string\\\"}}]}}"}}]}'
{
    "docs" : [
        {
          "doc" : {
            "_index" : "_index",
            "_type" : "_doc",
            "_id" : "_id",
            "_source" : {
              "pipeline" : {
                "description" : "JSON 处理器",
                "processors" : [
                  {
                    "json" : {
                      "field" : "json_string"
                    }
                  }
                ]
              },
              "json_string" : "{\"pipeline\":{\"description\":\"JSON 处理器\", \"pro
cessors\":[{\"json\":{\"field\":\"json_string\"}}]}}"
            },
            ……
        }
```

10.2.25 键值对处理器

键值对（kv）处理器可以将参数 field 指定的文本字段 X 的内容，按照参数 field_split 和参数 value_split 所指定的正则表达式进行匹配，匹配成功的部分文本，将被替换成键值对。文本字段 X 的内容应该包含一个或多个类似键值对的字符串，通过处理后，可以成功转换成键值对。键值对处理器提供参数 trim_field 可从键中修剪，而参数 trim_value 可从值中修剪。若参数 target_field 不存在，则将键值对存储在文档的根级别。若参数 target_field 存在并指定字段 Y，则创建字段 Y 并设定其值。若字段 Y 存在，则字段 Y 的值将重置。JSON 处理器还支持可选的参数有 ignore_missing、if、on_failure、ignore_failure 和 tag。

以下示例文本字段 kv_string 的内容，里面有 5 个等式。

```
{searchType=QUERY_THEN_FETCH, indices=[fund_basic], batchedReduceSize=512,
preFilterShardSize=128, allowPartialSearchResults=true}
```

如下面的例子可以设定参数 field_split 为逗号，而参数 value_split 为等号来提取键值对。

```
"field_split":",", "value_split":"=","
```

以下示例如同上述设定，测试键值对处理器。结果显示各个新字段按照提取的键值对产生于文档的顶层。

```
    curl   --request POST http://localhost:9200/_ingest/pipeline/_simulate?pretty=true
--header
    "Content-Type: application/json" --data $'{"pipeline":{"description":"键
值对处理器", "processors":[{"kv":{"field":"kv_string", "field_split":",", "value_
split":"="}}]}, "docs":[{"_source":{"kv_string":" {searchType=QUERY_THEN_FETCH,
indices=[fund_basic],
```

```
batchedReduceSize=512, preFilterShardSize=128, allowPartialSearchResults=
true}"}}]}}"}}]}'
{
  "docs" : [
    {
      "doc" : {
        "_index" : "_index",
        "_type" : "_doc",
        "_id" : "_id",
        "_source" : {
          " {searchType" : "QUERY_THEN_FETCH",
          " preFilterShardSize" : "128",
          " allowPartialSearchResults" : "true}",
          " batchedReduceSize" : "512",
          " indices" : "[fund_basic]",
          "kv_string" : " {searchType=QUERY_THEN_FETCH, indices=[fund_basic],
batchedReduceSize=512, preFilterShardSize=128, allowPartialSearchResults=true}"
        },
        ......
}
```

然而结果还需要进一步修剪,例如字段" {searchType"开头的空字符和字段"allowPartialSearchResults"的值"true}"含有右括号。以下在上面的示例加上微调的修剪设置。结果显示各个字段正确无误。

```
    curl --request POST http://localhost:9200/_ingest/pipeline/_simulate?pretty=true
--header
    "Content-Type: application/json" --data $'{"pipeline":{"description":"键值对处理器",
"processors":
    [{"kv":{"field":"kv_string", "field_split":",", "value_split":"=","trim_key":" {, ",
    "trim_value":" }"}}]}, "docs":[{"_source":{"kv_string":" {searchType=QUERY_THEN_
FETCH,
    indices=[fund_basic], batchedReduceSize=512, preFilterShardSize=128, allowPartialS
earchResults=true}"
    }}]}}"}}]}'
{
    ......
    "_source" : {
        "indices" : "[fund_basic]",
        "allowPartialSearchResults" : "true",
        "preFilterShardSize" : "128",
        "searchType" : "QUERY_THEN_FETCH",
        "batchedReduceSize" : "512",
         "kv_string" : " {searchType=QUERY_THEN_FETCH, indices=[fund_basic],
batchedReduceSize=512, preFilterShardSize=128, allowPartialSearchResults=true}"
    },
    ......
}
```

其他对键与值做修饰的参数描述如表 10-3 所示。

表 10-3 可修饰键与值的参数

参数	描述
include_keys	指定要包括的键名列表,默认包括所有匹配成功的键名
exclude_keys	指定要排除在外的键名列表
trim_key	指定要从初步匹配成功的键名,进行修剪的字符串
trim_value	指定要从初步匹配成功的键值对的值,进行修剪的字符串
prefix	指定要附加在初步匹配成功的键名的字首

参　数	描　述
strip_brackets	布尔数据类型参数，默认为 false。如果设置为 true，它将从键值对的值去除括号，包括小括号、中括号、大括号、角括号、单引号和双引号等

10.2.26　用户代理处理器

浏览器的用户代理（user agent）字符串可帮助识别使用的浏览器，版本和操作系统。用户代理（user_agent）处理器将参数 field 指定的字段 X（其内容是浏览器的用户代理字符串），提取例如浏览器名称、浏览器主要版本、浏览器次要版本、浏览器版本补丁、浏览器内部构建版本、操作系统、操作系统名称、操作系统主要版本、操作系统次要版本和设备等详细信息。若只需要部分内容，可使用可选参数 properties 指定要存储的内容。还提供了可选参数 regex_file 以指定在 config/ingest-user-agent 文件夹下自定义的 YAML 文件名称，内有正则表达式用来解析用户代理的信息。若参数 target_field 不存在，则将键值对存储在 user_agent 字段。若参数 target_field 存在并指定字段 Y，则创建字段 Y 并设定其值。若字段 Y 存在，则字段 Y 的值将重置。另外还支持可选的参数 ignore_missing。

以下示例如同上述设定，测试用户代理处理器，并使用删除处理器删除原字段 ua_string。结果显示各个新字段按照提取的信息产生，并且创建字段 user_agent 存储内容。原字段 ua_string 已删除。

```
    curl    --request POST http://localhost:9200/_ingest/pipeline/_simulate?pretty=true
--header
     "Content-Type: application/json" --data $'{"pipeline":{"description":"用户代理处理器
", "processors":[{"user_agent":{"field":"ua_string"}},{"remove":{"field":"ua_string"}}]},
"docs":[{"_source":{"ua_string":"Mozilla/5.0 (iPhone; CPU iPhone OS 10_3 like Mac OS X)
    AppleWebKit/602.1.50 (KHTML, like Gecko) CriOS/56.0.2924.75 Mobile/14E5239e
Safari/602.1"}}]}'
{
    "docs" : [
      {
        "doc" : {
          "_index" : "_index",
          "_type" : "_doc",
          "_id" : "_id",
          "_source" : {
            "user_agent" : {
              "name" : "Chrome Mobile iOS",
              "original" : "Mozilla/5.0 (iPhone; CPU iPhone OS 10_3 like Mac OS X)
    AppleWebKit/602.1.50 (KHTML, like Gecko) CriOS/56.0.2924.75 Mobile/14E5239e
Safari/602.1",
              "os" : {
                "name" : "iOS",
                "version" : "10.3",
                "full" : "iOS 10.3"
              },
              "device" : {
                "name" : "iPhone"
              },
              "version" : "56.0.2924.75"
            }
         },
     ……
    }
```

10.2.27 排序处理器

排序（sort）处理器将参数 field 指定数据类型为数组的字段 X，数组内的元素可以是数字或是字符串。可使用参数 order 指定排序方向 asc 和 desc，默认为 asc（升序）。默认情况下覆盖原来字段内容。若参数 target_field 存在并指定字段 Y，则创建字段 Y 并设定其值为解码操作后的值。若字段 Y 存在，则字段 Y 的值将重置。排序处理器还支持可选的参数有 if、on_failure、ignore_failure 和 tag。

以下示例设定测试文档的字段 field 为数组 sort_strings，并设定参数 target_field 为 sorted_strings。先使用排序处理器后，再使用删除处理器删除原字段 sort_strings。排序方向为默认的 asc。结果显示文档的字段 sorted_strings 存储排序后的内容，原始内容已被覆盖。

```
curl   --request POST http://localhost:9200/_ingest/pipeline/_simulate?pretty=true
--header
 "Content-Type: application/json" --data $'{"pipeline":{"description":"排序
处理器", "processors":[{"sort":{"field":"sort_strings", "target_field":"sorted_
strings"}}, {"remove":{"field":"sort_strings"}}]}, "docs":[{"_source":{"sort_
strings":["b","c","a"]}}]}'
{
  "docs" : [
    {
      "doc" : {
        "_index" : "_index",
        "_type" : "_doc",
        "_id" : "_id",
        "_source" : {
          "sorted_strings" : [
            "a",
            "b",
            "c"
          ]
        },
        ……
    }
}
```

10.2.28 点扩展器处理器

如果文档存在的字段包含点符号，例如 os.name、os.version 及 os.full，这些字段不能被其他处理器使用。使用点扩展器（dot_expander）处理器将参数 field 指定的字段 X，而字段 X 为含有点符号的字符串，可以转换带点的字段 X 扩展为对象字段。当字段 X 的内容只涉及原来字段的部分路径时，可使用可选参数 path 来指定字段的父路径。点扩展器处理器还支持可选的参数有 if、on_failure、ignore_failure 和 tag。

以下示例设定测试点扩展器的文档如以下的字段 top_level：

```
"top_level":{"user_agent.os.name":"iOS"}
```

需要点扩展的字符串为 user_agent.os.name，所以字段 field 设置为字符串 user_agent.os.name，并设定参数 path 为字段 top_level，这是因为字符串 user_agent.os.name 属于对象 top_level。结果显示字符串 user_agent.os.name 扩展为文档的字段 top_level 的嵌套子对象。

```
curl   --request POST http://localhost:9200/_ingest/pipeline/_simulate?pretty=true
--header
```

```
"Content-Type: application/json" --data $'{"pipeline":{"description":"点扩展器处理器
", "processors":[{"dot_expander":{"field":"user_agent.os.name","path":"top_level"}}]},
"docs":[{"_source":{"top_level":{"user_agent.os.name":"iOS"}}}]}'
{
    "docs" : [
      {
        "doc" : {
          "_index" : "_index",
          "_type" : "_doc",
          "_id" : "_id",
          "_source" : {
            "top_level" : {
              "user_agent" : {
                "os" : {
                  "name" : "iOS"
                }
              }
            }
          },
      ......
}
```

10.2.29 丰富处理器

丰富（enrich）处理器按照参数 policy_name 指定的丰富数据策略，将策略的参数 match_field 指定为字段 ts_code，若当前索引文档字段 ts_code 的内容匹配策略的参数 indices 指定的索引内文档字段 ts_code 的内容，则按照策略中数据类型为数组的参数 enrich_fields，提取各个指定字段的数据来丰富当前文档。参数 target_field 必须存在，若指定字段 Y 而字段 Y 不存在，则创建字段 Y 并设定其值为包含 match_field 和 enrich_fields 指定字段的内容。若字段 Y 存在，则字段 Y 的值将重置。另外支持可选的参数 max_matches 和参数 override。参数 max_matches 为最大匹配文档数，默认为 1。如果大于 1，参数 target_field 指定的字段 Y 将设定为 json 数组。参数 override 默认为 true，将覆盖参数 target_field 指定的字段 Y 的内容。丰富处理器还支持可选的参数有 ignore_missing、if、on_failure、ignore_failure 和 tag。

以下示例需要遵循几个的步骤，从而获得结果。

（1）创建丰富数据的策略（enrich policy）my_policy。

对 _enrich 接口使用 PUT 请求创建策略 my_policy，策略为将参数 match_field 指定的文档字段 shortname，与索引 fund_company 的文档字段 shortname 内容进行匹配。若匹配，将参数 enrich_fields 指定的数组内所有字段，从匹配的文档中提取内容，添加到当前文档中。

```
curl --request PUT http://localhost:9200/_enrich/policy/my_policy?pretty=true --header
"Content-Type: application/json" --data $'{"match":{"indices": "fund_company",
"match_field": "shortname", "enrich_fields": ["reg_capital"]}}'
{
  "acknowledged" : true
}
```

（2）执行 my_policy 创建索引 enrich。

```
curl --request POST http://localhost:9200/_enrich/policy/my_policy/_execute?pretty=true
{
    "status" : {
      "phase" : "COMPLETE"
```

```
    }
  }
```

（3）以下示例设定测试文档的字段 field 为 shortname，结果显示策略 my_policy 里的字段 enrich_fields 指定的数组内字段 reg_capital 已添加到测试文档中。

```
    curl    --request POST http://localhost:9200/_ingest/pipeline/_simulate?pretty=true
--header
    "Content-Type: application/json" --data $'{"pipeline":{"description":"丰富处理器
", "processors":[ {"enrich":{"policy_name":"my_policy",    "field":"shortname",
    "target_field":"company_info"}}]}, "docs":[{"_source": {\"ts_code\":\"159809.SZ\",
\"name\":\"恒生湾区 \",\"shortname\":\"博时基金\"}}]}'
{
  "docs" : [
    {
      "doc" : {
        "_index" : "_index",
        "_type" : "_doc",
        "_id" : "_id",
        "_source" : {
          "ts_code" : "159809.SZ",
          "name" : " 恒生湾区 ",
          "company_info" : {
            "reg_capital" : 25000,
            "shortname" : " 博时基金 "
          },
          "shortname" : " 博时基金 "
        },
        ……
}
```

10.2.30 日期索引名称处理器

日期索引名称（date_index_name）处理器按照参数 field 指定为日期格式的字段 X，按照参数 date_rounding 指定的格式以 y（年）、M（月）、w（周）、d（天）、h（小时）、m（分钟）或 s（秒）进行四舍五入，然后设定元字段 _index 为四舍五入后的值。若参数 date_formats 存在，其值将为字段 X 期望的日期读取格式，默认为 yyyy-MM-dd'T'HH:mm:ss.SSSXX。若参数 index_name_prefix 存在，其值将设为元字段 _index 的字首值。若参数 index_name_format 存在，其值将为元字段 _index 的日期输出格式，默认为 yyyy-MM-dd。另外还支持两个可选参数，时区 timezone 和区域 locale。参数 timezone 默认为 UTC 而参数 locale 默认为 ENGLISH。处理器还支持可选的参数有 if、on_failure、ignore_failure 和 tag。

以下示例设定测试文档的参数 field 指定的字段 date 如下：

```
    "date" : "2020-08-06T09:38:58.911Z"
```

另外参数 index_name_prefix 设定为字符串 Log-，产生的元字段 _index 字首会设定其值。结果显示 _index 转换成如下：

```
    "_index" : "<Log-{2020-08-06||/d{yyyy-MM-dd|UTC}}>"
```

提示：必须按照 1.6 节接口用法约定说明中日期的数学表示方式解读。可以理解为以字首 Log- 开始，接着以双竖线 || 结尾的日期字符串 2020-08-06，并且以四舍五入到最近一天。日期格式为 yyyy-MM-dd，而时区偏移量为 UTC。

```
    curl    --request POST http://localhost:9200/_ingest/pipeline/_
simulate?pretty=true   --header
```

```
    "Content-Type: application/json" --data $'{"pipeline":{"description":"日期索引名
称处理器", "processors":[{"date_index_name":{"field":"date", "date_rounding":"d",
"index_name_prefix":"Log-"}}]}, "docs":[{"_source":{"date":"2020-08-
06T09:38:58.911Z",
"message":"low disk watermark [85%] exceeded"}}]}'
{
  "docs" : [
    {
      "doc" : {
        "_index" : "<Log-{2020-08-06||/d{yyyy-MM-dd|UTC}}>",
        "_type" : "_doc",
        "_id" : "_id",
        "_source" : {
          "date" : "2020-08-06T09:38:58.911Z",
          "message" : "low disk watermark [85%] exceeded"
        },
        ......
    }
```

10.3 处理管道中的故障

当一系列依次执行的处理器其中一个发生异常时，如果没有特殊的设置，整个过程将突然的停止。也许服务失败中断时最佳的补救方式是优雅地失败，并提供足够的调试信息。当然，如果一个处理器的操作失败可以忽略，则可以设置参数 ingore_failure 为 true，以静默方式去忽略该故障并继续执行下一个处理器。如果一个处理器的操作失败无法忽略时，则可以设定参数 on_failure 来捕获异常并执行一些清理工作和编辑失败原因的消息。响应中返回的请求主体在元数据 _ingest.on_failure_message 显示该错误的代码和失败原因。有时候也可以使用故障处理器检查当前情况是否已经满足失败条件。若已经满足失败条件，则立刻跳过下一个处理器的处理过程，直接执行失败补救处理程序。在 10.2.14 节故障处理器曾测试文档不存在 open 字段下，故障处理器的处理方式。

1. 测试异常环境

以下示例测试当文档不存在字段 open 时，结果显示发生 null_pointer_exception 异常，返回的消息没有显示失败所在的真正原因。

```
    curl  --request POST http://localhost:9200/_ingest/pipeline/_simulate?pretty=true
--header
    "Content-Type: application/json" --data $'{"pipeline":{"description":"处理管道中
的故障", "processors":[{"script":{"source":"ctx.ohlc=(ctx.open+ctx.high+ctx.low+ctx.
close)/4"}}]},
    "docs":[{"_source":{ "ts_code": "159801.SZ", "trade_date": 20200731, "pre_
close": 1.296,
    "high": 1.333, "low": 1.29, "close": 1.329, "change": 0.033, "pct_chg": 2.5463,
    "vol": 1743608.88, "amount": 229883.049 }}]}'
{
  "docs" : [
    {
      "error" : {
        "root_cause" : [
          {
            "type" : "script_exception",
            "reason" : "runtime error",
            "script_stack" : [
              "ctx.ohlc=(ctx.open+ctx.high+ctx.low+ctx.close)/4",
```

```
                  "                      ^---- HERE"
        ],
        "script" : "ctx.ohlc=(ctx.open+ctx.high+ctx.low+ctx.close)/4",
        "lang" : "painless"
      }
    ],
    "type" : "script_exception",
    "reason" : "runtime error",
    "script_stack" : [
      "ctx.ohlc=(ctx.open+ctx.high+ctx.low+ctx.close)/4",
      "                      ^---- HERE"
    ],
    "script" : "ctx.ohlc=(ctx.open+ctx.high+ctx.low+ctx.close)/4",
    "lang" : "painless",
    "caused_by" : {
      "type" : "null_pointer_exception",
      "reason" : null
    }
  ......
}
```

2. 测试异常环境下设置 on_failure 参数

以下示例使用相同文档，测试参数 on_failure。失败补救处理程序为设置参数 if 的条件为测试字段 open 是否为 null。若字段 open 为真，则在 error_message 指出失败所在的真正原因。结果显示没有异常发生，而字段 error_message 指出脚本处理器的失败所在的真正原因为字段 open 的值为空。

```
    curl --request POST http://localhost:9200/_ingest/pipeline/_simulate?pretty=true
--header
    "Content-Type: application/json" --data $'{"pipeline":{"description":"处理管道中
的 故 障 ", "processors":[{"script":{"source":"ctx.ohlc=(ctx.open+ctx.high+ctx.low+ctx.
close)/4", "on_failure":[{"set":{"if":"ctx.open==null", "field":"error_message",
"value":"字段 open 值为空"}}]
    }}]}, "docs":[{"_source":{ "ts_code": "159801.SZ", "trade_date": 20200731, "pre_
close": 1.296,
    "high": 1.333, "low": 1.29, "close": 1.329, "change": 0.033, "pct_chg": 2.5463,
    "vol": 1743608.88, "amount": 229883.049 }}]}'
{
  "docs" : [
    {
      "doc" : {
        "_index" : "_index",
        "_type" : "_doc",
        "_id" : "_id",
        "_source" : {
          "error_message" : "字段 open 值为空",
          "amount" : 229883.049,
          "change" : 0.033,
          "trade_date" : 20200731,
          "pre_close" : 1.296,
          "high" : 1.333,
          "vol" : 1743608.88,
          "ts_code" : "159801.SZ",
          "low" : 1.29,
          "close" : 1.329,
          "pct_chg" : 2.5463
        },
      ......
    }
```

第 11 章　使用 Elasticsearch 进行探索性数据分析

根据美国国家标准与技术研究院（NIST）信息技术实验室（ITL）的定义，探索性数据分析（EDA）是一种通过揭示数据底层结构和模型来进行数据分析的方法。要进行探索性数据分析，首先要遵循几个理念：
- 深入了解数据集从而发现底层结构
- 从数据集中提取重要变量，并检测异常值
- 建立初始模型，然后测试基本假设
- 最后确定最佳设定

11.1　数据处理

为了在本章使用中更好的示例，在这里引进在第二章中提到由 Tushare 提供的公募基金场内行情数据。并随机挑选了基金代码 159801.SZ 及 159995.SZ 两支基金用于测试。由于场内基金日线行情 fund_daily 已由 Tushare 清理干净，数据底层结构已经很清楚。假设在文档索引过程中创建新字段 ohlc 及 avg_price，其定义如下：

$$ohlc = \frac{open + high + low + close}{4}$$

$$avg\_price = \frac{amount * 1000}{vol * 100}$$

其中，字段 ohlc 是如何其使用脚本处理器建立它，在第 10.1 节中已经说明清楚。在这里，字段 avg_price 代表该基金当日平均价格，计算公式是成交额（amount）除以成交量（vol）。成交额的单位为 1000 元，而成交量为手。1 手代表 100 份。

11.1.1　日线行情显式映射

文件 fund_daily_mappings.json 内定义了 fund_daily 索引的显式映射。字段与表 2-4 日线行情数据列表一一对应，另外还添加了两个字段，ohlc 和 avg_price。采用的显式映射内容如下：

```
{
        "mappings" : {"dynamic": false, "properties": {
          "ts_code":{"type":"text", "fields":{"keyword":{"type":"keyword"}}},
          "trade_date": {"type": "date", "format": "yyyyMMdd"},
          "open": {"type": "float"},
```

```
           "high": {"type": "float"},
           "low": {"type": "float"},
           "close": {"type": "float"},
           "pre_close": {"type": "float"},
           "change": {"type": "float"},
           "pct_chg": {"type": "float"},
           "vol": {"type": "float"},
           "amount": {"type": "float"},
           "ohlc": {"type": "float"},
           "avg_price": {"type": "float"}}}}
```

11.1.2　创建 ohlc_avg_price_pipeline 摄取节点管道

由于在索引 fund_daily 的文档索引之前需要进行预处理，亦即是计算和添加 ohlc 和 avg_price 这两个字段。我们定义了 ohlc_avg_price_pipeline 摄取节点管道，里面有两个脚本处理器，按照上面的公式逐一计算。

以下指令创建摄取节点管道 ohlc_avg_price_pipeline：

```
curl --request PUT http://localhost:9200/_ingest/pipeline/ohlc_avg_price_
pipeline?pretty=true
 --header "Content-Type: application/json" --data $'{"description":
"ohlc_avg_price_pipeline 摄取节点管道", "processors":[{"script":
{"source": "ctx.ohlc=(ctx.open+ctx.high+ctx.low+ctx.close)/4"}},
{"script":{"source":"ctx.avg_price=(ctx.amount*1000)/(ctx.vol*100)"}}]}'
{
  "acknowledged" : true
}
```

11.1.3　批量处理索引文档

在 4.1.3 节认识批量多文档接口功能中，已经介绍过如何使用 _bulk 接口为文档建立索引。下面列出如何在批量文档接口索引请求主体中，添加文档索引前要执行的摄取节点管道。请求语法如下：

```
{"index": { "pipeline":"摄取节点管道名称"}} \newline
{请求主体} \newline
...
```

以下使用文件 fund_daily_bulk.json 内编写好的批量处理文档之一指令举个例子：

```
{"index":{"pipeline":"ohlc_avg_price_pipeline"}}
 { "ts_code": "159801.SZ", "trade_date": 20200731, "pre_close": 1.296, "open":
1.295, "high": 1.333, "low": 1.29, "close": 1.329, "change": 0.033, "pct_chg":
2.5463, "vol": 1743608.88, "amount": 229883.049 }
```

11.1.4　公募基金交易行情文档索引操作

本章提供了 3 个文件用于基金交易行情文档索引操作，包括 fund_daily_bulk.json、fund_daily_bulk_index.sh 和 fund_daily_mappings.json。文件 fund_daily_bulk_index.sh 是个 bash 执行档，运行执行以下任务：

（1）使用在 fund_daily_mappings.json 内定义的设定和分析器创建 fund_daily 索引。

（2）使用 fund_daily_bulk.json 内编写好的批量处理文档接口指令，对 159801.SZ 及 159995.SZ 两支基金的日线行情数据进行文档索引。基金代码 159801.SZ 的数据时间段是在 2020 年 02 月 18 日至 2020 年 07 月 31 日之间。而基金代码 159995.SZ 的数据时间段是在

2020年02月10日至2020年07月31日之间。执行 fund_daily_bulk_index.sh 后的部分索引结果显示如下：

```
./fund_daily_bulk_index.sh
{"acknowledged":true,"shards_acknowledged":true,"index":"fund_daily"}
{
  "took" : 126,
  "errors" : false,
  "items" : [
    {
      "index" : {
        "_index" : "fund_daily",
        "_type" : "_doc",
        "_id" : "bTKcsHMBTmRt5jA40H-n",
        "_version" : 1,
        "result" : "created",
        "_shards" : {
          "total" : 2,
          "successful" : 1,
          "failed" : 0
        },
        "_seq_no" : 0,
        "_primary_term" : 1,
        "status" : 201
      }
    },
    ......
  ]
}
```

（3）搜索索引 fund_daily 内的文档，设置仅返回一个（hits）文档进行检查。检查管道 ohlc_avg_price_pipeline 中两个处理器是否已正常执行。结果显示搜索成功，返回的响应内容已经包括 ohlc、avg_price 和 ohlc_avg_price 两个字段。

```
curl --request POST http://localhost:9200/fund_daily/_search?pretty=true --header
"Content-Type: application/json" --data $'{"size":1}'
{
  ......
  "hits" : {
    "total" : {
      "value" : 232,
      ......
    "hits" : [
      {
        "_index" : "fund_daily",
        "_type" : "_doc",
        "_id" : "AFYI5nMBus1K8YrAbMwE",
        "_score" : 1.0,
        "_source" : {
          "ohlc" : 1.31175,
          "avg_price" : 1.3184324284927937,
          ......
      }
```

11.2 指标数据分析

假设 ohlc 和 avg_price 这两个字段被用作指标，以下可使用第9章的聚合框架来执行这些指标的数据分析。数据集是由 113 交易日和前面提到的两个基金组成。

11.2.1 执行扩展统计聚合

以下使用 ts_code 为 159801.SZ 对 ohlc 和 avg_price 这两个指标进行扩展统计（extended_stats）聚合，统计报告见表 11-1。在表 11-1 中加入方差系数值（coefficient of variation, cv）用于度量概率分布的离散度。基本上 CV<10% 表示数据分布得非常好，而 10%-%20 表示很好。方差系数计算公式如下：

$$cv = \frac{std\_deviation}{avg} \times 100$$

结果显示指标 ohlc 和 avg_price 的方差系数值约 13%，算做好等级。

```
curl --request POST http://localhost:9200/fund_daily/_search?pretty=true
--header
 "Content-Type: application/json" --data $'{"query":{"term":{"ts_code.keyword":"159801.SZ"}},
 "aggs": {"ohlc_extended_stats": {"extended_stats": {"field":"ohlc"}}}, "size":0}'

curl --request POST http://localhost:9200/fund_daily/_search?pretty=true
--header
 "Content-Type: application/json" --data $'{"query":{"term":{"ts_code.keyword":"159801.SZ"}},
 "aggs": {"avg_price_extended_stats": {"extended_stats": {"field":"avg_price"}}}, "size":0}'
```

表 11-1 执行 ohlc 和 avg_price 扩展统计聚合结果列表

统计指标名称	ohlc	avg_price
min	0.8855000138282776	0.8787862658500671
max	1.469249963760376	1.4562231302261353
avg	1.0945265509385977	1.095093605792628
variance	0.01957052780236542	0.019537603921325558
std_deviation	0.13989470255290376	0.13977697922521276
std_dev upper	1.3743159560444052	1.3746475642430536
std_dev lower	0.8147371458327902	0.8155396473422024
cv	12.78%	12.76%

11.2.2 执行矩阵统计聚合

在金融市场中，当变量表现出相似的行为时，协方差（covariance）为正。然而，线性关系的强度不能通过协方差值的大小来解释。如果需要相关强度的量度时，则使用相关性（correlation）。根据 Elasticsearch 官网上对相关性的定义，当相关性值为 1 时表示两个变量始终沿相同方向移动。当相关性值为 -1 时表示相反方向移动。当相关性值为 0 时表示两个变量是独立的。

以下使用 ts_code 为 159801.SZ 对指标 ohlc 和指标 avg_price 这两个指标进行矩阵统计（matrix_stats）聚合，统计报告见表 11-2。结果显示指标 ohlc 和指标 avg_price 的协方差为正，相关性为非常正相关。

```
curl --request POST http://localhost:9200/fund_daily/_search?pretty=true
--header
```

```
     "Content-Type: application/json" --data $'{"query":{"term":{"ts_code.
keyword":"159801.SZ"}},
     "aggs": {"ohlc_avg_price_matrix_stats": {"matrix_stats": {"fields":["ohlc",
"avg_price"]}}},
     "size":0}'
```

表 11-2 执行 ohlc 和 avg_price 矩阵统计聚合结果列表

统计指标名称	ohlc	avg_price
mean	1.0945265509385977	1.0950936057926275
variance	0.019745264657743806	0.019712046813480228
skewness	0.7363670953714218	0.7231021665436252
kurtosis	2.7771396753157584	2.723807982890492
covariance/ohlc	0.019745264657743806	0.019711026667456732
covariance/avg_price	0.019711026667456732	0.019712046813480228
correlation/ohlc	1.0	0.9991067773002028
correlation/avg_price	0.9991067773002028	1.0

11.2.3 执行百分位聚合和百分位等级聚合

以下使用 ts_code 为 159801.SZ 对 ohlc 和 avg_price 这两个指标进行百分位（percentiles）聚合，百分位数分布情况。测试结果见表 11-3。

```
     curl --request POST http://localhost:9200/fund_daily/_search?pretty=true --header
     "Content-Type: application/json" --data $'{"query":{"term":{"ts_code.
keyword":"159801.SZ"}},
     "aggs": {"ohlc_percentiles": {"percentiles":{"field":"ohlc"}}}, "size":0}'

     curl  --request  POST  http://localhost:9200/fund_daily/_search?pretty=true
--header
     "Content-Type: application/json" --data $'{"query":{"term":{"ts_code.
keyword":"159801.SZ"}},
     "aggs": {" avg_price_percentiles": {"percentiles":{"field":"avg_price"}}},
"size":0}'
```

表 11-3 执行 ohlc 和 avg_price 百分位聚合结果列表

百分位（%）	ohlc	avg_price
1.0	0.8877050042152405	0.8869015091657639
5.0	0.9095625221729279	0.908040189743042
25.0	0.9956874549388885	0.9949130415916443
50.0	1.0682499408721924	1.0700398683547974
75.0	1.1720000505447388	1.180947721004486
95.0	1.3464124858379363	1.3590025186538695
99.0	1.4601149499416353	1.4505003690719604

以下使用 ts_code 为 159801.SZ 对 ohlc 和 avg_price 这两个指标进行百分位等级（percentile_rank）聚合，测试其值为 1 时的百分位数。测试结果见表 11-4。

表 11-4 执行 ohlc 和 avg_price 百分位聚合结果列表

指标值	ohlc（%）	avg_price（%）
1	25.43591002139316	25.680655699306428

11.2.4 执行导数聚合

以下使用 ts_code 为 159801.SZ 对 ohlc 和 avg_price 这两个指标进行导数（derivative）聚合，测试其一阶和二阶导数，测试结果经编撰后，使用 Kibana 汇集于图 11-1。

```
curl --request POST http://localhost:9200/fund_daily/_search?pretty=true
--header
  "Content-Type: application/json" --data $'{"query":{"term": {"ts_code.
keyword":{"value":
  "159801.SZ"}}}, "aggs":{"daily_derivative_report":{"date_
histogram":{"field":"trade_date",
  "calendar_interval":"1d", "min_doc_count":"1"},"aggs":{"ohlc":{"sum":{"field":"
ohlc"}},
  "ohlc_1st_derivative":{"derivative":{"buckets_path":"ohlc"}},"ohlc_2nd_
derivative":{
  "derivative":{"buckets_path":"ohlc_1st_derivative"}},"avg_
price":{"sum":{"field":
  "avg_price"}},"avg_price_1st_derivative":{"derivative":{"buckets_path":"avg_
price"}},
  "avg_price_2nd_derivative":{"derivative":{"buckets_path":"avg_price_1st_
derivative"}}}}},
  "size":0}'
{
  ......
          "key_as_string" : "20200220",
          "key" : 1582156800000,
          "doc_count" : 1,
          "ohlc" : {
            "value" : 1.1927499771118164
          },
          "avg_price" : {
            "value" : 1.1873430013656616
          },
          "ohlc_1st_derivative" : {
            "value" : 0.022499918937683105
          },
          "ohlc_2nd_derivative" : {
            "value" : -0.019250154495239258
          },
          "avg_price_1st_derivative" : {
            "value" : 0.008053064346313477
          },
          "avg_price_2nd_derivative" : {
            "value" : -0.04696249961853027
          }
  ......
}
```

11.2.5 执行移动函数聚合

以下使用 ts_code 为 159801.SZ 对 ohlc 这个指标进行简单移动平均聚合，测试结果与 11.2.4 节执行导数聚合的结果经编撰后，使用 Kibana 汇集于图 11-1。简单移动平均函数 ewma 的参数设置为 window=5 执行指令如下：

```
curl --request POST http://localhost:9200/fund_daily/_search?pretty=true --header
  "Content-Type: application/json" -data $'{"query":{"term":{"ts_code.
keyword":{"value":
  "159801.SZ"}}}, "aggs":{"daily_report":{"date_histogram":{"field":"trade_date",
  "calendar_interval":"1d","min_doc_count":"1"},"aggs":{"ohlc":{"max":{"field":"
ohlc"}},
```

```
"moving_avg":{"moving_fn":{"buckets_path": "ohlc", "window":5,
"script":"MovingFunctions.unweightedAvg(values)"}}}},"size":0}'
```

图 11-1 显示 ohlc 和 avg_price 这两个指标及其一阶和二阶导数的曲线几乎重叠，它们的含义几乎相同。

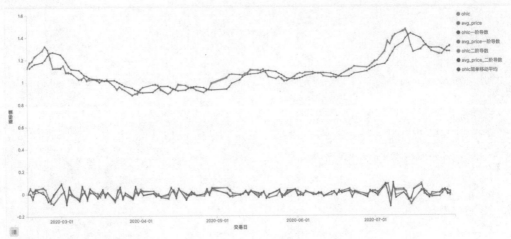

图 11-1　指标 ohlc 和指标 avg_price 及其导数聚合和移动函数聚合

11.3　投资组合

1981 诺贝尔经济学奖得主詹姆斯·托宾（James Tobin）曾经说过投资组合（portfolio）的基本原理—不要把你所有的鸡蛋都放在一个篮子里，但也不要放在太多的篮子里。在第 9 章曾介绍过 Tushare 的公募基金持仓数据，数据属于季度更新。这些数据，可以使投资者了解基金的资产配置结构、投资思路、投资风格等重要信息的最佳渠道。

投资组合季度清单

以下在索引 fund_portfolio 中使用 ts_code 为 515850.SH 首先按照截止日期进行日期直方图聚合，然后对股票代码进行词条聚合，并按照股票市值比大小进行桶排序聚合，执行指令如下，测试结果先依照截止日期按月份排序，然后在每个月份里按照股票市值比大小排序，所有结果见表 11-5，并汇集于图 11-2。

```
curl --request POST http://localhost:9200/fund_portfolio/_search?pretty=true --header
    "Content-Type: application/json" --data $'{"query":{"term":{"ts_code.
keyword":{"value":
    "515850.SH"}}}, "aggs":{"end_date_portfolio":{"date_histogram": {"field":"end_date",
"min_doc_count":1,"calendar_interval":"1M"}, "aggs":{"portfolio":{"terms":{"field":
    "symbol.keyword", "size":30},"aggs":{"terms_stk_mkv_ratio":{"max":{"field":"stk_
mkv_ratio"}},
    "stk_mkv_ratio_bucket_sort":{"bucket_sort": {"sort": [{"terms_stk_mkv_ratio ":
{"order": "desc"
    }}]}}}},"st_mkv_ratio_sum":{"sum_bucket":{"buckets_path":"portfolio>terms_stk_
mkv_ratio"}}}},
    "size":0}'
    {
        ……
```

第 11 章 使用 Elasticsearch 进行探索性数据分析

```
"aggregations" : {
  "end_date_portfolio" : {
    "buckets" : [
      {
        "key_as_string" : "20200201",
        "key" : 1580515200000,
        "doc_count" : 10,
        "portfolio" : {
          "doc_count_error_upper_bound" : 0,
          "sum_other_doc_count" : 0,
          "buckets" : [
            {
              "key" : "600030.SH",
              "doc_count" : 1,
              "terms_stk_mkv_ratio" : {
                "value" : 9.819999694824219
              }
            },
            ......
            {
              "key_as_string" : "20200301",
              "key" : 1583020800000,
              "doc_count" : 15,
              "portfolio" : {
                "doc_count_error_upper_bound" : 0,
                "sum_other_doc_count" : 0,
                "buckets" : [
                  {
                    "key" : "600030.SH",
                    "doc_count" : 1,
                    "terms_stk_mkv_ratio" : {
                      "value" : 14.619999885559082
                    }
                  },
                  ......
}
```

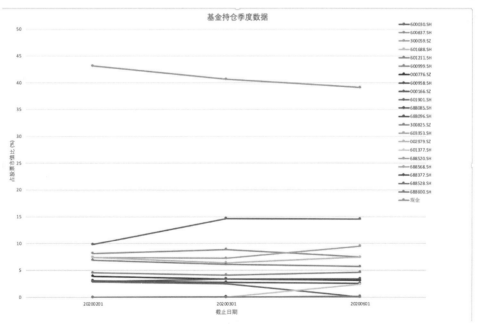

图 11-2　ts_code 为 515850.SH 的投资组合季度图表

另一种方式是先对股票代码进行词条聚合，然后按照截止日期进行日期直方图聚合，执行指令如下，测试结果先依照股票代码划分成组，然后按照截止日期依照月份排序。执行指令如下，聚合结果与上面的例子相同。

```
    curl --request POST http://localhost:9200/fund_portfolio/_search?pretty=true
--header "Content-Type: application/json" --data $'{"query":{"term":{"ts_code.
keyword":{"value":"515850.SH"}}},"aggs":{"portfolio":{"terms":{"field":"symb
ol.keyword"},"aggs":{ "end_date_portfolio":{"date_histogram": {"field":"end_
date", "min_doc_count":1,"calendar_interval":"1M"},"aggs":{"terms_stk_mkv_
ratio":{"max":{"field":"stk_mkv_ratio"}}}}}}}, "size":0}'
    {
      ......
      "aggregations" : {
        "portfolio" : {
          "doc_count_error_upper_bound" : 0,
          "sum_other_doc_count" : 11,
          "buckets" : [
            {
              "key" : "000166.SZ",
              "doc_count" : 3,
              "end_date_portfolio" : {
                "buckets" : [
                  {
                    "key_as_string" : "20200201",
                    "key" : 1580515200000,
                    "doc_count" : 1,
                    "terms_stk_mkv_ratio" : {
                      "value" : 2.9200000762939453
                    }
                  },
            ......
            {
              "key" : "000776.SZ",
              "doc_count" : 3,
              "end_date_portfolio" : {
                "buckets" : [
                  {
                    "key_as_string" : "20200201",
                    "key" : 1580515200000,
                    "doc_count" : 1,
                    "terms_stk_mkv_ratio" : {
                      "value" : 3.859999895095825
                    }
                  },
            ......
    }
```

表 11-5 ts_code 为 515850.SH 的投资组合股票市值比季度清单

股票代号\截止日期	20200201	20200301	20200601
600030.SH	9.819999695	14.61999989	14.56000042
600837.SH	8.109999657	8.840000153	7.429999828
300059.SZ	7.389999866	7.239999771	9.479999542
601688.SH	7.389999866	6.369999886	7.429999828
601211.SH	6.889999866	6.079999924	5.679999828
600999.SH	4.550000191	4.079999924	4.599999905

续表

股票代号\截止日期	20200201	20200301	20200601
000776.SZ	3.859999895	3.380000114	3.109999895
600958.SH	3.059999943	2.730000019	2.519999981
000166.SZ	2.920000076	3.380000114	3.359999895
601901.SH	2.869999886	2.49000001	
688085.SH		0.059999999	
688096.SH		0.029999999	
300825.SZ		0.01	
603353.SH		0.01	
002979.SZ		2.06452E-07	
601377.SH			2.390000105
688520.SH			0.150000006
688568.SH			0.059999999
688377.SH			0.050000001
688528.SH			0.039999999
688600.SH			0.039999999
总计 (%)	56.85999894	59.32	60.89999923
现金 (%)	43.14000106	40.68	39.10000077
基金数量	10	15	15

第三篇 Java 和 Python 客户端编程介绍

第 12 章 Java 客户端编程

Java REST 客户端可以分为高级别及低级别。区别在于，低级别 REST 客户端允许直接通过 http 和 Elasticsearch 集群进行通信。高级别 REST 客户端则建构与封装在低级别 REST 客户端之上，并且提供特定的接口方法。这些方法接受请求对象作为参数并返回响应对象，而请求和响应对象之间的序列化和反序列化的操作则完全由 Elasticsearch 系统负责。不过调用者需要了解和熟悉许多细节才能充分利用这些接口。如果选用低级别 REST 客户端，序列化和反序列的操作则需要自行处理，但是可以不必了解过多的接口关系与细节。两种 REST 客户端都可以同步调用或者异步调用。

12.1 Elasticsearch Java REST 客户端概览

许多编程语言都支持 Elasticsearch 客户端，但实际上官方真正仅支持两种协议，HTTP（通过 RESTful API）和原生（native）协议。使用 Java 语言时，传输客户端（transport）是首选方法。但是在 Elasticsearch 8.0 中将删除这个功能。所以使用 Java 编程语言时，高级别 REST 客户端应该是首选。

为了演示 Java REST 客户端，Spring Boot 2.3 被用来构造 Java 程序。Spring Tools 4 IDE 被用来创建一个 Spring Boot 启动程序项目称为 java_restful_client，其基本目录如图 12-1 所示。

在目录中可以找到 pom.xml 文件，文件中描述了如何使用 Maven 存储库配置 elasticsearch-rest-client 依存关系（dependencies）的方法。使用的 Spring Boot 版本是 2.3.4，默认使用的 Elasticsearch 客户端软件包 7.6.2，需要添加 7.5.1 版本号到 pom.xml 文件替换默认的版本：

图 12-1 Spring Boot 项目 java_restful_client 的基本目录

```
<properties>
    ......
    <elasticsearch.version>7.5.1</elasticsearch.version>
</properties>
```

另外，添加 Java 高级别及低级别 REST 客户端的依存关系项：

```
<dependencies>
    ......
    <dependency>
        <groupId>org.elasticsearch.client</groupId>
        <artifactId>elasticsearch-rest-client</artifactId>
    </dependency>
    <dependency>
        <groupId>org.elasticsearch.client</groupId>
        <artifactId>elasticsearch-rest-high-level-client</artifactId>
    </dependency>
    ......
```

12.2 Java 低级别 REST 客户端

Java 低级别 REST 客户端基于 Apache HTTP 异步客户端发送 http 请求。低级别意味着它对用户建立请求或解析响应的支持较少。低级别 REST 客户端定义于 org.elasticsearch.client.RestClient Java 类。

12.2.1 Java 低级别 REST 客户端操作流程

操作流程基本上涉及四个主要步骤：
（1）REST 客户端初始化。
（2）执行请求。
（3）处理响应。
（4）完成所有请求后关闭 REST 客户端。

12.2.2 REST 客户端初始化

客户端初始化的方法如图 12-2 所示。首先使用 RestClient 类的静态方法 builder（HttpHost[]）创建一个 RestClientBuilder 对象，然后再从该 RestClientBuilder 对象调用 build（）方法创建一个 RestClient 对象。RestClientBuilder 提供了 setDefaultHeaders（Header[]）方法来设置每个 Http 请求的默认标头。此外还提供 setFailureListener（FailureListener）方法来设置当请求失败时要通知的处理程序。以下列出了各方法中如何设置参数：

（1）builder（HttpHost[]）方法

org.apache.http.HttpHost 类提供数个对象构造函数指定协议的类型，这个测试使用其中一个方法构建 HttpHost 对象。需要设定的参数包括设为 localhost 的主机名称（hostname）、设为 9200 的连接埠（port）和设为 http 的通信协议（scheme）。以下是 HttpHost 的方法名称和签名。

```
HttpHost(String hostname, int port, String scheme)
```

（2）setDefaultHeaders（Header[]）方法

org.apache.http.message.BasicHeader 类实现了 org.apache.http.Header 接口，并提供一个对象构造函数生成 HTTP 协议的标头。需要设定的参数包括设为名称（name）及其值（value）。以下是 BasicHeader 的方法名称和签名。测试中设定两个标头，分别为 accept 和 content-type。用于设定请求和响应内容的媒体类型，设定为 application/json。

```
BasicHeader(String name, String value)
```

（3）setFailureListener（FailureListener）方法

RestClient 类包含了静态类 FailureListenerFailureListener，它的默认构造函数可用于创建对象。

```
static class FailureListener{public void onFailure(Node node) {}}
```

在 com.tdpress.restclient.config Java 程序包内的 LLRestClientConfig.java 文档，

LLRestClient() 方法封装了低级别 REST 客户端 RestClient 对象组件，便于创建和用作 Spring Bean，用于获取低级别的 Elasticsearch Rest 客户端。

```java
@Bean
// 此方法可用于获取低级别 REST 客户端
public RestClient LLRestClient() {
    return RestClient.builder(new HttpHost("localhost", 9200, "http"))
        .setDefaultHeaders(new Header[] {
            new BasicHeader("accept","application/json"),
            new BasicHeader("content-type","application/json")})
        .setFailureListener(new RestClient.FailureListener() {
            // 创建失败时的回调方法
            public void onFailure(Node node) {
                logger.error("Low level Rest Client Failure on node " + node.getName());
            }
        })
        .build();
}
```

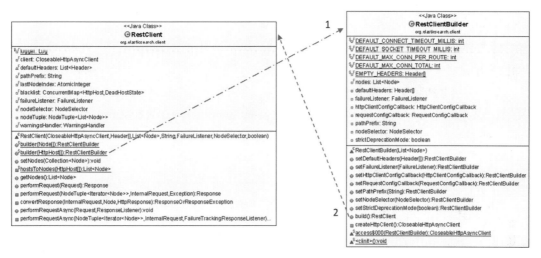

图 12-2 RestClient 和 RestClientBuilder 的 UML 类图

12.2.3 执行 REST 客户端请求

执行低级别 REST 客户端请求时可以同步 (performRequest) 或异步 (performRequestAsync) 发送请求。在同步模式下，调用线程被阻塞，直到收到响应结果之后才会返回，否则将引发异常而失败。事实上，同步模式是通过执行异步调用并等待结果来实现。在异步模式下，调用线程不被阻塞而会立刻返回，当响应结果准备好后，该程序会进行回调方法。无论使用哪一个模式，都需要在请求参数内使用 org.elasticsearch.client.Request 对象。在 org.elasticsearch.client Java 程序包内定义了 Request 对象。里面有一个构造函数，需要两个 String 类型参数。一个用于指定 HTTP 方法例如 POST，而另一个则用于指定 HTTP 端点例如 URL。以下是 Request 构造函数的名称和签名代码：

```
Request(java.lang.String method, java.lang.String endpoint)
```

以下是同步及异步请求方法的名称和签名代码：

```
Response performRequest(Request request) throws java.io.IOException
```

```
Cancellable performRequestAsync(Request request, ResponseListener
responseListener)
```

在异步请求方法 performRequestAsync 中包含 Cancellable 对象，可以用它来取消请求操作。参数 ResponseListener 使用匿名内部类 (Anonymous Inner class) 并设置两个回调方法 onSuccess 和 onFailure。当请求成功产生响应时回调方法 onSuccess。当请求失败时回调方法 onFailure。

12.2.4 处理 REST 客户端响应

无论由同步 performRequest 方法返回的 Response 对象或作为请求成功时的回调方法 onSuccess 中的参数 Response 对象，都是 org.elasticsearch.client.Response 数据类型，它包装了 http 客户端返回的响应。以下列出 Response 对象用于测试的一些方法：

（1）调用 Response 对象的 getStatusLine 方法，返回当前响应的状态行 (org.apache.http.StatusLine)，再从状态行支持的方法 getStatusCode 获取状态码。

（2）调用 Response 对象的 getEntity 方法，返回可用的响应主体 (org.apache.http.HttpEntity)。

（3）调用 Response 对象的 getHeaders 方法，返回响应的所有标头。

如果发送请求后发生任何通信异常，将引发 IOException。如果错误源自 HTTP 请求或响应，则需要处理异常 org.elasticsearch.client.ResponseException。

12.2.5 关闭 REST 客户端

调用 RestClient 对象的 close 方法，关闭 REST 客户端。

12.2.6 封装低级别 REST 客户端请求与处理其响应

在 com.tdpress.restclient.service Java 程序包内的 LLRestClientServiceImpl.java 文档，LLRestClientServiceImpl 类提供两个方法，分别封装了低级别 REST 客户端的同步 (performSyncRequest) 及异步 (performAsyncRequest) 请求方法，并简单地处理异常。返回响应结果为 Response 对象。自定义的返回方式是先从响应结果中提取数据，组成 RestClientResponse 对象，再转换为映射 (Map<String, Object>) 返回。

以下是测试程序的 performSyncRequest 方法：

```
//此方法调用低级别 REST 客户端的同步请求方法
public Map<String, Object> performSyncRequest(Request request) {
  RestClientResponse clientResponse = new RestClientResponse();
  try {
    Response response = restClient.performRequest(request);
    clientResponse.setStatusCode(response.getStatusLine().getStatusCode());
    clientResponse.setResponseBody(EntityUtils.toString(response.getEntity()));
    clientResponse.setHeaders(response.getHeaders());
  } catch (Exception ex) {
    clientResponse.setStatusCode(500);
    if (!ex.getMessage().isEmpty()) {
      clientResponse.setErrMessage(ex.getMessage());
    }
```

```
    }
    Map<String, Object> convertValue =
      (Map<String, Object>) (new ObjectMapper()).convertValue(clientResponse, Map.class);
    return convertValue;
}
```

以下是测试程序的 performAsyncRequest 方法,并简单地在其参数 ResponseListener 使用匿名内部类 (Anonymous Inner class),并设置两个回调方法 onSuccess 和 onFailure。返回响应结果为 Response 对象,示例简单地从屏幕输出结果。

```
// 此方法调用低级别 REST 客户端的异步请求方法
public Map<String, Object> performAsyncRequest(Request request) {
  RestClientResponse clientResponse = new RestClientResponse();
  restClient.performRequestAsync(request, new ResponseListener() {
    RestClientResponse clientResponse = new RestClientResponse();
    @Override
    // 请求成功产生响应时的回调方法
    public void onSuccess(Response response) {
      clientResponse.setStatusCode(response.getStatusLine().getStatusCode());
      try {
        clientResponse.setResponseBody(EntityUtils.toString(response.getEntity()));
        clientResponse.setHeaders(response.getHeaders());
        try {
          String jsonStr = (new ObjectMapper()). writeValueAsString(clientResponse);
          // 从屏幕输出验证结果
          System.out.println(jsonStr);
        }
        catch (IOException e) {
          e.printStackTrace();
        }
      } catch (Exception ex) {
        clientResponse.setStatusCode(500);
        if (!ex.getMessage().isEmpty()) {
          clientResponse.setErrMessage(ex.getMessage());
        }
      }
    }

    @Override
    // 请求失败时的回调方法
    public void onFailure(Exception exception) {
      clientResponse.setStatusCode(500);
      if (!exception.getMessage().isEmpty()) {
        clientResponse.setErrMessage(exception.getMessage());
      }
      try {
        String jsonStr = (new ObjectMapper()).writeValueAsString(clientResponse);
        // 从屏幕输出失败原因
        System.out.println(jsonStr);
      }
      catch (IOException e) {
        e.printStackTrace();
      }
    }
  });
  clientResponse.setStatusCode(200);
  // 从响应结果中提取数据,转换为映射
  Map<String, Object> convertValue =
    (Map<String, Object>) (new ObjectMapper()).convertValue(clientResponse, Map.class);
  return convertValue;
}
```

12.2.7　调用自定义的 performSyncRequest 和 performAsyncRequest 方法

无论同步或是异步发送低级别 REST 客户端请求，正如 12.2.3 节执行 REST 客户端请求讨论的，必须首先构造一个查询请求 Request 对象。

在 com.tdpress.restclient.controller Java 程序包内的 LLRestClientController.java 文档，LLRestClientController 类提供两个方法，同步 POST 请求方法 (performSyncPostRequest) 及异步 POST 请求方法 (performAsyncPostRequest)。分别调用 performSyncRequest 及 performAsyncRequest 方法，发送 Http POST 请求至 url 端点，请求主体设置为 requestBody。

```
@Autowired
private LLRestClientService llRestClient;
……
@ApiOperation("测试 Java 低级别 REST 客户端同步 POST 请求")
……
@RequestMapping(value="/sync_post", method=RequestMethod.POST)
public ResponseEntity<Map<String, Object>> performSyncPostRequest(
    @RequestParam(value = "url") String url,
    @RequestBody String requestBody) throws Exception {
    Request request = new Request("POST", url);
    request.setJsonEntity(requestBody);
    Map<String,Object> response = llRestClient.performSyncRequest(request);
    return new ResponseEntity<Map<String, Object>>(response, HttpStatus.OK);
}

@ApiOperation("测试 Java 低级别 REST 客户端异步 POST 请求")
……
@RequestMapping(value="/async_post", method=RequestMethod.POST)
public ResponseEntity<Map<String, Object>> performAsyncPostRequest(
    @RequestParam(value = "url") String url,
    @RequestBody String requestBody) throws Exception {
    Request request = new Request("POST", url);
    request.setJsonEntity(requestBody);
    Map<String,Object> response = llRestClient.performAsyncRequest(request);
    return new ResponseEntity<Map<String, Object>>(response, HttpStatus.OK);
}
```

12.3　使用 Swagger UI 测试低级别 REST 客户端

为了方便用户，本章测试程序已集成 Swagger UI，从已开发的低级别 REST 客户端程序生成可视化的测试接口。

1. 准备 Swagger UI 测试环境

以下分步说明如何使用 Maven 指令在 java_restful_client 目录下，运行清理、编译、构建和执行程序项目：

（1）运行清理项目

```
mvn clean
……
[INFO] BUILD SUCCESS
[INFO] ------------------------------------------------------------------------
[INFO] Total time:  1.989 s
[INFO] Finished at: 2020-08-21T10:55:34-07:00
[INFO] ------------------------------------------------------------------------
```

（2）运行编译和构建项目

```
mvn package
......
[INFO] -----------------------------------------------------------------
[INFO] BUILD SUCCESS
[INFO] -----------------------------------------------------------------
[INFO] Total time:  17.577 s
[INFO] Finished at: 2020-08-21T10:59:11-07:00
[INFO] -----------------------------------------------------------------
```

（3）执行程序项目

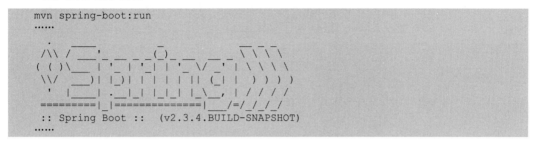

2.Java 低级别 REST 客户端项目的 Swagger UI 页面

使用网页浏览器进行互动测试访问，URL 为 http://localhost:10010/swagger-ui.html。然后按 ll-rest-client-controller 请求栏以展开面板，屏幕截图如图 12-3 所示。

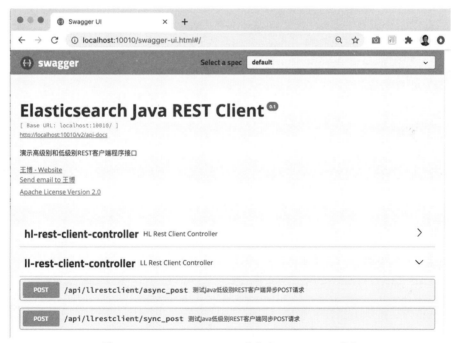

图 12-3　java_restful_client 项目的 Swagger UI 页面

3.Java 低级别 REST 客户端同步 POST 请求测试

点击"测试 Java 低级别 REST 客户端同步 POST 请求"一栏以展开面板，屏幕如图 12-4 所示。

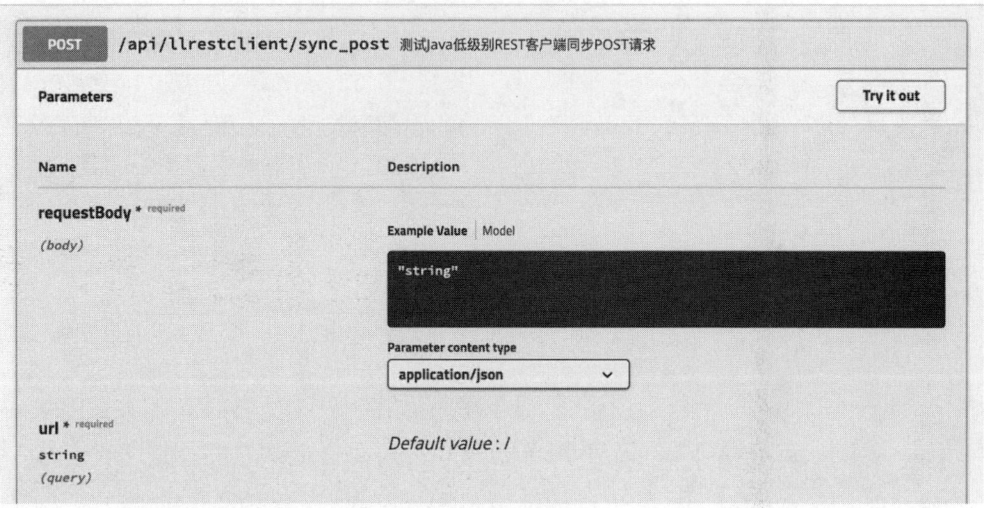

图 12-4　展开 Swagger UI 页面中低级别 REST 客户端同步 POST 请求一栏

点击"Try it out"按钮时，面板将更改为输入模式。为方便起见，使用 7.3.1 节认识查询领域特定语言的测试用例，键入 URL 为"/fund_basic/_search"和查询语句为在基金简称文本中匹配字符串"中银 ETF"。查询语句如下所示，如图 12-5 所示。

```
{"query": {"match" : {"name": {"query" : "中银ETF", "operator":"AND"}}}, "_source":["name","ts_code"]}
```

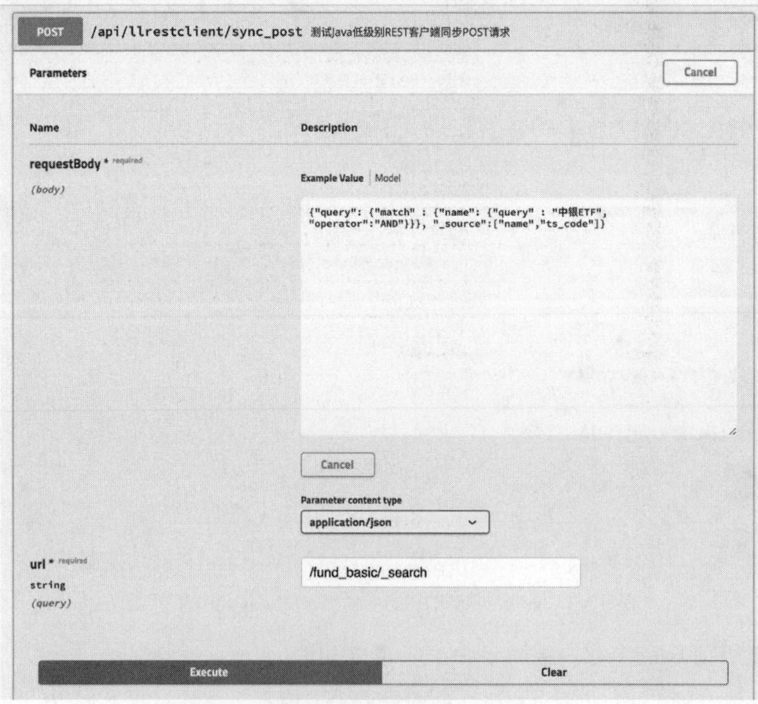

图 12-5　键入请求主体的查询语句和 URL

点击"Execute"按钮，向下滑动屏幕后可以看到响应结果。获取的状态码 (statusCode) 为 200，返回的响应结果 (responseBody) 为键值对形式的数据结构。屏幕截图如图 12-6 所示。

第 12 章 Java 客户端编程

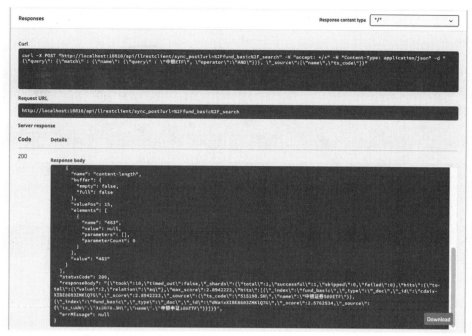

图 12-6 低级别 REST 客户端同步 POST 请求返回的响应结果

4.Java 低级别 REST 客户端异步 POST 请求测试

异步 POST 请求测试类似于同步测试，点击"测试 Java 低级别 REST 客户端异步 POST 请求"一栏以展开面板。点击"Try it out"按钮并键入请求主体的查询语句和 URL。查询语句如同步测试，如图 12-7 所示。

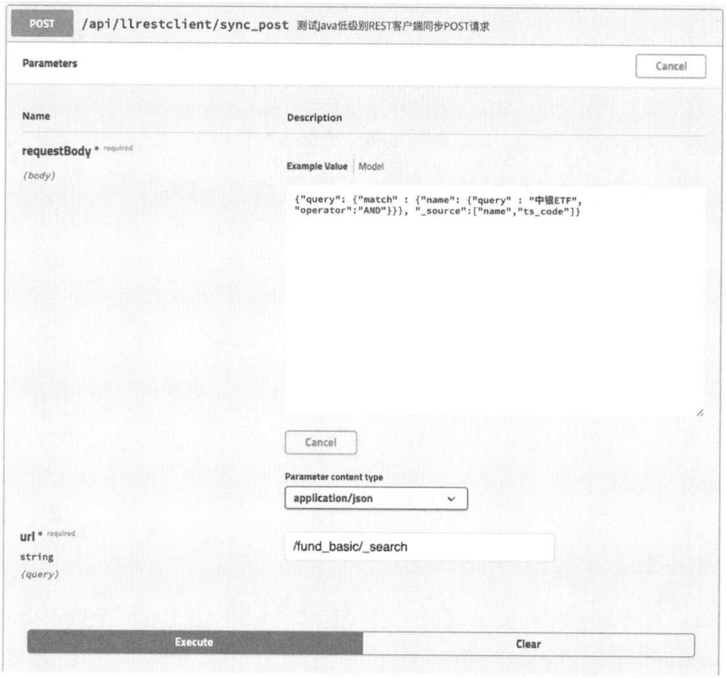

图 12-7 键入请求主体的查询语句和 URL

点击"Execute"按钮，向下滑动屏幕后可以看到请求发送结果。获取的状态码(statusCode)为 200，如图 12-8 所示。

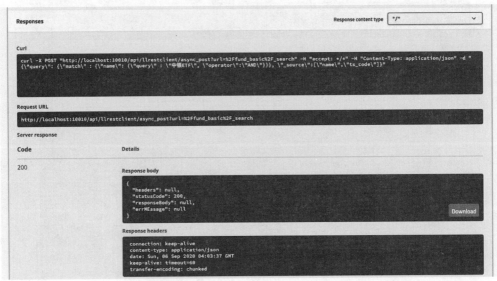

图 12-8　低级别 REST 客户端异步 POST 请求返回的响应结果

由于响应结果会在稍后发送到回调方法 onSuccess，因此这测试只将其打印在屏幕上。返回的响应结果 (responseBody) 为匹配查询字符串中银 ETF 的基金简称，包括中银证券500ETF 和中银中证 100ETF。以下是运行 Elasticsearch 服务器的屏幕出现的结果。

```
"statusCode":200,"responseBody":"{\"took\":5,\"timed_out\":false,\"_
shards\":{\"total\":1,
\"successful\":1,\"skipped\":0,\"failed\":0},\"hits\":{\"total\":{\"value\":2,
\"relation\":\"eq\"},\"max_score\":2.8942223,\"hits\":[{\"_index\":\"fund_basic\",
\"_type\":\"_doc\",\"_id\":\"cdaixXIBE8603ZMKlQ7G\",\"_score\":2.8942223,\"_
source\":{
    \"ts_code\":\"515190.SH\",\"name\":\"中银证券500ETF\"}}, {\"_index\":\"fund_basic\",
\"_type\":\"_doc\", \"_id\":\"dNaixXIBE8603ZMKlQ7G\",\"_score\":2.5762534,\"_
source\":{
    \"ts_code\":\"515670.SH\", \"name\":\"中银中证100ETF\"}}]}}", "errMEssage":null}
```

12.4　Java 高级别 REST 客户端

Java 高级别 REST 客户端建立在低级别 REST 客户端之上，并负责请求和响应对象的序列化和反序列化。高级别 REST 客户端定义于 org.elasticsearch.client.RestHighLevelClient Java 类。主要目标是提供各种特定接口的调用方法，有些方法可以直接从 REST 客户端调用，另一些方法则需要间接调用。

12.4.1　封装高级别 REST 客户端

在 com.tdpress.restclient.config Java 程 序 包 内 的 HLRestClientConfig.java 文 档，HLRestClient() 方法封装了高级别 REST 客户端 RestHighLevelClient 对象组件。如同前面提到 Java 高级别 REST 客户端建立在低级别 REST 客户端之上，RestHighLevelClient 构造函

数需要使用低级别 RestClient 对象作为参数。创建低级别 RestClient 对象的方法与 12.2.2 节 REST 客户端初始化相同。

```
@Bean
// 此方法可用于获取高级别 REST 客户端
public RestHighLevelClient HLRestClient() {
    return new RestHighLevelClient(
        RestClient.builder(new HttpHost("localhost", 9200, "http"))
            .setDefaultHeaders(new Header[] {
                new BasicHeader("accept","application/json"),
                new BasicHeader("content-type","application/json")})
            .setFailureListener(new RestClient.FailureListener() {
                // 创建失败时的回调方法
                public void onFailure(Node node) {
                    logger.error("build High level REST Client Failure on node " +
                        node.getName());
                }
            }));
}
```

12.4.2 提供间接调用方法的 RestHighLevelClient 成员

间接调用意味着首先必须获得来自 RestHighLevelClient 内的成员对象，然后从其成员对象直接调用其请求方法。提供间接调用方法的成员编撰汇集于表 12-1。

表 12-1　RestHighLevelClient 对象内提供间接调用方法的成员列表

获得提供间接调用方法的成员的方法	简述
CcrClient ccr()	用来访问跨集群复制接口
ClusterClient cluster()	用来访问集群接口
EnrichClient enrich()	用来访问 enrich 接口，丰富文档的字段内容
GraphClient graph()	用来访问 _graph/explore 接口，探索数据中存在的相关联系
IndexLifecycleClient indexLifecycle()	用来访问 _ilm(索引生命周期管理)接口，管理索引的 Hot、Warm、Cold 和 Delete 4 个阶段
IndicesClient indices()	用来访问索引接口
IngestClient ingest()	用来访问摄取节点接口
LicenseClient license()	用来访问软件许可接口
MachineLearningClient machineLearning()	用来访问机器学习接口
MigrationClient migration()	用来访问版本之间的迁移接口
RollupClient rollup()	用来访问数据上卷接口
SecurityClient security()	用来访问安全配置接口
SnapshotClient snapshot()	用来访问快照备份与恢复接口
TasksClient tasks()	用来访问任务接口
TransformClient transform()	用来访问 TransformClient 接口，管理通过转换源索引中数据到目标索引的操作
WatcherClient watcher()	用来访问告警接口，提供警报和通知
XPackClient xpack()	用来访问 X-Pack 接口

12.4.3 间接调用方法

由于提供间接调用方法的成员太多而没有办法一一展示，在这小节只详细描述 IndicesClient 作为一个例子。一般来说，IndicesClient 为 RestHighLevelClient 提供间接调用方法提供相关索引操作，可以同步或异步发送请求。这些请求相当于 12.2.3 节执行 REST 客户端请求所介绍的。调用方法编撰汇集于表 12-2。

表 12-2　通过 IndicesClient 提供间接调用方法列表

方法名称、签名和返回类型	简述
CreateIndexResponse create(CreateIndexRequest createIndexRequest, RequestOptions options)	创建索引
Cancellable createAsync(CreateIndexRequest createIndexRequest, RequestOptions options, ActionListener<CreateIndexResponse> listener)	
GetIndexResponse get(GetIndexRequest getIndexRequest, RequestOptions options)	检索索引的信息
Cancellable getAsync(GetRequest getRequest, RequestOptions options, ActionListener<GetResponse> listener)	
AcknowledgedResponse delete(DeleteIndexRequest deleteIndexRequest, RequestOptions options)	删除索引
Cancellable deleteAsync(DeleteIndexRequest deleteIndexRequest, RequestOptions options, ActionListener<AcknowledgedResponse> listener)	
ResizeResponse clone(ResizeRequest resizeRequest, RequestOptions options)	克隆索引
Cancellable cloneAsync(ResizeRequest resizeRequest, RequestOptions options, ActionListener<ResizeResponse> listener)	
boolean exists(GetIndexRequest request, RequestOptions options)	检查索引是否存在
Cancellable existsAsync(GetIndexRequest request, RequestOptions options, ActionListener<java.lang.Boolean> listener)	
OpenIndexResponse open(OpenIndexRequest openIndexRequest, RequestOptions options)	打开索引
Cancellable openAsync(OpenIndexRequest openIndexRequest, RequestOptions options, ActionListener<OpenIndexResponse> listener)	
CloseIndexResponse close(CloseIndexRequest closeIndexRequest, RequestOptions options)	关闭索引
Cancellable closeAsync(CloseIndexRequest closeIndexRequest, RequestOptions options, ActionListener<CloseIndexResponse> listener)	
AcknowledgedResponse putSettings(UpdateSettingsRequest updateSettingsRequest, RequestOptions options)	更新索引的设置
Cancellable putSettingsAsync(UpdateSettingsRequest updateSettingsRequest, RequestOptions options, ActionListener<AcknowledgedResponse> listener)	
GetSettingsResponse getSettings(GetSettingsRequest getSettingsRequest, RequestOptions options)	检索索引的设置
Cancellable getSettingsAsync(GetSettingsRequest getSettingsRequest, RequestOptions options, ActionListener<GetSettingsResponse> listener)	
AcknowledgedResponse putMapping(PutMappingRequest putMappingRequest, RequestOptions options)	更新索引的映射
Cancellable putMappingAsync(PutMappingRequest putMappingRequest, RequestOptions options, ActionListener<AcknowledgedResponse> listener)	
GetMappingsResponse getMapping(GetMappingsRequest getMappingsRequest, RequestOptions options)	检索索引的映射
Cancellable getMappingAsync(GetMappingsRequest getMappingsRequest, RequestOptions options, ActionListener<GetMappingsResponse> listener)	
AcknowledgedResponse putTemplate(PutIndexTemplateRequest putIndexTemplateRequest, RequestOptions options)	更新索引的模板
Cancellable putTemplateAsync(PutIndexTemplateRequest putIndexTemplateRequest, RequestOptions options, ActionListener<AcknowledgedResponse> listener)	

方法名称、签名和返回类型	简述
GetIndexTemplatesResponse getIndexTemplate(GetIndexTemplatesRequest getIndexTemplatesRequest, RequestOptions options)	检索索引的模板
Cancellable getIndexTemplateAsync(GetIndexTemplatesRequest getIndexTemplatesRequest, RequestOptions options, ActionListener<GetIndexTemplatesResponse> listener)	
AcknowledgedResponse deleteTemplate(DeleteIndexTemplateRequest request, RequestOptions options)	删除索引的模板
Cancellable deleteTemplateAsync(DeleteIndexTemplateRequest request, RequestOptions options, ActionListener<AcknowledgedResponse> listener)	
boolean existsTemplate(IndexTemplatesExistRequest indexTemplatesRequest, RequestOptions options)	检查索引的模板是否存在
Cancellable existsTemplateAsync(IndexTemplatesExistRequest indexTemplatesExistRequest, RequestOptions options, ActionListener<java.lang.Boolean> listener)	
GetAliasesResponse getAlias(GetAliasesRequest getAliasesRequest, RequestOptions options)	检索索引的别名
Cancellable getAliasAsync(GetAliasesRequest getAliasesRequest, RequestOptions options, ActionListener<GetAliasesResponse> listener)	
boolean existsAlias(GetAliasesRequest getAliasesRequest, RequestOptions options)	检查索引的别名是否存在
Cancellable existsAliasAsync(GetAliasesRequest getAliasesRequest, RequestOptions options, ActionListener<java.lang.Boolean> listener)	
AcknowledgedResponse updateAliases(IndicesAliasesRequest indicesAliasesRequest, RequestOptions options)	更新索引的别名
Cancellable updateAliasesAsync(IndicesAliasesRequest indicesAliasesRequest, RequestOptions options, ActionListener<AcknowledgedResponse> listener)	
GetFieldMappingsResponse getFieldMapping(GetFieldMappingsRequest getFieldMappingsRequest, RequestOptions options)	检索索引的字段映射
Cancellable getFieldMappingAsync(GetFieldMappingsRequest getFieldMappingsRequest, RequestOptions options, ActionListener<GetFieldMappingsResponse> listener)	
AnalyzeResponse analyze(AnalyzeRequest request, RequestOptions options)	文本分析
Cancellable analyzeAsync(AnalyzeRequest request, RequestOptions options, ActionListener<AnalyzeResponse> listener)	
ReloadAnalyzersResponse reloadAnalyzers(ReloadAnalyzersRequest request, RequestOptions options)	重新加载搜索分析器以同步同义词文件的更新
Cancellable reloadAnalyzersAsync(ReloadAnalyzersRequest request, RequestOptions options, ActionListener<ReloadAnalyzersResponse> listener)	
FlushResponse flush(FlushRequest flushRequest, RequestOptions options)	刷新到硬盘
Cancellable flushAsync(FlushRequest flushRequest, RequestOptions options, ActionListener<FlushResponse> listener)	
SyncedFlushResponse flushSynced(SyncedFlushRequest syncedFlushRequest, RequestOptions options)	同步刷新到硬盘
Cancellable flushSyncedAsync(SyncedFlushRequest syncedFlushRequest, RequestOptions options, ActionListener<SyncedFlushResponse> listener)	
ForceMergeResponse forcemerge(ForceMergeRequest forceMergeRequest, RequestOptions options)	强制合并索引
Cancellable forcemergeAsync(ForceMergeRequest forceMergeRequest, RequestOptions options, ActionListener<ForceMergeResponse> listener)	

续表

方法名称、签名和返回类型	简述
ShardsAcknowledgedResponse freeze(FreezeIndexRequest request, RequestOptions options)	冻结索引
Cancellable freezeAsync(FreezeIndexRequest request, RequestOptions options, ActionListener<ShardsAcknowledgedResponse> listener)	
ShardsAcknowledgedResponse unfreeze(UnfreezeIndexRequest request, RequestOptions options)	解冻索引
Cancellable unfreezeAsync(UnfreezeIndexRequest request, RequestOptions options, ActionListener<ShardsAcknowledgedResponse> listener)	
RefreshResponse refresh(RefreshRequest refreshRequest, RequestOptions options)	刷新索引相关的操作以用于搜索
Cancellable refreshAsync(RefreshRequest refreshRequest, RequestOptions options, ActionListener<RefreshResponse> listener)	
RolloverResponse rollover(RolloverRequest rolloverRequest, RequestOptions options)	滚动索引
Cancellable rolloverAsync(RolloverRequest rolloverRequest, RequestOptions options, ActionListener<RolloverResponse> listener)	
ResizeResponse shrink(ResizeRequest resizeRequest, RequestOptions options)	缩小现有索引以使用较少主分片
Cancellable shrinkAsync(ResizeRequest resizeRequest, RequestOptions options, ActionListener<ResizeResponse> listener)	
ResizeResponse split(ResizeRequest resizeRequest, RequestOptions options)	拆分索引
Cancellable splitAsync(ResizeRequest resizeRequest, RequestOptions options, ActionListener<ResizeResponse> listener)	
ValidateQueryResponse validateQuery(ValidateQueryRequest validateQueryRequest, RequestOptions options)	验证查询语句是否正确
Cancellable validateQueryAsync(ValidateQueryRequest validateQueryRequest, RequestOptions options, ActionListener<ValidateQueryResponse> listener)	
ClearIndicesCacheResponse clearCache(ClearIndicesCacheRequest clearIndicesCacheRequest, RequestOptions options)	清理索引缓存
Cancellable clearCacheAsync(ClearIndicesCacheRequest clearIndicesCacheRequest, RequestOptions options, ActionListener<ClearIndicesCacheResponse> listener)	

12.4.4 直接调用方法

RestHighLevelClient类提供的直接调用方法,可以针对接口类型,粗略地细分为索引文档、搜索查询、脚本、批量请求、重置节流和其他接口。分别编撰汇集于表 12-3—表 12-8。

表 12-3 RestHighLevelClient 提供有关索引文档的方法列表

方法名称、签名和返回类型	简述
IndexResponse index(IndexRequest indexRequest, RequestOptions options)	索引文档
Cancellable indexAsync(IndexRequest indexRequest, RequestOptions options, ActionListener<IndexResponse> listener)	
GetResponse get(GetRequest getRequest, RequestOptions options)	检索索引文档
Cancellable getAsync(GetRequest getRequest, RequestOptions options, ActionListener<GetResponse> listener)	
MultiGetResponse mget(MultiGetRequest multiGetRequest, RequestOptions options)	检索多个索引文档
Cancellable mgetAsync(MultiGetRequest multiGetRequest, RequestOptions options, ActionListener<MultiGetResponse> listener)	

续表

方法名称、签名和返回类型	简述
UpdateResponse update(UpdateRequest updateRequest, RequestOptions options)	更新索引文档内容
Cancellable updateAsync(UpdateRequest updateRequest, RequestOptions options, ActionListener <UpdateResponse> listener)	
DeleteResponse delete(DeleteRequest deleteRequest, RequestOptions options)	删除索引文档
Cancellable deleteAsync(DeleteRequest deleteRequest, RequestOptions options, ActionListener <DeleteResponse> listener)	
boolean exists(GetRequest getRequest, RequestOptions options)	检查索引文档是否存在
Cancellable existsAsync(GetRequest getRequest, RequestOptions options, ActionListener<java.lang.Boolean> listener)	
boolean existsSource(GetRequest getRequest, RequestOptions options)	检查索引文档的_source 字段是否存在
boolean existsSourceAsync(GetRequest getRequest, RequestOptions options, ActionListener<java.lang.Boolean> listener)	
MultiTermVectorsResponse mtermvectors(MultiTermVectorsRequest request, RequestOptions options)	检索多个文档的词条向量信息
Cancellable mtermvectorsAsync(MultiTermVectorsRequest request, RequestOptions options, ActionListener<MultiTermVectorsResponse> listener)	

表 12-4　RestHighLevelClient 提供有关搜索查询的方法列表

方法名称、签名和返回类型	简述
SearchResponse search(SearchRequest searchRequest, RequestOptions options)	查询文档
Cancellable searchAsync(SearchRequest searchRequest, RequestOptions options, ActionListener <SearchResponse> listener)	
SearchResponse scroll(SearchScrollRequest searchScrollRequest, RequestOptions options)	游标查询文档
Cancellable scrollAsync(SearchScrollRequest searchScrollRequest, RequestOptions options, ActionListener<SearchResponse> listener)	
ClearScrollResponse clearScroll(ClearScrollRequest clearScrollRequest, RequestOptions options)	关闭游标查询
Cancellable clearScrollAsync(ClearScrollRequest clearScrollRequest, RequestOptions options, ActionListener<ClearScrollResponse> listener)	
SearchTemplateResponse searchTemplate(SearchTemplateRequest searchTemplateRequest, RequestOptions options)	按照查询模板执行请求
Cancellable searchTemplateAsync(SearchTemplateRequest searchTemplateRequest, RequestOptions options, ActionListener<SearchTemplateResponse> listener)	
MultiSearchResponse msearch(MultiSearchRequest multiSearchRequest, RequestOptions options)	多重目标查询文档
Cancellable msearchAsync(MultiSearchRequest searchRequest, RequestOptions options, ActionListener<MultiSearchResponse> listener)	
MultiSearchTemplateResponse msearchTemplate(MultiSearchTemplateRequest multiSearchTemplateRequest, RequestOptions options)	按照多重目标查询模板执行请求
Cancellable msearchTemplateAsync(MultiSearchTemplateRequest multiSearchTemplateRequest, RequestOptions options, ActionListener<MultiSearchTemplateResponse> listener)	
CountResponse count(CountRequest countRequest, RequestOptions options)	搜索查询结果的匹配数
Cancellable countAsync(CountRequest countRequest, RequestOptions options, ActionListener <CountResponse> listener)	
FieldCapabilitiesResponse fieldCaps(FieldCapabilitiesRequest fieldCapabilitiesRequest, RequestOptions options)	检查字段功能
Cancellable fieldCapsAsync(FieldCapabilitiesRequest fieldCapabilitiesRequest, RequestOptions options, ActionListener<FieldCapabilitiesResponse> listener)	

方法名称、签名和返回类型	简述
ExplainResponse explain(ExplainRequest explainRequest, RequestOptions options)	解释为何匹配和得此分数
Cancellable explainAsync(ExplainRequest explainRequest, RequestOptions options, ActionListener<ExplainResponse> listener)	
RankEvalResponse rankEval(RankEvalRequest rankEvalRequest, RequestOptions options)	评估搜索查询结果中排名的质素
Cancellable rankEvalAsync(RankEvalRequest rankEvalRequest, RequestOptions options, ActionListener<RankEvalResponse> listener)	

表 12-5　RestHighLevelClient 提供有关脚本的方法列表

方法名称、签名和返回类型	简述
AcknowledgedResponse putScript(PutStoredScriptRequest putStoredScriptRequest, RequestOptions options)	存储脚本
Cancellable putScriptAsync(PutStoredScriptRequest putStoredScriptRequest, RequestOptions options, ActionListener<AcknowledgedResponse> listener)	
GetStoredScriptResponse getScript(GetStoredScriptRequest request, RequestOptions options)	检索脚本
Cancellable getScriptAsync(GetStoredScriptRequest request, RequestOptions options, ActionListener<GetStoredScriptResponse> listener)	
AcknowledgedResponse deleteScript(DeleteStoredScriptRequest request, RequestOptions options)	删除脚本
Cancellable deleteScriptAsync(DeleteStoredScriptRequest request, RequestOptions options, ActionListener<AcknowledgedResponse> listener)	

表 12-6　RestHighLevelClient 提供有关批量请求的方法列表

方法名称、签名和返回类型	简述
BulkResponse bulk(BulkRequest bulkRequest, RequestOptions options)	执行批量请求
Cancellable bulkAsync(BulkRequest bulkRequest, RequestOptions options, ActionListener<BulkResponse> listener)	
BulkByScrollResponse updateByQuery(UpdateByQueryRequest updateByQueryRequest, RequestOptions options)	根据请求执行批量更新
Cancellable updateByQueryAsync(UpdateByQueryRequest updateByQueryRequest, RequestOptions options, ActionListener<BulkByScrollResponse> listener)	
BulkByScrollResponse deleteByQuery(DeleteByQueryRequest deleteByQueryRequest, RequestOptions options)	根据请求执行批量删除
Cancellable deleteByQueryAsync(DeleteByQueryRequest deleteByQueryRequest, RequestOptions options, ActionListener<BulkByScrollResponse> listener)	
BulkByScrollResponse reindex(ReindexRequest reindexRequest, RequestOptions options)	重新索引
Cancellable reindexAsync(ReindexRequest reindexRequest, RequestOptions options, ActionListener<BulkByScrollResponse> listener)	

表 12-7　RestHighLevelClient 提供有关任务的方法列表

方法名称、签名和返回类型	简述
ListTasksResponse updateByQueryRethrottle(RethrottleRequest rethrottleRequest, RequestOptions options)	重置正在运行的查询更新任务当前每秒请求数的限制
Cancellable updateByQueryRethrottleAsync(RethrottleRequest rethrottleRequest, RequestOptions options, ActionListener<ListTasksResponse> listener)	
ListTasksResponse deleteByQueryRethrottle(RethrottleRequest rethrottleRequest, RequestOptions options)	重置正在运行的查询删除任务当前每秒请求数的限制
Cancellable deleteByQueryRethrottleAsync(RethrottleRequest rethrottleRequest, RequestOptions options, ActionListener<ListTasksResponse> listener)	

续表

方法名称、签名和返回类型	简述
ListTasksResponse reindexRethrottle(RethrottleRequest rethrottleRequest, RequestOptions options)	重置正在运行的重新索引任务当前每秒请求数的限制
Cancellable reindexRethrottleAsync(RethrottleRequest rethrottleRequest, RequestOptions options, ActionListener<ListTasksResponse> listener)	
TaskSubmissionResponse submitDeleteByQueryTask(DeleteByQueryRequest deleteByQueryRequest, RequestOptions options)	提交查询删除任务
TaskSubmissionResponse submitReindexTask(ReindexRequest reindexRequest, RequestOptions options)	提交重新索引任务

表 12-8 RestHighLevelClient 提供其他的方法接口列表

方法名称、签名和返回类型	简述
void close()	关闭高级别 REST 客户端
MainResponse info(RequestOptions options)	获取集群信息
java.util.Optional<Resp> performRequestAndParseOptionalEntity(Req request, CheckedFunction <Req,Request,java.io.IOException> requestConverter, RequestOptions options, CheckedFunction<XContentParser,Resp,java.io.IOException> entityParser)	提交请求，如果返回响应主体，则解析它
Cancellable performRequestAsyncAndParseEntity(Req request, CheckedFunction<Req,Request,java.io.IOException> requestConverter, RequestOptions options, CheckedFunction<XContentParser,Resp,java.io.IOException> entityParser, ActionListener<Resp> listener, java.util.Set<java.lang.Integer> ignores)	
Resp parseEntity(org.apache.http.HttpEntity entity, CheckedFunction<XContentParser,Resp,java.io.IOException> entityParser)	解析返回的响应主体

12.4.5 构造查询请求

各类型调用方法，无论是直接或是间接，同步或是异步发送请求，必须首先构造一个查询请求。从表 12-2 到表 12-8 可以看到不同类型的查询请求。下面的两个例子用来说明构造查询请求：

（1）使用 RestHighLevelClient 的成员 IndicesClient，间接调用检索索引设置的方法

♦ 无论是同步（getSettings）或是异步（getSettingsAsync）发送请求，都需要查询请求参数 GetSettingsRequest

```
    GetSettingsResponse getSettings(GetSettingsRequest getSettingsRequest,
RequestOptions options)
    Cancellable getSettingsAsync(GetSettingsRequest getSettingsRequest,
RequestOptions options, ActionListener<GetSettingsResponse> listener)
```

♦ 构造一个 GetSettingsRequest 查询请求对象作为调用方法的参数，此测试使用无参数构造函数，再设置索引名称为 indexName

```
    GetSettingsRequest request = new GetSettingsRequest();
        request.indices(indexName);
```

（2）使用 RestHighLevelClient 直接调用查询文档的方法

♦ 无论是同步（search）或是异步（searchAsync）发送请求，都需要查询请求参数 SearchRequest

```
    SearchResponse search(SearchRequest searchRequest, RequestOptions options)
    Cancellable searchAsync(SearchRequest searchRequest, RequestOptions options,
ActionListener<SearchResponse> listener)
```

- 构造一个 SearchRequest 查询请求对象作为调用方法的参数，需要用到 SearchRequest、SearchSourceBuilder 以及 QueryBuilder 类。SearchRequest 对象提供了一个名为 source 的方法，该方法带有 SearchSourceBuilder 参数，可存储 QueryBuilder 对象的查询语句。此外，SearchSourceBuilder 类提供 from、size 和 sort 等方法。以下为根据前述方式对索引 indexName 构建匹配查询（match query）的 SearchRequest 对象。

提示：在 com.tdpress.restclient.controller Java 程序包内的 HLRestClientController.java 文档，HLRestClientController 类提供两个方法，searchRequest 及 asyncSearchRequest。

```
SearchRequest request = new SearchRequest(indexName);
SearchSourceBuilder sourceBuilder = new SearchSourceBuilder();
sourceBuilder.from(from);
sourceBuilder.size(size);
MatchQueryBuilder queryBuilder = QueryBuilders.matchQuery(fieldName, fieldValue);
request.source(sourceBuilder.query(queryBuilder));
```

- 另外，每种查询请求对象必须利用由相对应的 QueryBuilder 构造函数构建的 SearchSourceBuilder 对象，各类构造函数编撰汇集于表 12-9。

表 12-9 各类型 QueryBuilder 对象构造函数列表

简述	方法名称和返回类型
对应全文搜索	
匹配所有查询	MatchAllQueryBuilder matchAllQuery()
标准匹配查询	MatchQueryBuilder matchQuery()
短语匹配查询	MatchPhraseQueryBuilder matchPhraseQuery()
布尔与前缀匹配查询	MatchBoolPrefixQueryBuilder matchBoolPrefixQuery()
短语前缀匹配查询	MatchPhrasePrefixQueryBuilder matchPhrasePrefixQuery()
多字段匹配	MultiMatchQueryBuilder multiMatchQuery()
简单查询字符串	SimpleQueryStringBuilder simpleQueryStringQuery()
查询语句查询	QueryStringQueryBuilder queryStringQuery()
对应词条级别搜索	
存在查询	ExistsQueryBuilder existsQuery()
模糊查询	FuzzyQueryBuilder fuzzyQuery()
文档标识符查询	IdsQueryBuilder idsQuery()
前缀查询	PrefixQueryBuilder prefixQuery()
范围查询	RangeQueryBuilder rangeQuery()
正则查询	RegexpQueryBuilder regexpQuery()
词条查询	TermQueryBuilder termQuery()
多个词条查询	TermsQueryBuilder termsQuery()
通配符查询	WildcardQueryBuilder wildcardQuery()
对应复合搜索	
布尔查询	BoolQueryBuilder boolQuery()
权重提升查询	BoostingQueryBuilder boostingQuery()
固定分数查询	ConstantScoreQueryBuilder constantScoreQuery()
分离最大化查询	DisMaxQueryBuilder disMaxQuery()
函数评分查询	functionScoreQuery functionScoreQuery()

续表

简述	方法名称和返回类型
对应地理搜索	
地理边框查询	GeoBoundingBoxQueryBuilder geoBoundingBoxQuery()
地理空间不重叠关系查询	GeoShapeQueryBuilder geoDisjointQuery()
地理距离查询	GeoDistanceQueryBuilder geoDistanceQuery()
地理空间重叠关系查询	GeoShapeQueryBuilder geoIntersectionQuery()
地理多边形查询	GeoPolygonQueryBuilder geoPolygonQuery()
地理形状查询	GeoShapeQueryBuilder geoShapeQuery()
地理空间包含关系查询	GeoShapeQueryBuilder geoWithinQuery()
对应嵌套查询	
嵌套查询	NestedQueryBuilder nestedQuery()
对应脚本查询	
脚本查询	ScriptQueryBuilder scriptQuery()
对应跨度查询	
跨距匹配，容许之跨度描述在参数 big 的跨距匹配查询结果，并且包含从参数 small 的跨距匹配查询的结果	SpanContainingQueryBuilder spanContainingQuery()
跨度匹配，参数 end 控制从开始到最后位置的跨度	SpanFirstQueryBuilder spanFirstQuery()
等效与词条查询，用于跨度查询	SpanTermQueryBuilder spanTermQuery()
包装词条、范围、前缀、通配符、正则表达式或模糊等查询，用于跨度查询	SpanMultiTermQueryBuilder spanMultiTermQueryBuilder()
跨距匹配，参数 slop 控制允许不匹配位置的跨度	SpanNearQueryBuilder spanNearQuery()
跨距匹配，容许之跨度为包括在参数 include 所描述的跨距匹配查询结果，但不能被包括在参数 exclude 所描述的跨距匹配查询结果	SpanNotQueryBuilder spanNotQuery ()
跨距匹配，跨度为属于所有子句的并集	SpanOrQueryBuilder spanOrQuery()
跨距匹配，容许之跨度描述在参数 small 的跨距匹配查询结果，并且在参数 big 的跨距匹配查询结果范围内	SpanWithinQueryBuilder spanWithinQuery()
跨域字段屏蔽查询，适用于单个字段同时使用多个分析器分析，并进行文档索引，然后使用多字段一起搜索	FieldMaskingSpanQueryBuilder fieldMaskingSpanQuery()
专业查询	
对距离功能查询	DistanceFeatureQueryBuilder distanceFeatureQuery()
与文档集的文档匹配查询	MoreLikeThisQueryBuilder moreLikeThisQuery()
脚本评分查询	ScriptScoreQueryBuilder scriptScoreQuery()
包装查询	WrapperQueryBuilder wrapperQuery()

12.4.6 自定义 searchSync 与处理其响应

在 com.tdpress.restclient.service Java 程序包内的 HLRestClientServiceImpl.java 文档，HLRestClientServiceImpl 类提供其 searchSync 方法。通过自定义高级别 REST 客户端 HLRestClient，直接调用其 search 同步请求，发送搜索操作。返回响应结果为 SearchResponse 对象。通过调用其 getHits 方法可以获得类似 curl 命令行的 hits 数据结构。自定义的返回

方式是先从响应结果中提取数据,组成 RestClientResponse 对象,再转换为映射(Map<String, Object>)返回。其中,hits.hits 数组中的项的 sourceAsMap 数据结构与文档元数据字段 _source 相同。

```java
// 此方法直接调用高级别 REST 客户端的同步搜索请求方法
    public Map<String, Object> searchSync(SearchRequest request, RequestOptions options) {
        SearchResponse response;
        Map<String, Object> convertValue;
        try {
            // 直接调用 RestHighLevelClient 的同步搜索请求
            response = hlRestClient.search(request, options);
            convertValue = new HashMap<String, Object>();
            if (response.getTook() != null)
              convertValue.put("took", response.getTook().seconds());
            convertValue.put("timed_out", response.isTimedOut());
            if (response.getHits() != null)
              convertValue.put("hits", response.getHits());
        } catch (IOException e) {
            e.printStackTrace();
            RestClientResponse clientResponse = new RestClientResponse();
            clientResponse.setStatusCode(500);
            convertValue =
            (Map<String, Object>) (new ObjectMapper()).convertValue(clientResponse,
                RestClientResponse.class);
        }
        return convertValue;
    }
```

12.4.7 自定义 searchASync 与处理其响应

HLRestClientServiceImpl 类提供自定义 searchAsync 方法。通过自定义高级别 REST 客户端 HLRestClient,直接调用其 searchAsync 异步请求,发送搜索操作。并在其参数 ActionListener 使用匿名内部类(Anonymous Inner class),并设置两个回调方法 onSuccess 和 onFailure。当请求成功产生响应时回调方法 onSuccess。当请求失败时回调方法 onFailure。返回响应结果为 SearchResponse 对象。自定义的返回方式是通过调用其 getHits 方法可以获得类似使用 curl 命令行的 hits 数据结构。其中,hits.hits 数组中的项的 sourceAsMap 数据结构与文档元数据字段 _source 相同。示例简单地从屏幕输出结果。

```java
// 此方法直接调用高级别 REST 客户端的异步搜索请求方法
    public Map<String, Object> searchAsync(SearchRequest request, RequestOptions options) {
        // 直接调用 RestHighLevelClient 的异步搜索请求
        hlRestClient.searchAsync(request, options, new ActionListener<SearchResponse>() {
            RestClientResponse clientResponse = new RestClientResponse();
            @Override
            // 请求成功产生响应时的回调方法
            public void onResponse(SearchResponse response) {
                Map<String, Object> convertValue = new HashMap<String, Object>();
                if (response.getTook() != null)
                    convertValue.put("took", response.getTook().seconds());
                convertValue.put("timed_out", response.isTimedOut());
                if (response.getHits() != null)
                    convertValue.put("hits", response.getHits());
                if (response.getAggregations() != null)
                    convertValue.put("aggs", response.getAggregations());
                String jsonStr;
                try {
                    // 从屏幕输出验证结果
```

```java
            jsonStr = (new ObjectMapper()).writeValueAsString(convertValue);
            System.out.println(jsonStr);
        } catch (JsonProcessingException e) {
            e.printStackTrace();
        }
    }

    @Override
    //请求失败时的回调方法
    public void onFailure(Exception exception) {
        clientResponse.setStatusCode(500);
        if (!exception.getMessage().isEmpty()) {
            clientResponse.setErrMessage(exception.getMessage());
        }
        try {
            String jsonStr = (new ObjectMapper()).writeValueAsString(clientResponse);
            //从屏幕输出失败原因
            System.out.println(jsonStr);
        }
        catch (IOException e) {
            e.printStackTrace();
        }
    }
});
RestClientResponse clientResponse = new RestClientResponse();
clientResponse.setStatusCode(200);
@SuppressWarnings("unchecked")
Map<String, Object> convertValue =
    (Map<String, Object>) (new ObjectMapper()).convertValue(clientResponse, Map.class);
return convertValue;
}
```

12.4.8 自定义 getIndexSettingsSync 方法与处理其响应

HLRestClientServiceImpl 类提供 getIndexSettingsSync 方法。过程是先通过调用 HLRestClient 对象的 indices 方法，取得 IndicesClient 对象后，再间接调其 getSettings 同步请求，执行检索索引设置操作。返回响应结果为 GetSettingsResponse 对象，通过调用其 getIndexToSettings() 方法，可取得类似于 Map 的数据结构 ImmutableOpenMap<String, Settings>。给定索引名称 index 及使用其提供的 get 方法，可取得索引 index 的设置。以下是封装 getIndexSettings 请求并简单地处理响应结果的代码。

```java
//此方法间接调用高级别 REST 客户端的同步检索索引设置请求方法
public Map<String, Object> getIndexSettingsSync(GetSettingsRequest request,
RequestOptions options) {
    GetSettingsResponse response;
    Map<String, Object> convertValue = new HashMap<String, Object>();
    try {
        //间接调用 RestHighLevelClient 的同步检索索引设置请求
        response = hlRestClient.indices().getSettings(request, options);
        String index = request.indices()[0];
        org.elasticsearch.common.settings.Settings settings =
            response.getIndexToSettings().get(index);
        Set<String>keys = settings.keySet();
        for (String key : keys) {
            convertValue.put(key, response.getSetting(index, key));
        }
    } catch (IOException e) {
        e.printStackTrace();
        RestClientResponse clientResponse = new RestClientResponse();
```

```
        clientResponse.setStatusCode(500);
        convertValue = (Map<String, Object>) (
          new ObjectMapper()).convertValue(clientResponse, RestClientResponse.class);
    }
    return convertValue;
}
```

12.4.9 自定义 getIndexSettingsAsync 方法与处理其响应

HLRestClientServiceImpl 类提供 getIndexSettingsAsync 方法。过程是先通过调用 HLRestClient 对象的 indices 方法，取得 IndicesClient 对象后，再间接调用其 getSettingsAsync 异步请求，执行检索索引设置操作。类似于 12.4.7 节自定义 searchASync 与处理其响应方式处理。当请求成功产生响应时回调方法 onSuccess，返回响应结果为 GetSettingsResponse 对象。当请求失败时回调方法 onFailure。示例简单地从屏幕输出结果。

```
//此方法调用高级别 REST 客户端的异步检索索引设置请求方法
public Map<String, Object> getIndexSettingsAsync(GetSettingsRequest request,
    RequestOptions options) {
    //间接调用 RestHighLevelClient 的异步检索索引设置请求
    hlRestClient.indices().getSettingsAsync(request, options,
      new ActionListener<GetSettingsResponse>() {
        RestClientResponse clientResponse = new RestClientResponse();
        @Override
        //请求成功产生响应时的回调方法
        public void onResponse(GetSettingsResponse response) {
            Map<String, Object> convertValue = new HashMap<String, Object>();
            String index = request.indices()[0];
            org.elasticsearch.common.settings.Settings settings =
                response.getIndexToSettings().get(index);
            ……
        }
        //请求失败时的回调方法
        public void onFailure(Exception exception) {
            clientResponse.setStatusCode(500);
            if (!exception.getMessage().isEmpty()) {
                clientResponse.setErrMessage(exception.getMessage());
            }
            ……
        }
    }
}
```

12.5 使用 Swagger UI 测试高级别 REST 客户端

高级别 REST 客户端代码与低级别 REST 客户端代码属于同一个开源项目，同在 java_restful_client 目录下。如果 Swagger UI 测试的用户界面还在，可以直接使用它。如何使用 Maven 指令运行清理、编译、构建和执行程序项目，与 12.3 节使用 Swagger UI 测试低级别 REST 客户端相同。

1. Java 高级别 REST 客户端项目 Swagger UI 页面

以下使用网页浏览器进行互动测试，访问项目的 Swagger UI 页面，URL 为 http://localhost:10010/swagger-ui.html。然后按 hl-rest-client-controller 请求栏以展开面板，屏幕截图如图 12-9 所示。

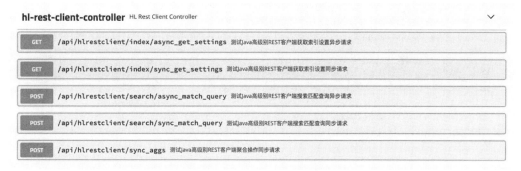

图 12-9　Swagger UI 页面中，高级别 REST 客户端请求一栏

2.Java 高级别 REST 客户端获取索引设置同步请求测试

点击"测试 Java 高级别 REST 客户端获取索引设置同步请求"一栏以展开面板，屏幕截图如图 12-10 所示。

图 12-10　Swagger UI 页面中高级别 REST 客户端获取索引设置同步请求一栏

点击"Try it out"按钮时，面板将更改为输入模式。键入索引名称 IndexName 为 fund_basic，屏幕截图如图 12-11 所示。

图 12-11　键入索引名称 fund_basic

点击"Execute"按钮，向下滑动屏幕后可以看到响应结果。获取的状态码 (statusCode) 为 200，返回的响应结果 (responseBody) 列出 fund_basic 的所有索引设置。屏幕截图如图 12-12 所示。

图 12-12 测试返回的结果，列出 fund_basic 的所有索引设置

3. Java 高级别 REST 客户端搜索匹配查询同步请求测试

类似于前面的测试，测试搜索匹配查询同步请求首先点击该测试面板，然后再点击"Try it out"按钮。为方便起见，使用与 12.3 节低级别 REST 客户端异步 POST 请求相同的测试用例。键入匹配字段 (fieldName) 为 name、匹配字符串 (fieldValue) 为"中银 ETF"、索引名称 (indexName) 为 fund_basic 和布尔运算符 (operator) 为 AND，屏幕截图如图 12-13 所示。

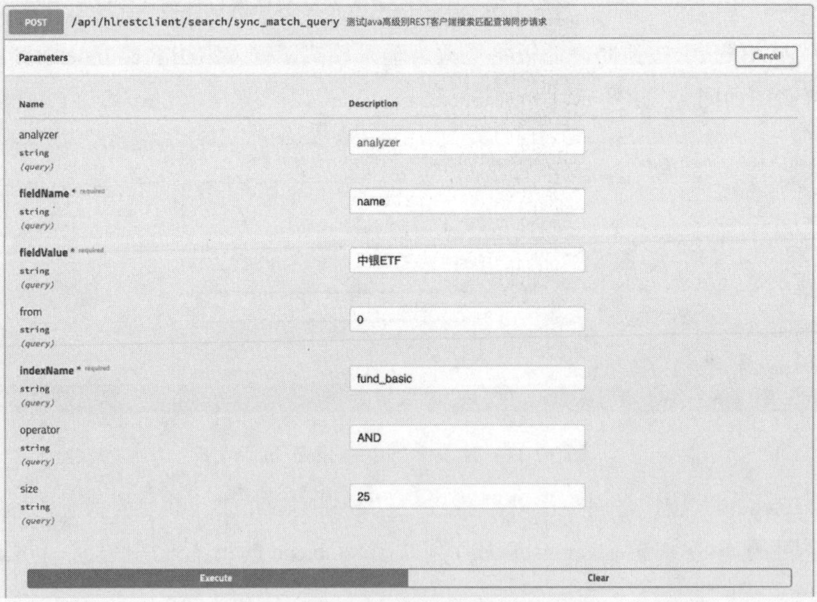

图 12-13 键入测试相对应的栏位内容

点击"Execute"按钮，向下滑动屏幕后可以看到返回的响应结果。匹配字符串"中银 ETF"的基金简称，包括中银证券 500ETF 和中银中证 100ETF。屏幕截图如图 12-14 所示。

图 12-14　测试返回的结果，包括中银证券 500ETF 和中银中证 100ETF

提示：感兴趣的读者可以进行 Java 高级别 REST 客户端异步请求测试。输入方式与同步请求测试相同，输出结果打印在运行 Elasticsearch 服务器的屏幕上。

12.6　Java 高级别 REST 客户端聚合操作简介

在 12.4.4 节构造查询请求曾介绍 SearchSourceBuilder 可存储 QueryBuilder 对象的查询语句。而且还提供了一个名为 aggregation 的方法，可存储聚合语句。aggregation 方法可以接受 AggregationBuilder 类或是 PipelineAggregationBuilder 类的对象作为参数。所有支持的聚合操作都继承这两个类。在 org.elasticsearch.search.aggregations 包内提供了辅助构建继承该两个类型的对象的类，这两个类的名称为 AggregationBuilders 和 PipelineAggregatorBuilders。分别提供构建继承 AggregationBuilder 类或 PipelineAggregationBuilder 类的对象的静态方法。如果需要使用子聚合，可以使用相应的 AggregationBuilder 类或 PipelineAggregationBuilder 类提供的 subAggregations 方法，把作为参数的子聚合对象链接在一起。表 12-10 汇集静态构建各种聚合的对象的方法。并且摘录开源项目 AggregationBuilder 类与子聚合相关的部分代码如下：

```
public abstract class AggregationBuilder
    implements NamedWriteable, ToXContentFragment, BaseAggregationBuilder {
    ……
    /** Add a sub aggregation to this builder. */
    public abstract AggregationBuilder subAggregation(AggregationBuilder aggregation);
    /** Add a sub aggregation to this builder. */
    public abstract AggregationBuilder subAggregation(PipelineAggregationBuilder aggregation);
    ……
}
```

表 12-10 构建各类型聚合的辅助方法列表

聚合家族	聚合类型	AggregationBuilders 提供的静态方法名称和返回数据类型
度量指标聚合 (metric)	平均值（avg）	AvgAggregationBuilder avg
	加权平均值（weighted_avg）	WeightedAvgAggregationBuilder weightedAvg
	基数（cardinality）	CardinalityAggregationBuilder cardinality
	扩展统计（extended_stats）	ExtendedStatsAggregationBuilder extendedStats
	最大值（max）	MaxAggregationBuilder max
	最小值（min）	MinAggregationBuilder min
	中位数绝对偏差（median_absolute_deviation）	MedianAbsoluteDeviationAggregationBuilder medianAbsoluteDeviation
	百分位（percentiles）	PercentilesAggregationBuilder percentiles
	百分位等级（percentile_ranks）	PercentileRanksAggregationBuilder percentileRanks
	脚本式度量指标（scripted_metric）	ScriptedMetricAggregationBuilder scriptedMetric
	统计（stats）	StatsAggregationBuilder stats
	总和（sum）	SumAggregationBuilder sum
	最热点（top_hits）	TopHitsAggregationBuilder topHits
	值计数（value_count）	ValueCountAggregationBuilder valueCount
	地理边界（geo_bounds）	GeoBoundsAggregationBuilder geoBounds
存储桶聚合 (bucket)	邻接矩阵（adjacency_matrix）	AdjacencyMatrixAggregationBuilder adjacencyMatrix
	日期直方图（date_histogram）	DateHistogramAggregationBuilder dateHistogram
	复合（composite）	CompositeAggregationBuilder composite
	日期范围（date_range）	DateRangeAggregationBuilder dateRange
	多元化采样器（diversified_sampler）	DiversifiedAggregationBuilder diversifiedSampler
	过滤器（filter）	FilterAggregationBuilder filter
	多过滤器（filters）	FiltersAggregationBuilder filters
	全局（global）	GlobalAggregationBuilder global
	直方图（histogram）	HistogramAggregationBuilder histogram
	IP 范围（ip_range）	IpRangeAggregationBuilder ipRange
	缺失字段（missing）	MissingAggregationBuilder missing
	嵌套（nested）	NestedAggregationBuilder nested
	范围（range）	RangeAggregationBuilder range
	反向嵌套（reserve_nested）	ReverseNestedAggregationBuilder reverseNested
	采样器（sampler）	SamplerAggregationBuilder sampler
	显著词条（significant_terms）	SignificantTermsAggregationBuilder significantTerms
	显著文本（significant_text）	SignificantTextAggregationBuilder significantText
	词条（terms）	TermsAggregationBuilder terms
	地理距离（geo_distance）	GeoDistanceAggregationBuilder geoDistance
	地理哈希网格（geohash_grid）	GeoHashGridAggregationBuilder geohashGrid
	地理瓦片网格（geotile_grid）	GeoTileGridAggregationBuilder geotileGrid
	地理重心（geo_centroid）	GeoCentroidAggregationBuilder geoCentroid

续表

聚合家族	聚合类型	AggregationBuilders 提供的静态方法名称和返回数据类型
管道聚合 (pipeline)	桶平均值 (avg_bucket)	AvgBucketPipelineAggregationBuilder avgBucket
	导数 (derivative)	DerivativePipelineAggregationBuilder derivative
	桶最大值 (max_bucket)	MaxBucketPipelineAggregationBuilder maxBucket
	桶最小值 (min_bucket)	MinBucketPipelineAggregationBuilder minBucket
	桶总和 (sum_bucket)	SumBucketPipelineAggregationBuilder sumBucket
	桶统计 (stats_bucket)	StatsBucketPipelineAggregationBuilder statsBucket
	桶扩展统计 (extended_stats_bucket)	ExtendedStatsBucketPipelineAggregationBuilder extendedStatsBucket
	桶百分位 (percentiles_bucket)	PercentilesBucketPipelineAggregationBuilder percentilesBucket
	移动函数 (moving_fn)	MovFnPipelineAggregationBuilder movingFunction
	累计总和 (cumulative_sum)	CumulativeSumPipelineAggregationBuilder cumulativeSum
	桶脚本 (bucket_script)	BucketScriptPipelineAggregationBuilder bucketScript
	桶选择器 (bucket_selector)	BucketSelectorPipelineAggregationBuilder bucketSelector
	桶排序 (bucket_sort)	BucketSortPipelineAggregationBuilder bucketSort
	串行差分 (serial_diff)	SerialDiffPipelineAggregationBuilder diff

以下例子使用与 9.6.14 节移动函数聚合的测试用例，用例使用了指数加权移动平均模型的单指数 (ewma)、二次指数 (holt) 和三次指数 (holtWinters) 聚合操作。这三个聚合都使用 MovFnPipelineAggregationBuildera 的 movingFunction 方法构建。其他所需的聚合操作，根据其类型按照表 12-10，使用相对应的聚合家族构建方法构建对象，然后使用 subAggregations 方法，把作为参数的子聚合对象链接在一起。在聚合过程中，需要使用日期直方图（date_histogram）聚合和最大值聚合找出每日的复权单位净值，供其他聚合操作之用。

```
@ApiOperation(" 测试 Java 高级别 REST 客户端聚合操作同步请求 ")
......
@RequestMapping(value="/sync_aggs", method=RequestMethod.POST)
  public ResponseEntity<Map<String, Object>> aggregationRequest(
    @RequestParam(value = "indexName", required=true) String indexName,
    @RequestParam(value = "ts_code", required=true) String ts_code,
    @RequestParam(value = "fieldName", required=true) String fieldName,
    @RequestParam(value = "alpha", required=true) double alpha,
    @RequestParam(value = "beta", required=true) double beta,
    @RequestParam(value = "gamma", required=true) double gamma,
    @RequestParam(value = "window", required=true) int window,
    @RequestParam(value = "period", required=true) int period) throws Exception {
// 构造 SearchRequest 查询请求
SearchRequest request = new SearchRequest(indexName);
SearchSourceBuilder sourceBuilder = new SearchSourceBuilder();
sourceBuilder.from(0);
sourceBuilder.size(0);
// 以基金代码字段进行词条聚合
TermQueryBuilder queryBuilder =
  QueryBuilders.termQuery("ts_code.keyword", ts_code);
// 以最大值聚合 daily_adj_nav 的结果进行单指数移动平均
MovFnPipelineAggregationBuilder ewma =
  PipelineAggregatorBuilders.movingFunction("ewma",
    new Script("MovingFunctions.ewma(values, " + alpha + ")"),
      "daily_adj_nav", 5);
ewma.setShift(1);
ewma.setWindow(5);
```

```
// 以最大值聚合daily_adj_nav的结果进行二次指数移动平均
MovFnPipelineAggregationBuilder holt =
    PipelineAggregatorBuilders.movingFunction("holt",
        new Script("MovingFunctions.holt(values, " + alpha + "," + beta + ")"),
        "daily_adj_nav", 5);
holt.setShift(1);
holt.setWindow(5);
// 以最大值聚合daily_adj_nav的结果进行三次指数移动平均
MovFnPipelineAggregationBuilder holtWinters =
    PipelineAggregatorBuilders.movingFunction("holtWinters",
        new Script("if (values.length>=5) MovingFunctions.holtWinters(values, " +
            alpha + "," + beta + "," + gamma + "," + period + ", false)"),
        "daily_adj_nav", 5);
        holtWinters.setShift(1);
        holtWinters.setWindow(5);
// 以复权单位净值截止日期（end_date）字段进行日期直方图聚合
AggregationBuilder aggs = AggregationBuilders.dateHistogram("daily_report")
    .calendarInterval(DateHistogramInterval.days(1)).field("end_date").minDocCount(1L)
    .subAggregation(AggregationBuilders.max("daily_adj_nav").field(fieldName))
    .subAggregation(ewma)
    .subAggregation(holt)
    .subAggregation(holtWinters);
sourceBuilder.aggregation(aggs);
request.source(sourceBuilder.query(queryBuilder));
RequestOptions options = RequestOptions.DEFAULT;
// 直接调用RestHighLevelClient的同步搜索请求
Map<String,Object> response = hlRestClient.searchSync(request, options);
return new ResponseEntity<Map<String, Object>>(response, HttpStatus.OK);
}
```

12.7 使用 Swagger UI 测试 Java 高级别 REST 客户端聚合操作

测试聚合操作的代码也在同一个项目，如果 Swagger UI 测试的用户界面还在，可以直接使用它。

1. Java 高级别 REST 客户端聚合操作

类似于前面的测试，测试指数加权移动平均模型聚合操作请求首先点击 "sync_aggs" 一栏以展开面板，屏幕截图如图 12-15 所示。

图 12-15　指数加权移动平均模型聚合操作测试相对应的栏位内容

点击"Try it out"按钮时，面板将更改为输入模式。键入每个变量值，包括基金代码、字段名称、移动函数的参数(alpha、beta、gamma、period 和 window)和索引名称，屏幕截图如图 12-16 所示。

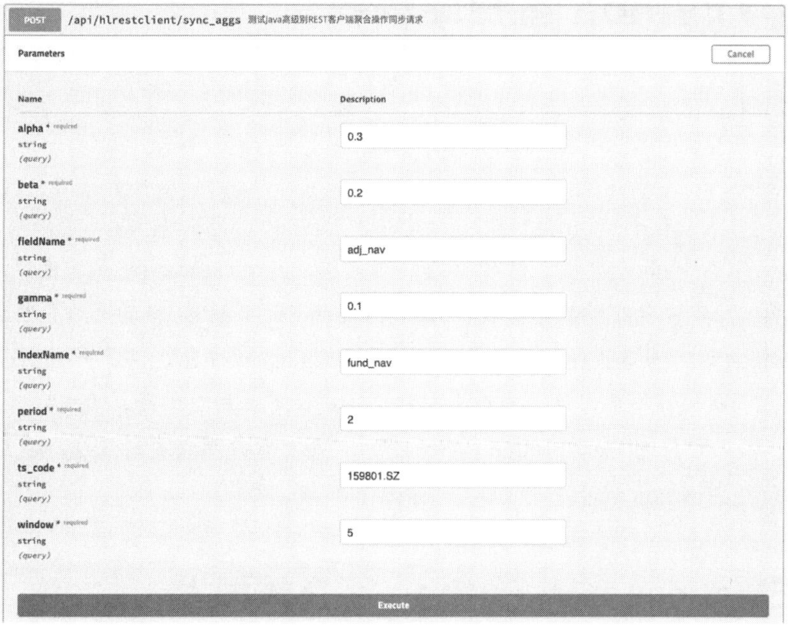

图 12-16　指数加权移动平均模型聚合操作测试输入模式

点击"Execute"按钮，向下滑动屏幕后可以看到指数加权移动平均模型的单指数、二次指数和三次指数聚合返回的响应结果。屏幕如图 12-17 所示。

图 12-17　测试指数加权移动平均模型的单指数、二次指数和三次指数聚合返回的结果

第 13 章

Python 客户端编程

Elasticsearch 为 Python 提供了两种类型的客户端。elasticsearch-py 软件包提供了官方的低级别客户端,此软件包非常轻量化包装 Elasticsearch REST API,目的是为 Python 中所有与 Elasticsearch 相关的代码提供共同基础。而 elasticsearch-dsl 软件包提供了一个在 elasticsearch-py 软件包之上构建的程序库,作为高级别客户端。提供了一种更方便,更惯用的方式来编写和操作查询,因为它靠近 Elasticsearch JSON DSL。两个客户端都支持从 2.x 到 7.x 的版本。本章将学习如何在 python 程序应用 Elasticsearch 客户端。

13.1 Elasticsearch Python 客户端概览

假设您具有适用于 Python 3.6 的工作环境,而 Python 的软件安装程序包 pip 也是 3.6 版本,请使用以下步骤执行指令:

(1)使用 pip 命令安装 Python elasticsearch-py 软件包 7.5.1 版本,代码如下所示:

```
pip install elasticsearch==7.5.1
Collecting elasticsearch==7.5.1
  Downloading elasticsearch-7.5.1-py2.py3-none-any.whl (86 kB)
     || 86 kB 2.1 MB/s
Requirement already satisfied: urllib3>=1.21.1 in /Library/Frameworks/Python.
framework/Versions/3.6/lib/python3.6/site-packages (from elasticsearch==7.5.1)
(1.24.1)
Installing collected packages: elasticsearch
  Attempting uninstall: elasticsearch
    Found existing installation: elasticsearch 7.0.0
    Uninstalling elasticsearch-7.0.0:
      Successfully uninstalled elasticsearch-7.0.0
Successfully installed elasticsearch-7.5.1
```

(2)由于当前(2020 年 8 月)elasticsearch-dsl 软件包支持的版本为 7.2.1,只能使用 pip 命令安装 7.2.1 版本,代码如下所示:

```
pip install elasticsearch-dsl==7.2.1
Collecting elasticsearch-dsl==7.2.1
  Downloading elasticsearch_dsl-7.2.1-py2.py3-none-any.whl (53 kB)
     || 53 kB 958 kB/s
Requirement already satisfied: elasticsearch<8.0.0,>=7.0.0 in /Library/
Frameworks/Python.framework/Versions/3.6/lib/python3.6/site-packages (from
elasticsearch-dsl==7.2.1) (7.5.1)
Requirement already satisfied: python-dateutil in /Library/Frameworks/Python.
framework/Versions/3.6/lib/python3.6/site-packages (from elasticsearch-dsl==7.2.1) (2.7.5)
```

```
Requirement already satisfied: six in /Library/Frameworks/Python.framework/
Versions/3.6/lib/python3.6/site-packages (from elasticsearch-dsl==7.2.1) (1.11.0)
  Requirement already satisfied: urllib3>=1.21.1 in /Library/
Frameworks/Python.framework/Versions/3.6/lib/python3.6/site-packages (from
elasticsearch<8.0.0,>=7.0.0->elasticsearch-dsl==7.2.1) (1.24.1)
Installing collected packages: elasticsearch-dsl
  Attempting uninstall: elasticsearch-dsl
    Found existing installation: elasticsearch-dsl 7.0.0
    Uninstalling elasticsearch-dsl-7.0.0:
      Successfully uninstalled elasticsearch-dsl-7.0.0
Successfully installed elasticsearch-dsl-7.2.1
```

（3）如果在项目上工作，建议的方法是在 setup.py 文件或 requirements.txt 中设置版本范围，假设项目正在使用 7.5.1，代码如下所示：

```
#Elasticsearch 7.5.1
elasticsearch>= 7.5.1, <8.0.0
elasticsearch-dsl>= 7.5.1, <8.0.0
```

（4）假设 Elasticsearch 服务器正在运行，可以使用 Python 脚本来测试连接，并使用 es.info()命令检索 Elasticsearch 实例的信息。代码如下所示：

```
python3.6
Python 3.6.7 (v3.6.7:6ec5cf24b7, Oct 20 2018, 03:02:14)
[GCC 4.2.1 Compatible Apple LLVM 6.0 (clang-600.0.57)] on darwin
Type "help", "copyright", "credits" or "license" for more information.
>>> from elasticsearch import Elasticsearch
>>> es = Elasticsearch()
>>> es.info()
{'name': 'WTW.attlocal.net', 'cluster_name': 'elasticsearch', 'cluster_uuid':
'c07VU4HPQomgXvky01ndbg', 'version': {'number': '7.5.1', 'build_flavor': 'default',
'build_type': 'tar', 'build_hash': '3ae9ac9a93c95bd0cdc054951cf95d88e1e18d96',
'build_date': '2019-12-16T22:57:37.835892Z', 'build_snapshot': False, 'lucene_
version': '8.3.0', 'minimum_wire_compatibility_version': '6.8.0', 'minimum_index_
compatibility_version': '6.0.0-beta1'}, 'tagline': 'You Know, for Search'}
```

13.2　elasticsearch-py 软件包

elasticsearch-py 软件包非常轻量地包装 Elasticsearch RESTful 接口以获得最大的灵活性。每个 Elasticsearch 对象都配备有单独的连接池，负责维护与 Elasticsearch 节点的连接，以支持持久连接。默认情况下每个节点允许打开 10 个连接。Elasticsearch 构造函数提供参数 maxsize 用于设置最大的连接数。Elasticsearch 对象是线程安全的，可以在多线程环境中使用。最佳实践是创建一个全局使用的单例。但是，对于并行处理，应在调用 fork 之后重新创建 Elasticsearch 对象。

elasticsearch-py 软件包提供的方法调用方式，非常类似于 Java 高级别 REST 客户端，有些方法可以直接调用，而有些方法则需要间接调用。

13.2.1　提供间接调用方法的成员

参考官方 elasticsearch-py 开源软件网站中的 elasticsearch/client 包内的 __init__() 方法，定义了 Elasticsearch 类，里面可以看到提供间接调用方法的成员，并且在表 13-1。

表 13-1　Elasticsearch 类提供间接调用方法的成员列表

提供间接调用方法的成员	成员的数据类型	简述
cat	CatClient	用来访问目录接口
ccr	CcrClient	用来访问跨集群复制接口
cluster	ClusterClient	用来访问集群接口
enrich	EnrichClient	用来访问 enrich 接口，丰富文档的字段内容
graph	GraphClient	用来访问 _graph xplore 接口，探索数据中存在的相关联系
ilm	IlmClient	用来访问 _ilm (索引生命周期管理) 接口，管理索引的 Hot、Warm、Cold 和 Delete 4 个阶段
indices	IndicesClient	用来访问索引接口
ingest	IngestClient	用来访问摄取节点接口
license	LicenseClient	用来访问软件许可接口
migration	MigrationClient	用来访问版本之间的迁移接口
ml	MlClient	用来访问机器学习接口
monitoring	MonitoringClient	用来访问监控接口
nodes	NodesClient	用来访问节点接口
remote	RemoteClient	用来访问远程集群信息接口
rollup	RollupClient	用来访问数据上卷接口
security	SecurityClient	用来访问安全配置接口
slm	SlmClient	用来访问快照备份与恢复接口
snapshot	SnapshotClient	用来访问快照生命周期管理接口
sql	SqlClient	用来访问 Elasticsearch SQL 接口
ssl	SslClient	用来访问 SSL 安全接口
tasks	TasksClient	用来访问任务接口
transform	TransformClient	用来访问 TransformClient 接口，管理通过转换源索引中数据到目标索引的操作
watcher	WatcherClient	用来访问告警接口，提供警报和通知

13.2.2　间接调用方法

由于提供间接调用方法的成员太多而没有办法一一展示，在这小节只详细描述 IndicesClient 类作一个例子。IndicesClient 提供间接调用方法，但只可以同步发送请求。通过 IndicesClient 间接调用方法见表 13-2。

提示：Elasticsearch 7.8.0 版本的 elasticsearch-py 软件包，开始支持异步请求，需要在本地安装 Python 3.6 或更高版本。7.5.1 版本不支持异步请求，感兴趣的读者请参考官网。

表 13-2　通过 IndicesClient 提供间接调用方法列表

方法名称和签名	简述
analyze(body=None, index=None, params=None)	对文本执行分析
create(index, body=None, params=None, headers=None):	创建索引
get(index, params=None, headers=None)	检索索引的信息
delete(index, params=None, headers=None)	删除索引
clone(index, target, body=None, params=None, headers=None)	克隆索引
exists(index, params=None, headers=None)	检查索引是否存在
open(index, params=None, headers=None)	打开索引

续表

方法名称和签名	简述
close(index, params=None, headers=None)	关闭索引
put_settings(body, index=None, params=None, headers=None)	更新索引的设置
get_settings(index=None, name=None, params=None, headers=None)	检索索引的设置
put_mapping(index, body, params=None, headers=None)	更新索引的映射
get_mapping(index=None, params=None, headers=None)	检索索引的映射
put_template(name, body, params=None, headers=None)	更新索引的模板
put_index_template(name, body, params=None, headers=None)	更新索引的模板
get_template(name=None, params=None, headers=None)	检索索引的模板
delete_template(name, params=None, headers=None)	删除索引的模板
exists_template(name, params=None, headers=None)	检查索引的模板是否存在
put_alias(index, name, body=None, params=None, headers=None)	创建或更新索引的别名
get_alias(index=None, name=None, params=None, headers=None)	检索索引的别名
exists_alias(name, index=None, params=None, headers=None)	检查索引的别名是否存在
update_aliases(body, params=None, headers=None)	更新索引的别名
delete_alias(index, name, params=None, headers=None)	删除索引的别名
get_field_mapping(fields, index=None, params=None, headers=None)	检索索引的字段映射
reload_search_analyzers(index, params=None, headers=None))	重新加载搜索分析器以同步同义词文件的更新
flush(index=None, params=None, headers=None)	刷新到硬盘
flush_synced(index=None, params=None)	同步刷新到硬盘
forcemerge(index=None, params=None, headers=None)	强制合并索引
freeze(index, params=None, headers=None)	冻结索引
unfreeze(index, params=None, headers=None)	解冻索引
refresh(index=None, params=None, headers=None)	刷新索引相关的操作以用于搜索
rollover(alias, body=None, new_index=None, params=None, headers=None)	滚动索引
shrink(index, target, body=None, params=None)	缩小现有索引以使用较少主分片
split(index, target, body=None, params=None, headers=None)	拆分索引
validate_query(body=None, index=None, doc_type=None, params=None, headers=None)	验证查询语句是否正确
clear_cache(index=None, params=None, headers=None)	清理索引缓存
recovery(index=None, params=None)	返回进行中的索引分片恢复的信息
segments(index=None, params=None)	提供有关索引的段的信息
shard_stores(index=None, params=None)	提供索引分片副本的信息
stats(index=None, metric=None, params=None)	提供有关索引操作的统计信息

13.2.3 直接调用方法

Elasticsearch 类的定义里，提供直接调用方法，可以针对接口类型，粗略地细分为索引、文档搜索查询、脚本、批量请求、重置节流和其他接口。分别见表 13-3~表 13-8。

表 13-3　Elasticsearch 类提供有关索引文档的方法列表

方法名称和签名	简　述
index(index, body, doc_type=None, id=None, params=None)	索引文档
get(index, id, doc_type=None, params=None)	检索索引文档
mget(body, index=None, doc_type=None, params=None)	检索多个索引文档
update(index, id, body, doc_type=None, params=None)	更新索引文档内容
delete(index, id, doc_type=None, params=None)	删除索引文档
exists(index, id, doc_type=None, params=None)	检查索引文档是否存在
exists_source(index, id, doc_type=None, params=None)	检查索引文档的 _source 字段是否存在
get_source(index, id, doc_type=None, params=None)	检索索引文档的 _source 字段
mtermvectors(body=None, index=None, doc_type=None, params=None)	检索多个文档的词条向量信息
termvectors(index, body=None, doc_type=None, id=None, params=None)	返回特定文档字段中词条的信息和统计信息

表 13-4　Elasticsearch 类提供有关搜索查询的方法列表

方法名称和签名	简　述
search(body=None, index=None, doc_type=None, params=None)	查询文档
search_shards(index=None, params=None)	返回有关执行搜索请求的索引和分片的信息
scroll(body=None, scroll_id=None, params=None)	游标查询文档
clear_scroll(body=None, scroll_id=None, params=None)	关闭游标查询
search_template(body, index=None, doc_type=None, params=None)	按照查询模板执行请求
render_search_template(body=None, id=None, params=None)	预先呈现搜索定义
msearch(body, index=None, doc_type=None, params=None)	多重目标查询文档
msearch_template(body, index=None, doc_type=None, params=None)	按照多重目标查询模板执行请求
count(body=None, index=None, doc_type=None, params=None)	搜索查询结果的匹配数
field_caps(index=None, params=None)	检查字段功能
explain(index, id, body=None, doc_type=None, params=None)	解释为何匹配和得此分数
rank_eval(body, index=None, params=None)	评估搜索查询结果中排名的质素

表 13-5　Elasticsearch 类提供有关脚本的方法列表

方法名称和签名	简　述
put_script(id, body, context=None, params=None)	存储脚本
get_script(id, params=None)	检索脚本
delete_script(id, params=None)	删除脚本
scripts_painless_execute(body=None, params=None)	允许执行任意脚本

表 13-6　Elasticsearch 类提供有关批量请求的方法列表

方法名称和签名	简　述
bulk(body, index=None, doc_type=None, params=None)	执行批量请求
update_by_query(index, body=None, doc_type=None, params=None)	根据请求执行批量更新
delete_by_query(index, body, doc_type=None, params=None)	根据请求执行批量删除
reindex(body, params=None)	重新索引

表 13-7　Elasticsearch 类提供有关任务的方法列表

方法名称和签名	简　述
update_by_query_rethrottle(task_id, params=None)	重置正在运行的查询更新任务当前每秒请求数的限制
delete_by_query_rethrottle(task_id, params=None)	重置正在运行的查询删除任务当前每秒请求数的限制
reindex_rethrottle(task_id, params=None)	重置正在运行的重新索引任务当前每秒请求数的限制

表 13-8 Elasticsearch 类提供其他的方法接口列表

方法名称和签名	简　述
void close()	关闭高级别 REST 客户端
info(params=None)	获取集群信息

13.2.4 操作流程

操作流程基本上涉及三个主要步骤：

（1）Elasticsearch 对象初始化

（2）执行请求与处理其响应

（3）完成所有请求后关闭 Elasticsearch 对象

1.Elasticsearch 对象初始化

在 com.tdpress.client.config 程序包内 LLClient.py 文档的 LLClient 类，使用单例设计模式 (Singleton Design Pattern)封装 get_instance 方法，提供单例 Elasticsearch 对象连接本地主机的 9200 埠。

```python
# 此 Python 类封装了低级别 REST 客户端并进行配置
    __es = None
    __es_lock = threading.Lock()

    @staticmethod
    # 使用单例设计模式封装方法，提供单例 Elasticsearch 对象
    def get_instance():
        if LLClient.__es is None:
            with LLClient.__es_lock:
                if LLClient.__es is None:
                    LLClient.__es = Elasticsearch(['localhost'], port=9200)
        return LLClient.__es

    def __init__(self):
        raise Exception("This class is a singleton!, use static method getInstance()")
```

2.执行请求方法与处理其响应

经由 LLClient 类的静态方法 get_instance 取得 Elasticsearch 对象后，间接调用请求方法需要先取得相对应的成员，例如 indices，再调用 indices 对象的方法，例如执行检索索引设置操作的 get_settings 方法。而直接调用请求方法在取得 Elasticsearch 对象后，可以直接调用其方法，例如 search 方法。返回的响应结果通过 Flask-Jsonpify 包的 jsonify 方法，用 JSON 表示形式创建一个 Response 代码如下所示：

```python
# 调用低级别 REST 客户端
ll_es = LLClient.get_instance()

#Flask API 的 Swagger 设置样式
@name_space_ll.route("/indexSettings/<index>")
@name_space_ll.doc(params={'index': '索引名称'})
# 此 Python 类提供低级别客户端同步检索索引设置方法
class LLIndexSettings(Resource):
    def get(self, index):
        # 通过 indices 对象，间接调用同步检索索引设置请求
        response = ll_es.indices.get_settings(index=index)
        if response is not None:
            return jsonify(response)
        else:
```

```
            return {}
@name_space_ll.route("/search/<index>")
@name_space_ll.doc(params={'index': '索引名称'}, body=search_fields)
# 此 Python 类提供低级别客户端同步搜索方法
class LLSearch(Resource):
    def post(self, index):
        payload = request.get_json()
        # 直接调用同步搜索请求
        response = ll_es.search(index=index, body=payload)
        if response['hits']['total']['value'] > 0:
            return jsonify(response)
        else:
            return {}
```

参看官方源代码存储库，无论是 get_settings 或是 search 方法，这两种方法中都是调用 org.elasticsearch.transport.Transport 类的 perform_request 方法发送请求。方法名称和签名如下：

```
perform_request(self, method, url, headers=None, params=None, body=None)
```

3. 关闭 Elasticsearch 对象

版本 7.5.1 尚未支持关闭功能，只能在应用程序关闭时关闭。从版本 7.8.0 开始 Elasticsearch 类支持调用 close 的方法。

13.3 使用 Swagger UI 测试调用方法

为了方便用户，本章中的测试将使用 flask，flask_restplus 和 swagger UI。两种调用方法代码同在 es-python-client 目录下。两个 elasticsearch-py 软件包的调用方法，get_settings 及 search，被封装成 RESTful 接口供调用，然后集成 Swagger UI 方便使用。请求成功时返回的响应，使用 flask_jsonpify.jsonify 对象，将响应结果数据序列化为 JSON 格式，并将其包装在具有 application/json MIME 媒体类型的响应对象中。

1. 准备测试环境

在目录 es_python_client 下，执行以下指令：

（1）设置虚拟环境

```
source venv/bin/activate
```

（2）设置 PYTHONPATH 路径环境

```
export PYTHONPATH=./:$PYTHONPATH
```

（3）执行主程序

```
python3.6 com/tdpress/flask/main.py
 * Serving Flask app "main" (lazy loading)
 * Environment: production
   WARNING: This is a development server. Do not use it in a production deployment.
   Use a production WSGI server instead.
 * Debug mode: off
 * Running on http://127.0.0.1:10010/ (Press CTRL+C to quit)
```

2. 测试 elasticsearch-py 软件包的 Swagger UI 页面

使用网页浏览器进行互动测试，访问项目的 Swagger UI 页面，URL 为 http://

localhost:10010。然后按 /api/llclient 请求栏以展开面板。GET /api/llclient/indexSettings 为间接调用，而 POST /api/llclient/search 为直接调用。屏幕截图如图 13-1 所示。

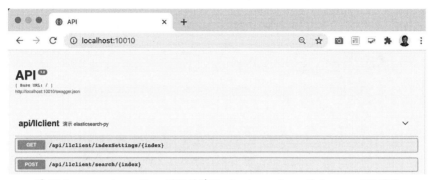

图 13-1　es_python_client 项目的演示 elasticsearch-py 的 Swagger UI 页面

3. 检索索引设置间接调用请求测试

点击"GET"一栏以展开面板，屏幕截图如图 13-2 所示。

图 13-2　展开 Swagger UI 页面中执行检索索引设置操作的 get_settings 方法一栏

点击"Try it out"按钮时，面板将更改为输入模式。为方便起见，使用检索索引设置操作相同的测试用例，输入 index 为 fund_basic，屏幕截图如图 13-3 所示。

图 13-3　输入检索索引设置请求主体的查询语句和 URL

点击"Execute"按钮,向下滑动屏幕后可以看到检索索引设置操作的响应结果。获取的状态码(statusCode)为200,返回的响应结果如图13-4所示。

图13-4 间接调用检索索引设置请求的响应结果

4. 文档搜索查询直接调用请求测试

直接调用测试类似于间接调用测试,点击"POST"一栏以展开面板。点击"Try it out"按钮并输入请求主体的查询语句和URL,输入索引名称index为fund_basic,查询语句为7.3.1节认识查询领域特定语言的测试用例,查询语句如下所示,屏幕如图13-5所示。

```
{"query": {"match" : {"name": {"query" : "中银ETF", "operator":"AND"}}}, "_source":["name","ts_code"]}
```

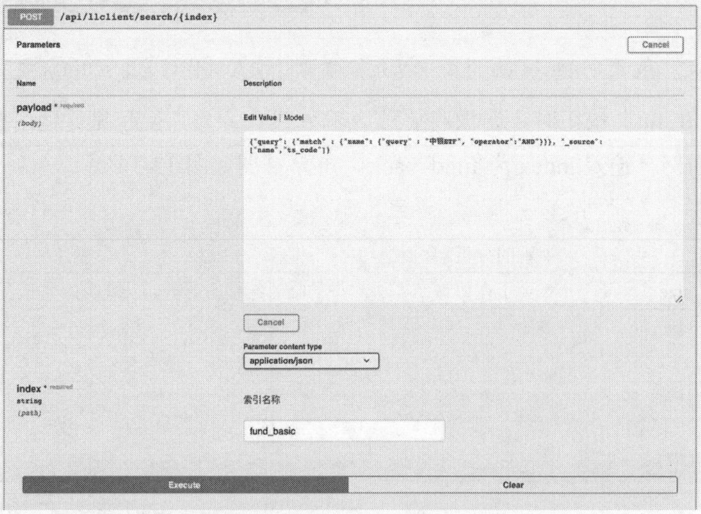

图13-5 输入请求主体的查询语句和索引名称index

点击"Execute"按钮,向下滑动屏幕后可以看到请求发送结果。获取的状态码(statusCode)为200,返回的search请求响应结果为匹配查询字符串中银ETF的基金简称,包括中银证券500ETF和中银中证100ETF。如图13-6所示。

图 13-6 直接调用 search 请求的响应结果

直接调用方法 search 除了在表 13-4 描述的参数外，还有其他约 40 个参数，见表 13-9。

表 13-9 search 方法的参数列表

参数名称	简 述
body	请求主体，DSL 查询语句
index	操作的索引名称，用逗号分隔的列表。若目标是所有索引，可用 _all 或空字符串
doc_type	映射类型，目前已弃用，将在 8.0.0 中完全删除
_source	输出文档的原始内容，字段名称可使用通配符
_source_excludes	过滤出文档的特定字段后再输出原始内容
_source_includes	输出文档的部分原始内容
allow_no_indices	是否当请求的索引为不可用（例如关闭、不存在）时，请求不会返回错误
allow_partial_search_results	是否在失败的情况下返回部分结果，默认为 true
analyze_wildcard	是否分析查询字符串中的通配符，默认为 false
analyzer	使用的分析器名称
batched_reduce_size	减少了在协调节点中收集的临时结果的数量，以减少内存使用量
ccs_minimize_roundtrips	跨集群搜索请求执行过程中，网络往返行程是否应最小化，默认为 true
default_operator	指定布尔运算符（默认为 or 或 |）对条件进行分组
df	指定要搜索的默认字段
docvalue_fields	在搜索结果中，是否返回指定的 doc_values 字段，以列表方式显示
expand_wildcards	通配符表达式可以扩展到什么状态的索引，选项为 all、open、closed 和 none
explain	是否在发生错误时响应将返回详细信息。默认为 false
from_	返回的搜索结果从第几项开始（默认为 0）
ignore_throttled	是否忽略线程的限速配置
ignore_unavailable	是否在响应中忽略丢失或闭合的索引。默认为 false
lenient	是否忽略搜索时和索引时的数据类型不匹配错误。默认为 false

续表

参数名称	简 述
max_concurrent_shard_requests	限制并发分片请求的数量,默认为 5
pre_filter_shard_size	预过滤搜索分片阈值。如果一个分片无法匹配任何文档,那么就无需参与基于查询重写的步骤。因此能够限制分片的操作数量,默认值为 128
preference	指定节点或分片以执行搜索请求
q	遵循 DSL 语法的查询句子
request_cache	缓存是否应用于此请求,默认为索引级别设置
rest_total_hits_as_int	在搜索响应中将 hits.total 呈现为整数还是对象,默认值为 false
routing	执行搜索时,可以通过提供路由参数来控制要搜索哪些分片
scroll	游标查询时,要保持索引的一致属性多长时间
search_type	定义搜索操作的方式。可选 query_then_fetch 或 dfs_query_then_fetch
seq_no_primary_term	用于返回文档的 _seq_no 和 _primary_term 字段
size	返回的搜索结果总项数(默认为 10)
sort	返回的搜索结果的字段排序,可用逗号分隔的指定键值对列表 <字段:asc|desc>
stats	用于日志记录和统计目的
stored_fields	检索在映射中标记为 stored 的那些字段。可用逗号分隔的指定字段列表
suggest_field	指定用于建议的字段
suggest_mode	指定建议模式。可用选项 always、missing(默认)和 popular
suggest_size	允许返回的建议总数
suggest_text	是否返回建议的源文本
terminate_after	指定在分片中允许收集的文档数,默认为无限制
timeout	指定完成搜索操作所允许的时间。默认值为无超时
track_scores	追踪评分,即使它们不用于排序。默认为 false
track_total_hits	追踪匹配文档的总数。可以使用 false 值禁用它。默认为 10000
typed_keys	是否在聚合名称和建议器名称使用各自的类型作为前缀
version	是否返回匹配文档的版本号

13.4 elasticsearch-dsl 软件包

elasticsearch-dsl 软件包提供了一个在 elasticsearch-py 客户端之上构建的高级程序库。虽然可以直接使用 Elasticsearch 对象在 elasticsearch-dsl 包提供的方法,如开源项目中显示的代码所示:

```
from elasticsearch import Elasticsearch
from elasticsearch_dsl import Search

client = Elasticsearch()
s = Search(using=client, index="my-index")
...
```

但是强烈建议使用 elasticsearch_dsl/connections.py 模块中的 create_connection 方法,因为这种方法提供了持久连接和连接池。以下的 HLClient 类,使用单例设计模式封装 get_instance 方法,提供单例 Elasticsearch 对象连接本地主机的 9200 埠。测试代码如下所示:

```
# 此Python类提供创建高级别客户端方法
```

```python
class HLClient:
    __conn = None
    __conn_lock = threading.Lock()

    @staticmethod
    # 使用单例设计模式封装 get_instance 方法
    def get_instance():
        if HLClient.__conn is None:
            with HLClient.__conn_lock:
                if HLClient.__conn is None:
                    # 使用 create_connection 方法，创建高级别 REST 客户端
                    HLClient.__conn = connections.create_connection( 'hlclient',
hosts=['localhost'], port=9200)
        return HLClient.__conn

    def __init__(self):
        raise Exception("This class is a singleton!, use static method getInstance()")
```

在原始开源文件中的 Connections 类提供的 create_connection 方法中，返回一个 Elasticsearch 对象作为高级别 REST 客户端，并且设置序列化程序（serializer），目的是将请求和响应对象正确序列化为 JSON 格式。

```python
def create_connection(self, alias="default", **kwargs):
    kwargs.setdefault("serializer", serializer)
    conn = self._conns[alias] = Elasticsearch(**kwargs)
    return conn
```

13.4.1 提供特定接口的类

elasticsearch-dsl 软件包提供特定接口的类（见表 13-10），并使用 elasticsearch-py 软件包的调用方法发送请求。举个例子，以下显示开源项目 elasticsearch_dsl/search.py 模块中的 Search 类，该类提供 execute 方法去执行请求，该方法调用了从 get_connection 方法返回的 Elasticsearch 对象所提供的 search 方法。

```python
def execute(self, ignore_cache=False):
    if ignore_cache or not hasattr(self, "_response"):
        es = get_connection(self._using)
        self._response = self._response_class(
            self, es.search(index=self._index, body=self.to_dict(), **self._params)
        )
    return self._response
```

表 13-10　elasticsearch-dsl 软件包提供特定接口的类列表

Elasticsearch 接口名称	对应接口的类名称	简述
_doc	Document	提供有关索引文档的方法
_index	Index	提供有关索引的方法
_search	Search	提供有关搜索查询的方法
_msearch	MultiSearch	提供有关多重目标查询文档的方法
_update_by_query	UpdateByQuery	提供有关批量更新的方法
_mapping	Mapping	提供有关索引映射的方法
_template	IndexTemplate	提供有关索引模板的方法

13.4.2　elasticsearch-dsl 软件包中 Index 类提供的调用方法

类似于 13.2.2 节间接调用方法介绍的 IndicesClient，开源项目 elasticsearch_dsl/index.py 模块中定义的 Index 类，提供的方法只可以同步发送请求。常见的方法如表 13-11 所示。

表 13-11　通过 Index 类提供对应接口的调用方法列表

方法名称和签名	简述
analyzer(*args, **kwargs)	对文本执行分析
create(using=None, **kwargs)	创建索引
get(using=None, **kwargs)	检索索引的信息
delete(using=None, **kwargs)	删除索引
clone(name=None, using=None)	克隆索引
exists(using=None, **kwargs)	检查索引是否存在
open(using=None, **kwargs)	打开索引
close(using=None, **kwargs)	关闭索引
save(using=None)	将索引定义与 elasticsearch 同步，如果不存在则创建索引，如果存在则更新其设置和映射
is_closed(using=None)	检查索引是否
put_settings(using=None, **kwargs)	更新索引的设置
get_settings(using=None, **kwargs)	检索索引的设置
put_mapping(using=None, **kwargs)	更新索引的映射
get_mapping(using=None, **kwargs)	检索索引的映射
put_alias(using=None, **kwargs)	创建或更新索引的别名
get_alias(using=None, **kwargs)	检索索引的别名
exists_alias(using=None, **kwargs)	检查索引的别名是否存在
delete_alias(using=None, **kwargs)	删除索引的别名
get_field_mapping(using=None, **kwargs)	检索索引的字段映射
flush(using=None, **kwargs)	刷新到硬盘
flush_synced(using=None, **kwargs)	同步刷新到硬盘
forcemerge(using=None, **kwargs)	强制合并索引
refresh(using=None, **kwargs)	刷新索引相关的操作以用于搜索
shrink(using=None, **kwargs)	缩小现有索引以使用较少主分片
validate_query(using=None, **kwargs)	验证查询语句是否正确
clear_cache(using=None, **kwargs)	清理索引缓存
recovery(using=None, **kwargs)	返回进行中的索引分片恢复的信息
segments(using=None, **kwargs)	提供有关索引的段的信息
shard_stores(using=None, **kwargs)	提供索引分片副本的信息
stats(using=None, **kwargs)	提供有关索引操作的统计信息
upgrade(using=None, **kwargs)	升级索引的操作
get_upgrade(using=None, **kwargs)	提供有关已经升级的索引

13.4.3　elasticsearch-dsl 软件包中 Search 类提供的调用方法

开源项目 elasticsearch_dsl/search.py 模块中定义的 Search 类所提供的方法，也是只可以同步发送请求。常见的方法如表 13-12 所示。

表 13-12　通过 Search 类提供对应接口的调用方法列表

方法名称和签名	简　述
count()	返回与查询和过滤器匹配的文档数
delete()	通过委派给 delete_by_query 方法执行查询
execute(ignore_cache=False)	执行搜索并返回数据
filter(*args, **kwargs)	添加滤波器
highlight(*fields, **kwargs)	要求高亮显示某些字段
scan()	将搜索方式变成扫描搜索
script_fields(**kwargs)	定义要在匹配中的脚本字段
sort(*keys)	将排序信息添加到搜索请求
source(fields=None, **kwargs)	选择 _source 字段的返回方式
suggest(name, text, **kwargs)	向搜索添加建议请求

13.4.4　构造查询请求

要使用 Search 对象 search 构造查询语句，可以通过以下三种方法进行：

（1）使用 Q 方法

在 elasticsearch_dsl/query.py 模块中定义的 Q 方法，是构造任何查询的简单辅助函数，调用 Q 方法可以获得查询对象，然后指定到 Search 对象的 query 变量，或是用作 Search 对象的 query 方法的参数。Q 方法的详细用法请参考官网。以下例子为 7.3.1 节认识查询领域特定语言的测试用例，现在使用 elasticsearch-dsl 的 Q 方法的查询语句如下所示：

```
search.query = Q('match', ** {field_name: {'query': field_value, 'operator': operator}})
```

（2）使用 Search 对象提供的 query 查询方法

query 查询方法有两种用法，第一种是采用 Q 方法同样的参数，而第二种则是把 Q 方法返回的结果当作参数。查询语句如下所示：

```
search = search.query('match', ** {field_name: {'query': field_value, 'operator': operator}})
search = search.query(Q('match', ** {field_name: {'query': field_value, 'operator': operator}}))
```

（3）使用合适的 Query 类构造函数

类似于 12.4.4 节构造查询请求介绍的 QueryBuilder，每种查询必须利用 Query 对象作参数，由相对应的构造函数构建，用法如 Q 方法。各类构造函数见表 13-13。

```
search = search.query(Match(field_name={'query': field_value, 'operator': operator}))
```

表 13-13　各类型 Query 类列表

简　述	类名称和继承类型
对应全文搜索	
匹配所有查询	MatchAll(Query)
不匹配查询	MatchNone(Query)
标准匹配查询	Match(Query)

续表

简　述	类名称和继承类型
短语匹配查询	MatchBoolPrefix(Query)
布尔与前缀匹配查询	MatchBoolPrefix(Query)
短语前缀匹配查询	MatchPhrasePrefix(Query)
多字段匹配	MultiMatch(Query)
简单查询字符串	SimpleQueryString(Query)
查询语句查询	QueryString(Query)
对应词条级别搜索	
存在查询	Exists(Query)
过滤查询	Filtered(Query)
模糊查询	Fuzzy(Query)
文档标识符查询	Ids(Query)
前缀查询	Prefix(Query)
范围查询	Range(Query)
正则查询	Regexp(Query)
词条查询	Term(Query)
多个词条查询	Terms(Query)
包含最少确切词条查询	TermsSet(Query)
通配符查询	Wildcard(Query)
对应复合搜索	
布尔查询	Bool(Query)
权重提升查询	Boosting(Query)
固定分数查询	ConstantScore(Query)
分离最大化查询	DisMax(Query)
函数评分查询	FunctionScore(Query)
对应地理搜索	
地理边框查询	GeoBoundingBox(Query)
地理距离查询	GeoDistance(Query)
地理距离范围查询	GeoDistanceRange(Query)
地理多边形查询	GeoPolygon(Query)
地理形状查询	GeoShape(Query)
对应嵌套查询	
嵌套查询	Nested(Query)
匹配子角色查询	HasChild(Query)
匹配父角色查询	HasParent(Query)
对应脚本查询	
脚本查询	Script(Query)
对应跨度查询	
跨距匹配，容许之跨度描述在参数 big 的跨距匹配查询结果，并且包含从参数 small 的跨距匹配查询的结果	SpanContainining(Query)
跨度匹配，参数 end 控制从开始到最后位置的跨度	SpanFirst(Query)
等效与词条查询，用于跨度查询	SpanTerm(Query)

简 述	类名称和继承类型
包装词条、范围、前缀、通配符、正则表达式或模糊等查询，用于跨度查询	SpanMulti(Query)
跨距匹配，参数 slop 控制允许不匹配位置的跨度	SpanNear(Query)
跨距匹配，容许之跨度为包括在参数 include 所描述的跨距匹配查询结果，但不能被包括在参数 exclude 所描述的跨距匹配查询结果	SpanNot(Query)
跨距匹配，跨度为属于所有子句的并集	SpanOr(Query)
跨距匹配，容许之跨度描述在参数 small 的跨距匹配查询结果，并且在参数 big 的跨距匹配查询结果范围内	SpanWithin(Query)
跨域字段屏蔽查询，适用于单个字段同时使用多个分析器分析，并进行文档索引，然后使用多字段一起搜索	FieldMaskingSpan(Query)
专业查询	
对距离功能查询	DistanceFeature(Query)
rank_feature 查询，可修改评分公式	RankFeature(Query)
与文档集的文档匹配查询	MoreLikeThis(Query)
从存储匹配查询句子的索引搜寻	Percolate(Query)

13.4.5 执行请求方法与处理其响应

经由自定义的 HLClient 类的 get_instance 方法的取得 Connection 的高级别 REST 客户端，再构建 elasticsearch-dsl 软件包特定接口的类的对象。然后再调用该对象的方法。例如 Index 对象提供检索索引设置操作的 get_settings 方法和 Search 对象提供搜寻查询的 search 方法。返回的响应结果首先通过 to_dict() 转换为 python 字典数据结构，然后再使用 Flask-Jsonpify 包的 jsonify 方法转换为 JSON 表示形式代码。执行请求方法与处理其响应如下所示：

（1）Index 对象的 get_settings 方法

```
#Flask API 的 Swagger 设置样式
@name_space_hl.route("/dsl_indexSettings/<index>")
@name_space_hl.doc(params={'index': '索引名称'})
# 此 Python 类提供高级别客户端间接调用同步检索索引设置请求方法
class HLIndexSettings(Resource):
    def get(self, index):
        # 此方法通过特定接口的 Index 对象，间接调用同步检索索引设置请求
        response = Index(name=index, using=hl_es).get_settings()
        if response is not None:
            return jsonify(response)
        else:
            return {}
```

（2）Search 对象的 search 方法

```
@name_space_hl.route("/dsl_search/<index>")
@name_space_hl.doc(params={'index': '索引名称'})
# 此 Python 类提供高级别客户端间接调用同步搜索请求方法
class HLSearch(Resource):
    @name_space_hl.doc(params={'field_name': 'field name', 'field_value': 'field value',
        'operator': 'AND|OR'})
    def post(self, index):
        field_name = request.args.get('field_name');
        field_value = request.args.get('field_value');
        operator = request.args.get('operator');
        # 通过特定接口的 Search 对象，间接调用同步搜索请求
```

```
            search = Search(index=index, using=hl_es)[0:10]
#  (1) 使用 Q 方法
#     search.query =
#         Q('match', **{field_name: {'query': field_value, 'operator': operator}})
#     search = search.query(
#.        Q('match', ** {field_name: {'query': field_value, 'operator': operator}}))
#  (2) 使用 Search 提供的 query 查询方法
#     search = search.query(
#.        'match', ** {field_name: {'query': field_value, 'operator': operator}})
#  (3) 使用合适的 Query 类构造函数
            search = search.query(Match(** {field_name: {'query': field_value,
'operator': operator}}))
            search = search.source(['name', 'ts_code'])
            esponse = search.execute()
            if response['hits']['total']['value'] > 0:
                return jsonify(response.to_dict())
            else:
                return {}
```

13.5 使用 Swagger UI 测试 elasticsearch-dsl 软件包

测试的 elasticsearch-dsl 代码与 elasticsearch-py 代码属于同一个项目，同在 es-python-client 目录下。两个 python 调用方法，get_settings 及 search，被封装成 RESTful 接口供调用，然后集成 Swagger UI 方便使用。如果 Swagger UI 测试的用户界面还在，可以直接使用它。如何使用 python 指令执行程序项目，与 13.3 节使用 Swagger UI 测试 elasticsearch-py 软件包相同。

1. 测试 elasticsearch-dsl 软件包的 Swagger UI 页面

以下使用网页浏览器进行互动测试，访问项目的 Swagger UI 页面，URL 为 http://localhost:10010。然后按 /api/hlclient 请求栏以展开面板，屏幕截图如图 13-7 所示。

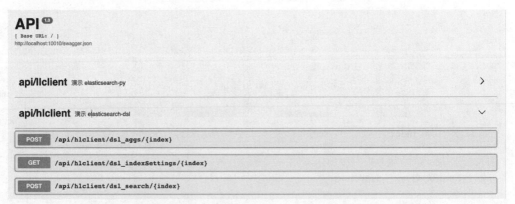

图 13-7　es_python_client 项目演示 elasticsearch-dsl 的 Swagger UI 页面

2.Index 类提供的检索索引设置调用方法测试

类似于前面的测试，测试 get_settings 请求首先点击"GET"一栏以展开面板，然后再点击"Try it out"按钮时，面板将更改为输入模式。输入索引名称 index 为 fund_basic，如图 13-8 所示。

图 13-8　使用 Index 对象的 get_settings 方法获取索引设置

点击"Execute"按钮，向下滑动屏幕后可以看到响应结果。获取的状态码 (statusCode) 为 200，返回的响应结果列出 fund_basic 的所有索引设置。屏幕截图如图 13-9 所示。

图 13-9　调用 Index 对象的 get_settings 方法返回的结果，列出 fund_basic 的所有索引设置

3. Search 类提供的文档搜索查询调用方法测试

类似于前面的测试，测试 Search 请求首先点击"POST"一栏以展开面板，然后再点击"Try it out"按钮，面板将更改为输入模式。输入匹配字段 (field_name) 为 name、匹配字符串 (field_value) 为"中银 ETF"、布尔运算符 (operator) 为 AND 和索引名称 (index) 为 fund_basic，如图 13-10 所示。

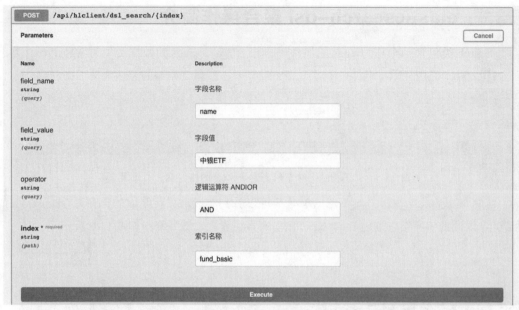

图 13-10 键入 elasticsearch-dsl Search 请求测试相对应的栏位内容

点击"Execute"按钮,向下滑动屏幕后可以看到返回的响应结果。匹配字符串"中银ETF"的基金简称,包括中银证券 500ETF 和中银中证 100ETF。屏幕如图 13-11 所示。

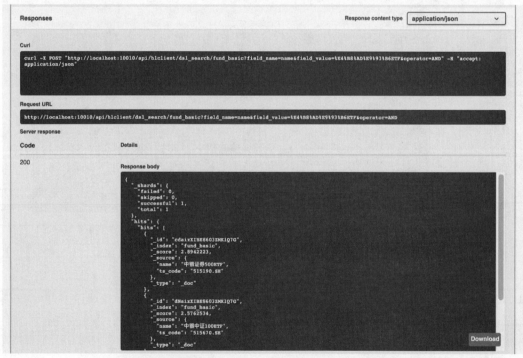

图 13-11 测试返回的结果,包括中银证券 500ETF 和中银中证 100ETF

13.6 elasticsearch-dsl 聚合操作简介

Search 类包含一个称为 aggs 的属性用于聚合操作。aggs 的数据类型为 AggsProxy 类，而 AggsProxy 类则继承了 AggBase 类，AggBase 类按照聚合框架的家族定义了构建聚合操作的方法，四个主要聚合家族中唯一没有支持的是矩阵聚合 (matrix)。其他三个家族就是度量指标聚合 (metric)、存储桶聚合 (bucket) 和管道聚合 (pipeline)。这三种方法间接使用类似于 Q 方法的 A 方法，其定义在 elasticsearch_dsl/aggs.py 模块中，可以构造任何聚合操作的简单辅助函数。由 metric、bucket 和 pipeline 方法构建的聚合可以链接在一起，与我们在 9.2 节聚合查询语法中介绍的子聚合的概念相同。摘录开源项目 AggBase 类部分的代码如下：

```python
class AggBase(object):
    ……
    def _agg(self, bucket, name, agg_type, *args, **params):
        agg = self[name] = A(agg_type, *args, **params)
        # For chaining - when creating new buckets return them...
        if bucket:
            return agg
        # otherwise return self._base so we can keep chaining
        else:
            return self._base
    ……
    def metric(self, name, agg_type, *args, **params):
        return self._agg(False, name, agg_type, *args, **params)

    def bucket(self, name, agg_type, *args, **params):
        return self._agg(True, name, agg_type, *args, **params)

    def pipeline(self, name, agg_type, *args, **params):
        return self._agg(False, name, agg_type, *args, **params)
    ……
```

A 方法的详细用法请参考官网。A 方法支持的各类型聚合操作见表 13-14。

表 13-14 A 方法支持的各类型聚合操作列表

聚合家族	聚合类型	Python 类名称
度量指标聚合 (metric)	平均值 (avg)	Avg
	加权平均值 (weighted_avg)	WeightedAvg
	基数（cardinality）	Cardinality
	扩展统计（extended_stats）	ExtendedStats
	最大值（max）	Max
	最小值（min）	Min
	百分位（percentiles）	Percentiles
	百分位等级（percentile_ranks）	PercentileRanks
	脚本式度量指标（scripted_metric）	ScriptedMetric
	统计（stats）	Stats
	总和（sum）	Sum
	最热点（top_hits）	TopHits
	值计数（value_count）	ValueCount
	地理边界（geo_bounds）	GeoBounds

续表

聚合家族	聚合类型	Python 类名称
存储桶聚合 (bucket)	子文档（children）	Children
	父文档（parent）	Parent
	日期直方图（date_histogram）	DateHistogram
	自动间隔日期直方图（auto_date_histogram）	AutoDateHistogram
	复合（composite）	Composite
	日期范围（date_range）	DateRange
	多元化采样器（diversified_sampler）	DiversifiedSampler
	过滤器（filter）	Filter
	多过滤器（filters）	Filters
	全局（global）	Global
	直方图（histogram）	Histogram
	IP 范围（ip_range）	IPRange
	缺失字段（missing）	Missing
	嵌套（nested）	Nested
	范围（range）	Range
	反向嵌套（reserve_nested）	ReverseNested
	采样器（sampler）	Sampler
	显著词条（significant_terms）	SignificantTerms
	显著文本（significant_text）	SignificantText
	词条（terms）	Terms
	地理距离（geo_distance）	GeoDistance
	地理哈希网格（geohash_grid）	GeohashGrid
	地理瓦片网格（geotile_grid）	GeotileGrid
	地理重心（geo_centroid）	GeoCentroid
管道聚合 (pipeline)	桶平均值（avg_bucket）	AvgBucket
	导数（derivative）	Derivative
	桶最大值（max_bucket）	MaxBucket
	桶最小值（min_bucket）	MinBucket
	桶总和（sum_bucket）	SumBucket
	桶统计（stats_bucket）	StatsBucket
	桶扩展统计（extended_stats_bucket）	ExtendedStatsBucket
	桶百分位（percentiles_bucket）	PercentilesBucket
	移动函数（moving_fn）	MovingFn
	累计总和（cumulative_sum）	CumulativeSum
	桶脚本（bucket_script）	BucketScript
	桶选择器（bucket_selector）	BucketSelector
	桶排序（bucket_sort）	BucketSort
	串行差分（serial_diff）	SerialDiff

以下例子使用与 9.6.14 节移动函数聚合的测试用例，用例使用了指数加权移动平均模型的单指数 (ewma)、二次指数 (holt) 和三次指数 (holtWinters) 聚合操作。这三个聚合都使用 A

方法构建。其他所需的聚合操作，根据其类型按照表 13-14 使用相对应的聚合家族构建方法指定到 Search 对象的 aggs 属性，例如日期直方图（date_histogram）是存储桶聚合 (bucket)，最大值（max）是度量指标聚合 (metric)，而移动函数（moving_fn）则是管道聚合 (pipeline)。在聚合过程中，需要使用日期直方图（date_histogram）聚合和最大值聚合找出每日的复权单位净值，供其他聚合操作之用。

```
@name_space_hl.route("/dsl_aggs/<index>")
@name_space_hl.doc(params={'index': 'index name'})
# 此 Python 类提供高级别客户端指定用例的聚合操作方法
class HLAggr(Resource):
    @name_space_hl.doc(params={'ts_code': '基金代码', 'field_name': '字段名称', 'alpha': '0-1',
     'beta': '0-1', 'gamma': '0-1', 'period': '2', 'window': '滑动窗口'})
    def post(self, index):
        ts_code = request.args.get('ts_code');
        field_name = request.args.get('field_name');
        alpha = request.args.get('alpha');
        beta = request.args.get('beta');
        gamma = request.args.get('gamma');
        period = request.args.get('period');
        window = request.args.get('window');
        # 通过特定接口的 Search 对象，间接调用同步搜索请求
        search = Search(index=index, using=hl_es)[0:0]
        search = search.filter('term', **{'ts_code.keyword': ts_code})
        daily_report = A(DateHistogram(field='end_date', calendar_interval='1d',
            min_doc_count=1))
        daily_adj_nav = A(Max(field='adj_nav'))
        # 指数加权移动平均模型的单指数、二次指数和三次指数聚合操作
        ewma = A(MovingFn(script='MovingFunctions.ewma(values, ' + alpha + ')',
            window=window, shift=1, buckets_path='daily_adj_nav'))
        holt = A(MovingFn(script='MovingFunctions.holt(values, ' + alpha + ',' + beta + ')',
            window=window, shift=1, buckets_path='daily_adj_nav'))
            holtWinters = A(MovingFn(script='if (values.length>=5) {
MovingFunctions.holtWinters(values,' + alpha + ',' + beta + ',' + gamma + ',' + 
period + ',false)}',
    window=window, shift=1, buckets_path='daily_adj_nav'))
        search.aggs.bucket('daily_report', daily_report).metric('daily_adj_nav',
daily_adj_nav) \
                .pipeline('ewma', ewma).pipeline('holt', holt).
pipeline('holtWinters', holtWinters)
        # 执行 Search 类提供 execute 的方法
        response = search.execute()
        if len(response['aggregations']['daily_report']['buckets']) > 0:
            return jsonify(response.to_dict())
        else:
            return {}
```

13.7 使用 Swagger UI 测试 elasticsearch-dsl 聚合操作

测试 elasticsearch-dsl 聚合操作代码也在同一个项目，如果 Swagger UI 测试的用户界面还在，可以直接使用它。以下使用网页浏览器进行指数加权移动平均模型的聚合操作互动测试步骤：

（1）类似于前面的测试，测试聚合操作请求首先点击 "dsl_aggs" 一栏以展开面板，然后再点击 "Try it out" 按钮时，面板将更改为输入模式。输入每个变量值，包括基金代码、

字段名称、移动函数的参数 (alpha、beta、gamma、period 和 window) 和索引名称，屏幕截图如图 13-12 所示。

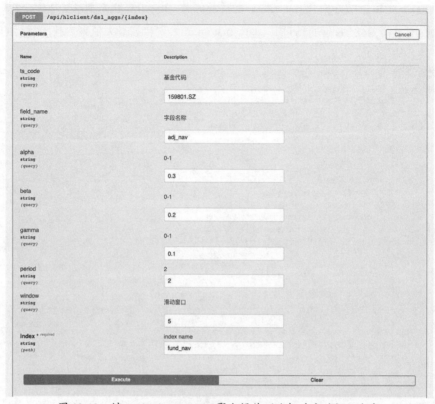

图 13-12　键入 elasticsearch-dsl 聚合操作测试相对应的栏位内容

（2）点击"Execute"按钮，向下滑动屏幕后可以看到指数加权移动平均模型的单指数、二次指数和三次指数聚合返回的响应结果。屏幕截图如图 13-13 所示。

图 13-13　测试指数加权移动平均模型的单指数、二次指数和三次指数聚合返回的结果

第四篇 进阶功能和数据分析实战

第 14 章 Elasticsearch 与金融舆情分析

根据百度百科的解释，舆情是"舆论情况"的简称。而舆论就是较多的群众对于特定事件或现象所持有的共同的情绪和倾向。基于情感语义分析是当前舆情监测的热门方法。目前，大多数使用 Elasticsearch 进行情感分析工作的开源项目，都只将 Elasticsearch 当作数据存储工具。而情感分析的工作则交由第三方软件处理。本章将介绍文本情感分析原理和情感分析插件，并讨论在 Elasticsearch 环境中进行股票分析和预测。

14.1 文本情感分析简介

文本情感分析是判断文本中要表达的观点、态度和情感。一般是根据原始资料使用上下文挖掘技术来识别和提取主观信息。然后判断所表达的潜在情绪是正面、负面还是中性。基本上，情感分析技术大致分为基于词典的方法、统计方法和混合方法。

- 基于词典的方法

预构建包含情感词极性、情感倾向程度和预设的数量组成的词典。这种方法计算效率高且可扩展，但是完全取决于语言规则对情感的认可。

- 统计方法

较多涉及机器学习和深度学习方法。需要标记训练数据来进行极性检测，基于深度学习模型的研究已经非常流行。一般而言，统计方法需要大量的训练数据，在语义上相对较弱。

- 混合方法

利用基于词典的方法和统计方法进行极性检测。基本目标是从统计方法继承高精准度的判断，并从基于词典的方法继承了语义上的稳定性。

14.2 文本情感分析软件服务

国内外已经有许多公司提供基于文本的情感倾向分析软件服务。国内三大互联网公司旗下的百度云、阿里云和腾讯云，都提供相关的自然语言处理软件服务。而在国外，有大量来自不同公司的情感倾向分析产品与服务。例如亚马逊的 AWS Comprehend、IBM 的 Watson NLU、Google 的 AutoML Natural Language、Microsoft Azure 的 Text Analytics API、SAS 的 Visual Text Analytics 以及一些来自相对较小型的公司，例如 Lexalytics、Rosette 等等。根据他们的介绍，依赖的主要技术是机器学习。

14.3 文本情感分析开源项目

由于文本情感分析被用于不同的业务场景都具有出色的表现，例如消费者决策、民意分析和个性化推荐等领域，已经越来越多开源项目被商业化或是被淘汰掉。再加上国情因素，目前已经剩下很少有维护良好的中文文本情感分析开源项目。在本节中将会讨论介绍一些值得关注的开源项目。

14.3.1 TextBlob

Textblob 是在情感分析开源项目中非常流行的工具，尤其是与 Twitter 的文本一起使用，最有可能的原因是 Textblob 非常简单和容易使用。以下是来自 Textblob 开源项目的一段 Python 代码段，可以说明执行其情感分析步骤的简易性。根据开源代码段的介绍，只需要使用几行代码就可以提供文本的极性值 (sentence.sentiment.polarity)。只可惜它不支持中文文本。

```
from textblob import TextBlob
text = 'This one tastes good'
blob = TextBlob(text)
for sentence in blob.sentences:
    print(sentence.sentiment.polarity)
```

TextBlob 开源项目是基于 NLTK 和 pattern 构建而成。近期，已经有学学开始研究在 NLTK 中采用中文的可行性。例如中南大学信息科学与工程学院李晨和刘卫国著的"基于 NLTK 的中文文本内容抽取方法"。如果该方法可以成功集成在 Textblob 开源项目，剩下要处理的问题就是中文文本的精确度。

实现 TextBlob 的情感分析技术默认依赖于词典的模式库 PatternAnalyzer。但是可以选择贝叶斯预测模型的 NaiveBayesAnalyzer。若是要创建一个自定义的情感分析器，需要准备贴好标签的数据，并对其进行机器学习训练。TextBlob 已经发布到 PyPI，最近的版本号为 0.15.3。最近在 github 的更新是 2020 年 3 月，维护良好。

14.3.2 SnowNLP

SnowNLP 是一个 python 程序库，是受到了 TextBlob 的启发而写的，目标是方便处理中文文本内容。SnowNLP 支持情感分析技术似乎依赖贝叶斯模型的预测。根据开源代码的介绍，也是只需要使用几行代码就可以提供文本的情感分析结果。

```
from snownlp import SnowNLP
text = u'这家味道还不错'
blob = SnowNLP(text)
```

```
for sentence in blob.sentences:
        print(SnowNLP(sentence).sentiments)
```

按照 SnowNLP 开源项目中的自述，自带训练好的字典主要是买卖东西时的评价，所以对于其他的场景，需要创建贴好标签的数据，并对其进行机器学习训练。SnowNLP 已经发布到 PyPI，最近的版本号为 0.12.3。最近在 github 的更新是 2017 年 5 月，似乎已经停止开发和维护。

14.3.3　BosonNLP

玻森数据是一家中文自然语言分析云服务提供商，拥有自我开发的中文语义 BosonNLP，也是一个 python 程序库，并且已经发布到 PyPI，最近的版本号为 0.11.1。最近在 github 的更新是 2018 年 11 月，似乎已经停止开发和维护。根据开源代码段的介绍，也是只需要使用几行代码就可以提供文本的情感分析结果。由于在开源项目内的代码没有看到支持的训练函数，似乎不支持定制的学习训练数据。

```
from bosonnlp import BosonNLP
nlp = BosonNLP('YOUR_API_TOKEN')
nlp.sentiment('这家味道还不错')
```

14.3.4　Stanford CoreNLP

Stanford CoreNLP 是由斯坦福大学自然语言处理小组开发，提供了一组用 Java 编写的自然语言分析工具。主要是针对英语文本，但是也支持多种人类语言。对于中文文本已经提供许多功能，目前（2020 年 9 月）版本为 4.1.0，情感分析功能仅适用于英语文本。当前版本的分析技术已发展到采用深度学习。Stanford CoreNLP 也有很多 python 程序接口经发布到 PyPI。Python 程序主要是与 Stanford CoreNLP Java 服务器连接并取得服务。来自 stanfordn/CoreNLP 开源项目的代码介绍，虽然不像其他产品那么简单，但也不复杂。最近在 github 的更新是 2020 年 8 月，开发和维护很好。使用 python 的读者可以参考 Lukas Frei 在 2019 年 2 月于 Towards Data Science 发布的一文"Natural Language Processing Using Stanford's CoreNLP"。

14.3.5　百度 Senta

百度的 AI 开放平台开放了情感分析服务，感兴趣的读者可以上网进行尝试。以下是在线服务的试用。测试语句来自证券时报的文章"一日售罄爆款再现 两只新基金一天大卖 300 亿如下"，测试句子比较正面倾向。显示如下：

> 9月7日，17只新基金携手首发，其中两只产品的募集规模超过了百亿。

测试结果为正面 84%，屏幕如图 14-1 所示。

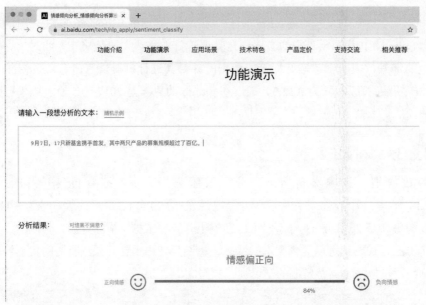

图 14-1 百度 AI 开放平台情感分析服务测试比较正面倾向的中文句子

另一个测试使用来自 Kaggle 数据集名为 "Sentiment Analysis for Financial News" 其中之一句正面倾向的英文句子，显示如下：

```
Operating profit totalled EUR 21.1 mn, up from EUR 18.6 mn in 2007, representing 9.7 % of net sales.
```

测试结果为正面 92%，屏幕如图 14-2 所示。

图 14-2 百度 AI 开放平台情感分析服务测试测试比较正面倾向的英文句子

百度除了开放 AI 平台外，还在 github 开放一个大规模挖掘而来的高质量情感数据和基于深度学习的语义训练模型 "Sentiment Knowledge Enhanced Pre-training"（SKEP）。对输入文本进行语义理解，并基于语义进行情感倾向的判断。开源项目名称为 baidu/Senta。最近的更新是 2020 年 6 月，开发和维护良好。Senta 是一个 Python 程序库，并且已经发布到 PyPI，最近的版本号为 2.0.0。

14.4 文本情感分析插件开源项目

插件是一种以自定义方式增强 Elasticsearch 功能的方法，例如添加映射类型、分析器和脚本等等。我们已经在第六章文本分析插件详细探讨。Elasticsearch 带有许多内置的核心插件。还有许多由不同社区贡献的自定义插件。根据我们当前的网络搜索，在 github 开放源码项目中并没有关于中文文本情感分析的 Elasticsearch 插件。所以在本节中使用一个名为 TechnocratSid/elastic-sentiment-analysis-plugin 的 github（在此简称为 ESAP）开源项目作为示例，以逐步说明制作情感分析 Elasticsearch 插件的过程。但因该项目基于 Stanford CoreNLP，正如在 14.3.4 节提到，其情感分析功能目前仅适用于英语文本。有兴趣的读者可以参考 TechnocratSid 在 2018 年 10 月发布在 technocratsid.com 的一文 "Elasticsearch plugin for Sentiment Analysis"。

14.4.1 ESAP 开源项目简介

此开源项目仅支持 Elasticsearch 6.4.1 版本。由于 Elasticsearch 服务器是用 Java 语言编写，因此编写插件也需要使用 Java。该项目使用的 Stanford CoreNLP 版本为 2017 年 6 月的 3.8.0，尽管比较旧版，也许会影响文本情感分析结果的准确性。然而，这是一个很好的实用示例供开发插件之用。读者可以同时参考 TechnocratSid 在 2018 年 10 月发布在 technocratsid.com 的另一篇文章 "How to create an Elasticsearch 6.4.1 Plugin" 及其开源项目，阅读这两篇文章可以清晰地了解如何编写文本情感分析 Elasticsearch 插件。以下的说明，摘录自该两篇文章及其开源项目，项目中的 src 目录下，所有子目录和文件如图 14-3 所示。

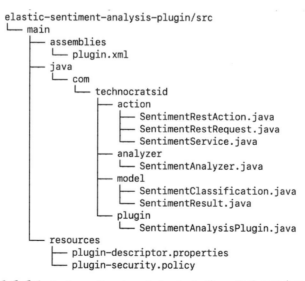

图 14-3 开源项目 elastic-sentiment-analysis-plugin 的 src 目录下所有子目录和文件

项目开发的简要说明如下：

（1）Elasticsearch 插件是一个 ZIP 文件，其中包含一个或多个带有已编译代码和资源的 JAR 文件。该项目提供的名为 elastic-sentiment-analyis-plugin-6.4.1.zip 的插件文件。

（2）使用的软件项目管理为 Maven，Maven 编译器插件（compiler.plugin.version）版本为 3.5.1。使用的 Java 版本为 1.8。

（3）插件必须包含一个名为 plugin-descriptor.properties 的文件。用于描述该插件所包含的 JAR 文件、脚本和配置文件。文件位置在项目中的 src/main/resources 目录。

（4）如果需要指定使用的安全权限，在 src/main/resources 目录中名为 plugin-security.policy 文件，编写权限的授予语句。默认为 java.security.AllPermission。

（5）在 src/main/assemblies 目录中，有一个名为 plugin.xml 文件，用于配置插件的打包方式。例如目标文件的格式、文件的位置并需要打包什么类型的文件等等。

（6）该项目创建了一个 Elasticsearch 插件，以提供一个名为 _sentiment 的自定义接口。用法是在请求正文中提供一个文本字段 text，分析结果给出 5 个不同级别的情感倾向分析结果。分别为非常正面、正面、中性、负面和非常负面。

（7）要创建一个插件，需要定义一个继承 org.elasticsearch.plugins.Plugin 类，并实现了 org.elasticsearch.plugins.ActionPlugin 接口的类。该项目创建了一个名为 SentimentAnalysisPlugin 的自定义插件类，主要目的是实现名为 getRestHandlers 接口方法，文件位置在项目中的 src/main/java/com/technocratsid/plugin 目录。此接口在 org.elasticsearch.plugins.ActionPlugin 文件中声明。实现接口的方法是创建一个 SentimentRestAction 类的自定义对象，并转换为 Collections.singletonList 数据类型返回。getRestHandlers 接口方法摘录如下：

```
public class SentimentAnalysisPlugin extends Plugin implements ActionPlugin {
  ……
  public List<RestHandler> getRestHandlers(…) {
    return Collections.singletonList(new SentimentRestAction(settings,
      restController));
  }
}
```

（8）自定义 SentimentRestAction 类主要目的是创建自定义接口 _sentiment 的端点，并实现 prepareRequest 方法以提供解决方案，给 _sentiment 端点的 GET 请求返回结果。以下代码段显示过程为创建一个 SentimentService 的自定义对象，并调用其 calculateSentiment 方法计算情感倾向。SentimentRestAction 类摘录如下：

```
public class SentimentRestAction extends BaseRestHandler {
  private static String NAME = "_sentiment";
  @Inject
  public SentimentRestAction(Settings settings, RestController
    restController) {
    super(settings);
    restController.registerHandler(RestRequest.Method.POST, "/" + NAME,
      this);
    restController.registerHandler(RestRequest.Method.GET, "/" + NAME,
      this);
  }
  @Override
  protected RestChannelConsumer prepareRequest(RestRequest request, NodeClient
```

```
      client) throws IOException {
      ……
      return channel -> {
         XContentBuilder builder = channel.newBuilder();
         AccessController.doPrivileged((PrivilegedAction) () -> {
         SentimentServiceservice = new
            SentimentService(restRequest);
         try {
            service.calculateSentiment(builder);
         }
         ……
   }
}
```

（9）自定义 SentimentService 类的主要目的是创建一个 SentimentAnalyzer 的自定义对象，并在其 calculateSentiment 方法中调用 SentimentAnalyzer 对象的 getSentimentResult 方法取得情感倾向分析结果。

```
public class SentimentService {
   private SentimentRestRequest request;
   private SentimentAnalyzer sentimentAnalyzer;
   ……
   public void calculateSentiment(XContentBuilder builder) throws IOException {
      if (builder != null) {
         synchronized (builder) {
            builder.startObject().
               field("sentiment_score", sentimentAnalyzer.
               getSentimentResult(request.getText()).getSentimentScore()).
               field("sentiment_type", sentimentAnalyzer.
               getSentimentResult(request.getText()).getSentimentType()).
               field("very_positive", sentimentAnalyzer.
               getSentimentResult(request.getText()).getSentimentClass().
               getVeryPositive()+"%")
            ……
         }
         ……
   }
}
```

（10）自定义 SentimentAnalyzer 类的主要目的是创建一个 StanfordCoreNLP 对象，并在其 getSentimentResult 方法中调用 StanfordCoreNLP 对象的 process 方法获得真正的情感分析处理过程。

```
public class SentimentAnalyzer {
   static Properties props;
   static StanfordCoreNLP pipeline;
   public void initialize() {
      ……
      props = new Properties();
      props.setProperty("annotators", "tokenize, ssplit, parse,
         sentiment");
      pipeline = new StanfordCoreNLP(props);
   }
   public SentimentResult getSentimentResult(String text) {
      ……
      Annotation annotation = pipeline.process(text);
      ……
   }
}
```

14.4.2 ESAP 开源项目安装与测试

ESAP 开源项目的安装、运行与测试步骤如下：

（1）安装 Elasticsearch 6.4.1 版本

◆ 安装

安装前，先按照 1.1 节准备环境关闭 Elasticsearch 7.5.1 进程，然后从官网下载和安装 Elasticsearch 6.4.1 版本，要执行的步骤没有区别。

◆ 运行

按照 1.2 节运行 Elasticsearch 6.4.1 节在指定的路径（elasticsearch-6.4.1 的文件夹）执行指令，运行 Elasticsearch 程序。

◆ 测试

按照 1.3 节与 Elasticsearch 6.4.1 节进行通信，在命令行中，使用"curl"命令发送 HTTP GET 指令，并通过默认的 9200 端口与 Elasticsearch 通信。在返回的结果中，检验正在运行的版本（Version.number）是 6.4.1。

（2）安装 elastic-sentiment-analysis-plugin 插件

使用如下 elasticsearch-plugin 命令和 elastic-sentiment-analysis-plugin 插件的 url，直接安装插件。插件安装完成后需要重启 Elasticsearch 服务器。

```
sudo bin/elasticsearch-plugin install https://github.com/TechnocratSid/elastic-sentiment-analysis-plugin/releases/download/6.4.1/elastic-sentiment-analyis-plugin-6.4.1.zip
Password:
-> Downloading https://github.com/TechnocratSid/elastic-sentiment-analysis-plugin/releases/download/6.4.1/elastic-sentiment-analyis-plugin-6.4.1.zip
[=================================================] 100%
@@@@@@@@@@@@@@@@@@@@@@@@@@@@@@@@@@@@@@@@@@@@@@@@@@
@     WARNING: plugin requires additional permissions     @
@@@@@@@@@@@@@@@@@@@@@@@@@@@@@@@@@@@@@@@@@@@@@@@@@@
* java.lang.RuntimePermission setIO
See http://docs.oracle.com/javase/8/docs/technotes/guides/security/permissions.html
for descriptions of what these permissions allow and the associated risks.
Continue with installation? [y/N]y
-> Installed sentiment-analysis-plugin
```

（3）测试 elastic-sentiment-analysis-plugin 插件

插件的接口为 _sentiment，需要使用 HTTP POST 方法。下面一些例子包括正面倾向和负面倾向的测试。

◆ 使用 Google 翻译工具将 14.3.5 节百度 Senta 的示例从中文翻译成英文，并使用 curl 将 POST 请求发送到 _sentiment 接口。测试结果与百度 Senta 的结果完全相反。

```
curl --include --request POST http://localhost:9200/_sentiment?pretty=true --header "Content-Type: application/json" --data $'{"text" : "On September 7, 17 new funds were launched first time, of which two products raised more than tens of billions."}'
{
  "sentiment_score" : 1.0,
  "sentiment_type" : "Negative",
  "very_positive" : "1.0%",
  "positive" : "2.0%",
```

```
    "neutral" : "20.0%",
    "negative" : "68.0%",
    "very_negative" : "10.0%"
}
```

- 使用与 14.3.5 节百度 Senta 来自 Kaggle 数据集的相同示例,测试结果与百度 Senta 的结果完全相反。

```
curl --include --request POST http://localhost:9200/_sentiment?pretty=true
--header "Content-Type: application/json" --data $'{"text" : "Operating profit
totalled EUR 21.1 mn , up from EUR 18.6 mn in 2007 , representing 9.7 % of net
sales."}'
{
    "sentiment_score" : 1.0,
    "sentiment_type" : "Negative",
    "very_positive" : "1.0%",
    "positive" : "1.0%",
    "neutral" : "10.0%",
    "negative" : "60.0%",
    "very_negative" : "29.0%"
}
```

- 使用来自 Kaggle 数据集中的一个中性倾向的句子,测试结果与百度 Senta 的结果不同。

```
curl --include --request POST http://localhost:9200/_sentiment?pretty=true
--header "Content-Type: application/json" --data $'{"text" : "According to Gran ,
the company has no plans to move all production to Russia , although that is where
the company is growing."}'
{
    "sentiment_score" : 1.0,
    "sentiment_type" : "Negative",
    "very_positive" : "1.0%",
    "positive" : "3.0%",
    "neutral" : "19.0%",
    "negative" : "64.0%",
    "very_negative" : "13.0%"
}
```

14.5 中文金融领域文本情感分析

在国内,有关于中文金融领域文本情感分析的研究和应用还真不少。在学术上有很多创新的相关研究发表论文。在专利申请上也可以看到一些已经公布。此外,还有公开比赛例如 CCF BDCI 2019 金融实体级情感分析大赛。在 github 中相关的中文金融领域项目也有。

无论文本情感分析是基于词典的方法、统计方法或是混合方法,都需要一本合适的情感词典作为训练使用或验证其有效性。每个领域都有其专业知识、语义和情感投入程度的区别。所以为金融领域建立情感词典以能提高情感分析的性能非常重要。以下介绍一些例子,也许可以基于此为中文金融领域文本开发一个 Elasticsearch 情感分析插件。

(1)金融新闻事件数据来源

Tushare 提供主流新闻网站的快讯新闻数据,包括新浪财经实时资讯、华尔街见闻快讯、同花顺财经新闻、东方财富财经新闻和云财经新闻。有足够丰富的信息,但是使用权限有积分的限制。

(2)开源项目

开源项目 MengLingchao/Chinese_financial_sentiment_dictionary 在 2020 年 4 月发布于 github。词典制作来源是刊登于金融学国际顶尖期刊(Journal of Financial Economics),由

Fuwei Jiang 等学者所著的"Manager Sentiment and Stock Returns"一文。此词典总共有消极词语 5890 词和积极词语 3338 词。词语来源是由 LM 词典中文翻译、Tsinghua 词典、知网词典、NTUSD 词典和 Word2vec 词典整合而来。

(3) 来自发表的期刊论文

- 在 2018 年 10 月刊登于数据分析与知识发现,由胡家珩等学者所著的"基于深度学习的领域情感词典自动构建——以金融领域"为例一文。提出一种通过基于词向量多层全连接神经网络,自动构建情感词典的方法,尤其是以金融领域作为主要发展重点。由于词典的内容是一天天地更新,因此自动创建词典是一种明智的方法。

- 在 2019 年 9 月刊登于 SSRN eLibrary,由 Shibo Bian1 等学者所著的"A New Chinese Financial Sentiment Dictionary for Textual Analysis in Accounting and Finance"。此词典总共有消极词语 1488 词和积极词语 1107 词。词语来源是基于 HOWNET、DLUTSD 和 NTUSD 这三本词典,再从路演信息添加 1 411 个词、从股票收益会议记录添加 7 138 个词、从 IPO 招股说明书添加 2 043 个词并从股票份年度报告添加 29 737 个词。经过算法和人工判断,使用多阶段剔除法来构建此金融词典。

14.6 应用 Elasticsearch 进行股票分析和预测

根据刊登于社会杂志 2017 年第 37 卷第 2 期,由陈云松等学者所著的"网络舆情是否影响股市行情?基于新浪微博大数据的 ARDL 模型边限分析"一文得出的结论,[在股市震荡期,早前三天内的"微博信心指数"有助于预测上证指数;"微博信心指数"和"上证指数"存在正向相关的均衡关系;在股市行情平稳期,以上的统计关联并不存在]。

分析互联网上的股市舆情是否影响股市行情是一个非常热门的话题,国内外已经有许多的研究和开源项目。在本节中,我们将讨论 Stocksight 开源项目。根据该项目的说明,Stocksight 是一个应用 Elasticsearch、Twitter、新闻头条、Python 自然语言处理和情感分析的股市分析器和预测器。Stocksight 使用的 Elasticsearch 版本为 5.x,但是该项目使用 docker 和 docker-compose 软件,在 docker 运行环境中自动安装 5.6.16 版的 elasticsearch 和 kibana。下面的安装、配置并测试的步骤,是从原始 github 网站中提取文本,翻译和重新编排。

14.6.1 安装与运行相关软件

假设开源项目 Stocksight、docker 和 docker-compose 已在系统中成功安装,以下步骤是运行相关命令及设置工作环境。

(1) 进入 Stocksight 项目目录。

```
cd stocksight
```

(2) 打开名为 config.py.sample 的文件,输入可用的 Twitter 帐户中的令牌访问密钥访问密钥。如果 sentiment.py 程序准备在容器内运行,将变量 elasticsearch_host 设置为容器的名称 stocksight_elasticsearch_1。如果 sentiment.py 程序准备在主机内运行,将变量 elasticsearch_host 设置为 localhost。

```
access_token = "abcxxx"
access_token_secret = "defxxx"
elasticsearch_host = "stocksight_elasticsearch_1"
```

（3）复制文件 config.py.sample 到名为 config.py。

```
cp config.py.sample config.py
```

（4）以下 docker-compose 命令，运行相关工作环境。

```
docker-compose build && docker-compose up
elasticsearch uses an image, skipping
kibana uses an image, skipping
Building stocksight
Step 1/11 : FROM python:3.6
……
```

（5）使用 docker 命令检查相关的工作环境，可以看到正在运行的 stocksight、kibana 和 elasticsearch 进程。

```
docker container list
  CONTAINER ID        IMAGE                    COMMAND                  CREATED
STATUS              PORTS                    NAMES
  dd68889dd25c        stocksight_stocksight    "bash startup.sh"        About
an hour ago     Up About an hour                                  stocksight_
stocksight_1
  01af49c79a84        kibana:5.6.16            "/docker-entrypoint.…"   About
an hour ago     Up About an hour    0.0.0.0:5601->5601/tcp        stocksight_
kibana_1
  fc9267fd031a        elasticsearch:5.6.16     "/docker-entrypoint.…"   About
an hour ago     Up About an hour    0.0.0.0:9200->9200/tcp, 9300/tcp  stocksight_
elasticsearch_1
```

（6）容器（containers）启动后，执行以下命令进入 stocksight_stocksight 容器内，容器的名称是 stocksight_stocksight_1。进入后可以看到命令提示符 dd68889dd25c:/app。dd68889dd25c 是 stocksight_stocksight 容器的标识符。

```
docker exec -it stocksight_stocksight_1 bash
dd68889dd25c:/app
```

（7）可以浏览当前目录或者检查文件。

```
dd68889dd25c:/app/ls
config.py     requirements.txt    sentiment.py    startup.sh    stockprice.py
twitteruserids.txt
```

（8）stocksight_stocksight 容器内已经安装好运行环境，例如 nltk 软件包。

```
dd68889dd25c:/app# ls /root/nltk_data
corpora    tokenizers
```

（9）执行以下命令，创建"stocksight"索引，并使用股票代号 TSLA 挖掘和分析 Tweets。

```
dd68889dd25c:/app# python sentiment.py -s TSLA -k 'Elon Musk',Musk,Tesla,SpaceX
--debug
……
  2020-11-20 02:48:38,506 [INFO][stocksight] Creating new Elasticsearch index or
using existing stocksight
……
```

（10）根据 stocksight 程序设计，有不同的选择来收集来自 twitter 的数据并进行挖掘和情感分析。

① 仅从推文收集数据。

```
dd68889dd25c:/app#python sentiment.py -s TSLA -k 'Elon Musk',Musk,Tesla,SpaceX --debug
```

② 从推文及推文的任何 URL 的链接网页收集数据。

```
dd68889dd25c:/app# python sentiment.py -s TSLA -k 'Elon Musk',Musk,Tesla,SpaceX -l --debug
```

③ 还有很多其他选择可以从中收集数据，感兴趣的读者可以参考官网。

（11）执行以下命令，将股票价格添加到 Elasticsearch 中的"stocksight"索引。

```
dd68889dd25c:/app#python stockprice.py -s TSLA --debug
```

（12）运行 Kibana（http://localhost: 5601），如图 14-4 所示，设置索引模式（index pattern）为"stocksight"，时间字段名称为 date。

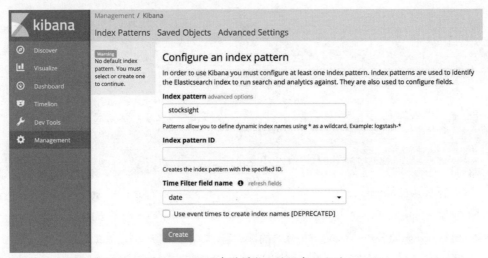

图 14-4 设置索引模式和时间字段名称

（13）按下创建（create）按钮创建索引模式，屏幕将显示如图 14-5 所示。

图 14-5 索引模式设置创建后的屏幕

（14）单击左侧菜单 Management 后，屏幕将显示如图 14-6 所示。单击顶部中间菜单 Saved Objects，单击导入顶部右侧按钮 Import，然后导入已保存的仪表板（export.json 文件）。导入后可以看到仪表板名称为 stocksight_dashboard。

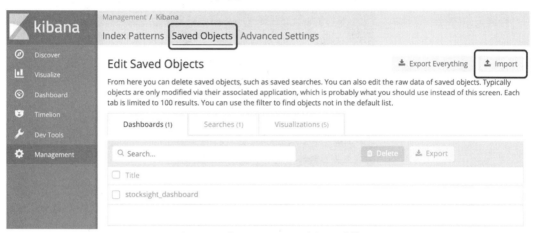

图 14-6　导入可视化/仪表板的屏幕

（15）单击表格中的项目 stocksight_dashboard，进入仪表板。屏幕将显示如图 14-7 所示。

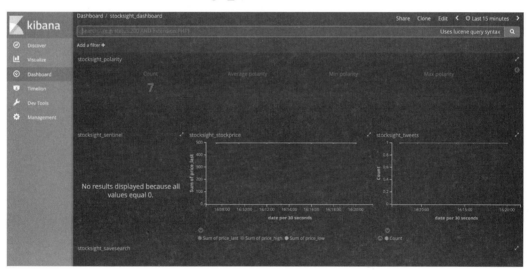

图 14-7　进入 stocksight_dashboard 仪表板

提示：由于 Twitter 帐户的安全设置问题，因此图 14-7 未完全显示数据。读者可以参考 twitter 开发人员指南的 apps.html 文件和 access-tokens.html 文件。

14.6.2　Stocksight 开源项目的 sentiment_analysis 程序

Stocksight 项目目录显示如图 14-8 所示，项目中程序文件 sentiment.py 包括股市舆情分析主要部分，以下显示该部分编程的摘要和重点。

图 14-8 Stocksight 项目目录

在函数 sentiment_analysis 里：

- 使用 14.3.1 节介绍的 TextBlob。
- 使用 Github 开源项目 cjhutto/vaderSentiment，VADER 是一种基于词典和规则的情感分析工具。
- 如果收集到达推文的任何 URL 的链接网页数据，则同时使用。否则，只使用推文内容。
- 按照 Stocksight 开源项目作者的策略，结合 TextBlob 和 VaderSentiment 的结果，调整文本的最终极性值（polarity）。

```python
def sentiment_analysis(text):
……
    # pass text into TextBlob
    text_tb = TextBlob(text)

    # pass text into VADER Sentiment
    analyzer = SentimentIntensityAnalyzer()
    text_vs = analyzer.polarity_scores(text)

    # determine sentiment from our sources
    if sentiment_url is None:
        if text_tb.sentiment.polarity < 0 and text_vs['compound'] <= -0.05:
            sentiment = "negative"
        elif text_tb.sentiment.polarity > 0 and text_vs['compound'] >= 0.05:
            sentiment = "positive"
        else:
            sentiment = "neutral"
    else:
         if text_tb.sentiment.polarity < 0 and text_vs['compound'] <= -0.05 and sentiment_url == "negative":
            sentiment = "negative"
         elif text_tb.sentiment.polarity > 0 and text_vs['compound'] >= 0.05 and sentiment_url == "positive":
            sentiment = "positive"
         else:
            sentiment = "neutral"

    # calculate average polarity from TextBlob and VADER
    polarity = (text_tb.sentiment.polarity + text_vs['compound']) / 2
……
```

第 15 章 使用 Elasticsearch 进行机器学习

机器学习被视为 Elasticsearch 中搜索和分析功能的自然扩展。Elasticsearch 支持的机器学习功能目前可以分为两类，无监督学习的异常检测（anomaly detection）和监督学习的数据框分析（data frame analytics），这两种技术都可以帮助分析时间序列数据。由于从下至上的方法来理解 Elasticsearch 机器学习操作方式相当不容易，所以使用自上而下的方式，由 Elastic Stack 的 Kibana 开始介绍如何创建机器学习作业，然后深入到 Elasticsearch 支持的接口。

15.1 Kibana 简介

Kibana 是一个开放的图形用户界面，能将 Elasticsearch 数据可视化，并可浏览 Elastic Stack 的组成元件，例如 Logstash 和 Beats。

15.1.1 准备环境和运行

以下是从官网下载、安装和运行 Kibana 7.5.1 版本的步骤：

（1）选择合适的操作系统的软件包，然后下载 7.5.1 版本。对于 Linux，文件名是 kibana-7.5.1-linux-x86_64.tar.gz。

（2）将压缩文件解压缩到目标目录，用以下命令生成一个名为 kibana-7.5.1 的文件夹：

```
tar -zxvf kibana-7.5.1-linux-86_64.tar.gz
```

（3）在指定的路径（kibana-7.5.1-linux-x86_64 目录），使用前台模式运行程序：

```
./bin/kibana
    log   [16:59:18.279] [info][plugins-system] Setting up [15] plugins: [security,
licensing,code,timelion,features,spaces,translations,uiActions,inspector,embeddable,
advancedUiActions,data,newsfeed,expressions,eui_utils]
    ......
    log   [16:59:29.685] [info][listening] Server running at http://localhost:5601
    log   [16:59:29.731] [info][server][Kibana][http] http server running at http://
localhost:5601
```

（4）如果要关闭 Kibana 进程，可以按【Ctrl＋C】组合键停止它

15.1.2 测试 Kibana

将浏览器地址指向 http://localhost:5601 与 Kibana 通信。Kibana 的主页如图 15-1 所示：

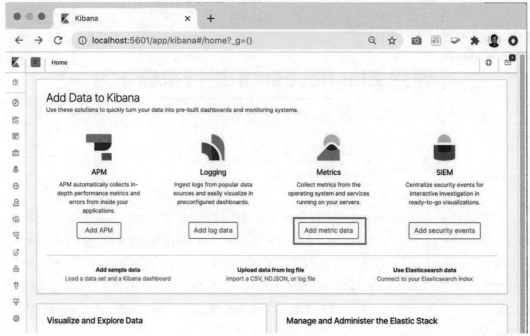

图 15-1 Kibana 的主页

为了证明运行的系统没有问题，请尝试使用官方网站提供的示例数据测试 Kibana：

（1）按下添加指标数据（Add metric data）按钮，可以看到以下屏幕如图 15-2 所示：

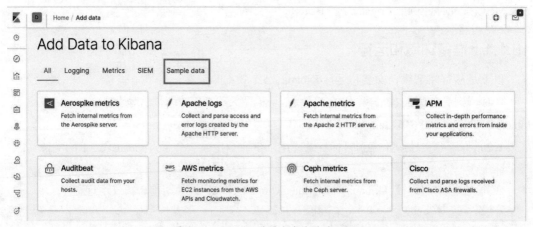

图 15-2 Kibana 的添加指标数据页面

（2）按下样本数据（Sample data）超连结，可以看到以下屏幕如图 15-3 所示：

第 15 章 使用 Elasticsearch 进行机器学习

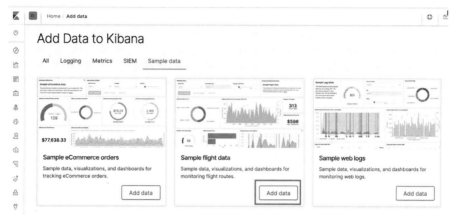

图 15-3 按下 Kibana 样本数据超连结后的页面

(3) 按下航班样本数据 Add data 按钮，可以看到以下屏幕如图 15-4 所示：

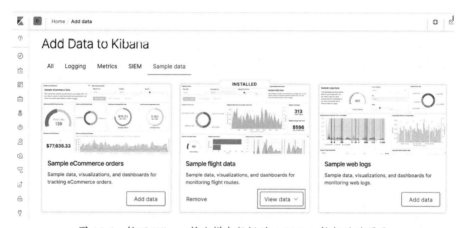

图 15-4 按下 Kibana 航班样本数据的 Add data 按钮后的页面

(4) 按下浏览数据（view data）按钮，可以看到以下屏幕如图 15-5 所示：

图 15-5 按下 Kibana 航班样本数据的 view data 按钮后的页面

（5）将鼠标移到左侧工具栏底部被框的按钮，按下后将进入管理（Management）界面，然后按索引管理（Index Management）超链接，可以看到以下屏幕如图 15-6 所示。此测试创建了 kibana_sample_data_flight 索引示例数据。其他索引是在之前的章节创建的。

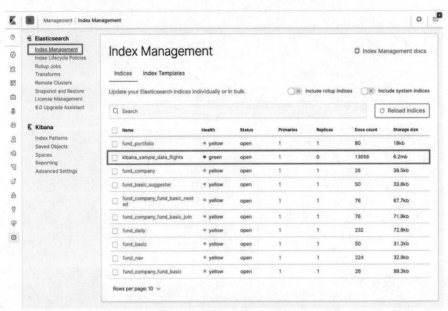

图 15-6　按下 Kibana 管理界面超链接后，再按 index management 超链接的页面

15.2　Kibana、Elasticsearch 与机器学习

根据 Elastic Stack 许可规则，Machine learning 功能需要 Platinum 许可。要运行有关程序，可以先安装试用许可证，并且激活 30 天试用许可证才能使用 Platinum 功能。

15.2.1　安装试用许可证

按照以下步骤启动 Elastic Stack 试用许可证：

（1）在管理界面左侧，有一个许可证管理（License Management）超链接，按下后可以看到许可证管理界面的开始试用 30 天按钮，屏幕如图 15-7 所示。

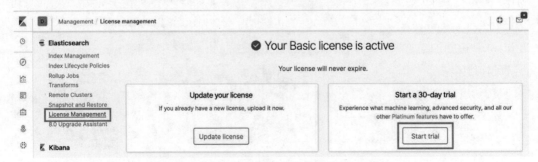

图 15-7　许可证管理界面的开始试用 30 天按钮

第 15 章　使用 Elasticsearch 进行机器学习

（2）按下开始试用 30 天按钮后，可以看到许可证说明弹出面板，屏幕如图 15-8 所示。

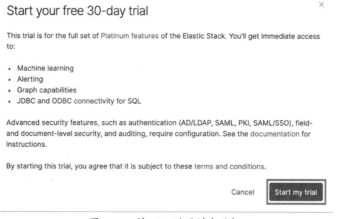

图 15-8　许可证说明弹出面板

（3）按下开始试用按钮后，可以看到屏幕一条通知，意思是 Elastic Stack 试用版许可证处于活动状态，并将在 30 天后过期。屏幕如图 15-9 所示。

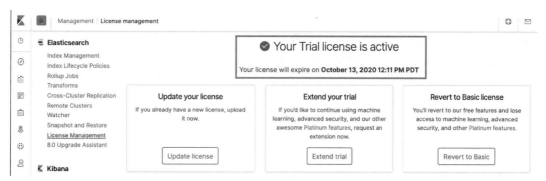

图 15-9　许可证管理界面显示试用版许可证处于活动状态页面

15.2.2　机器学习测试数据

为了机器学习测试有更好的结果，本章添加了新的数据集。类似 11.1.4 节公募基金交易行情文档索引操作，新创建名称为 510350_daily 的索引。并进行基金代码为 510350.SZ(工银沪深 300ETF) 的文档索引。由于此索引只储存工银沪深 300ETF 的文档，所以在显式映射文件 510350_mappings.json 中删除字段 ts_code，其他字段则使用相同的定义。在文档进行索引之前，会执行 11.1.2 节描述的 ohlc_avg_price_pipeline 摄取节点管道操作。请确认此摄取节点管道已经存在。

1. 创建相关索引

相关的索引操作文件包括 510350_bulk.json、510350_index.sh 和 510350_mappings.json。文件 510350_index.sh 是个 bash 执行档，运行执行以下任务：

（1）使用在 510350_mappings.json 内定义的设定和分析器创建 510350_daily 索引。

（2）使用 510350_bulk.json 内编写好的批量处理文档接口指令，对基金 510350.SZ 的日线行情数据进行文档索引。数据时间段是在 2019 年 08 月 16 日至 2020 年 09 月 11 日之间。执行 510350_index.sh 后的部分索引结果显示如下：

```
./510350_index.sh
{"acknowledged":true,"shards_acknowledged":true,"index":"510350_daily"}
{
  ......
  "items" : [
    {
      "index" : {
        "_index" : "510350_daily",
        "_type" : "_doc",
        "_id" : "CmsVinQBQAdw2QlpvHVp",
        "_version" : 1,
        "result" : "created",
        "_shards" : {
          "total" : 2,
          "successful" : 1,
          "failed" : 0
        },
        "_seq_no" : 0,
        "_primary_term" : 1,
      ......
}
```

（3）检查管道 ohlc_avg_price_pipeline 中两个处理器是否已正常执行。搜索索引 510350_daily 的部分结果显示如下，共有 263 文档，已经包含 ohlc 字段。

```
curl --request POST http://localhost:9200/510350_daily/_search?pretty=true
{
  ......
  "hits" : {
    "total" : {
      "value" : 263,
      "relation" : "eq"
    },
    "max_score" : 1.0,
    "hits" : [
      {
        "_index" : "510350_daily",
        "_type" : "_doc",
        "_id" : "AZTOM3UBGbpZEaNBtTcp",
        "_score" : 1.0,
        "_source" : {
          "amount" : 25217.033,
          "ohlc" : 4.6290000000000004,
          "change" : 0.029,
      ......
}
```

2. 创建 Kibana 索引模式

在 Kibana，必须先为索引创建索引模式（index pattern），然后才能使用于机器学习。创建索引模式 510350_daily* 步骤如下：

（1）如图 15-10 所示首先指定输入索引模式为 510350_daily*。

（2）如图 15-11 所示，指定字段名称 trade_date（交易日）为此索引模式的时间过滤器（time filter），然后单击创建索引模式按钮。

第 15 章　使用 Elasticsearch 进行机器学习

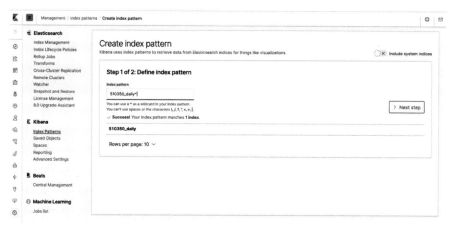

图 15-10　在 Kibana 创建索引模式 510350_daily*

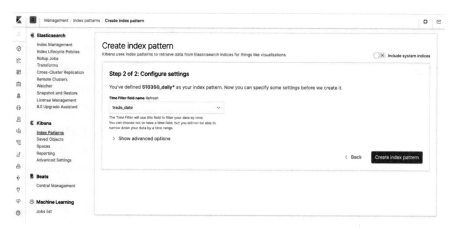

图 15-11　在 Kibana 指定字段名称 trade_date 为 time filter

（3）如图 15-12 所示，510350_daily* 索引模式创建成功后，会显示其各个字段的特征，如可用于搜索和可用于聚合等。

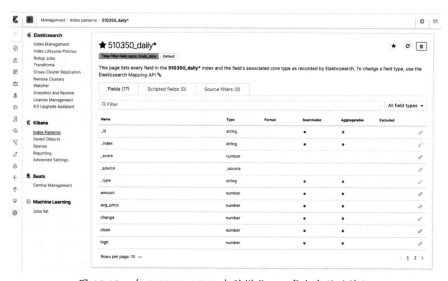

图 15-12　在 510350_daily* 索引模式下，每个字段的特征

15.2.3　Elastic 机器学习工作流程

按照 Kibana 机器学习使用主界面，目前支持的工作流程，正如屏幕如图 15-13 所示，可以分为两类，无监督学习的异常检测（anomaly detection）和监督学习的数据框分析（data frame analytics）。

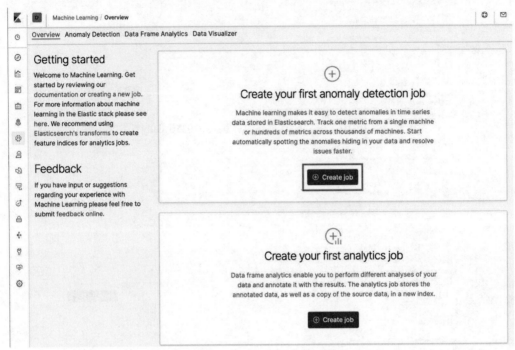

图 15-13　Kibana 机器学习主界面

以下使用两个示例来说明这两类机器学习的工作流程。

1. 无监督学习异常检测

按下 Create job 按钮去创建异常检测任务后，显示的是选择索引模式，如图 15-14 所示选择索引模式 510350_daily*。

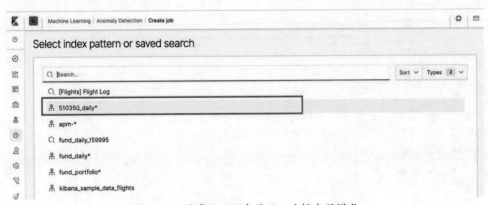

图 15-14　异常检测任务界面，选择索引模式

选择索引 510350_daily 索引模式后，显示异常检测任务四种子类型供选择。如图 15-15

所示，分别为单指标作业、多指标作业、数据分布行为作业和高级作业。还可以执行数据可视化。

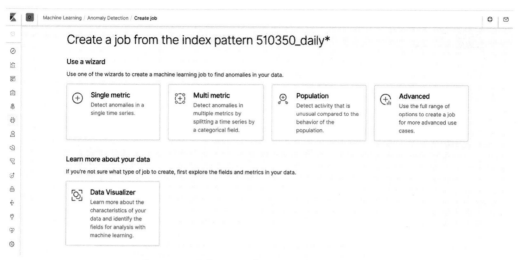

图 15-15　异常检测任务界面四种子类型供选择

- 单指标作业检测

如图 15-16 所示，单指标作业检测共需要四个步骤才能得出结果（summary）：

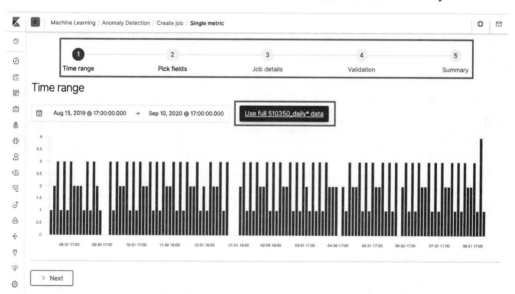

图 15-16　单指标作业的四种个步骤，并且选择使用完整数据

（1）选择使用完整数据（Use full … data）按钮，然后单击下一步（Next）按钮，将出现选择需要分析的字段，如图 15-17 所示。举个例子，选择字段 vol 的总和 Sum（vol）。桶间隔（Bucket span）为 1 天（1d），并且忽略空桶（Sparse data）。

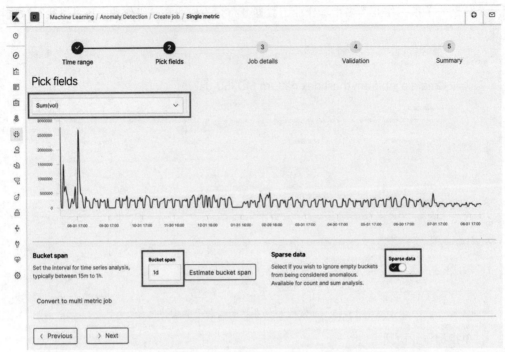

图 15-17　单指标作业中，选择需要分析的字段和指定选项步骤

（2）选择使用完整数据和指定其他选项后，单击下一步（Next）按钮。将出现单指标作业任务使用界面，需要填写任务名称标识和描述，如图 15-18 所示。本次测试任务名称为 510350_daily_1。

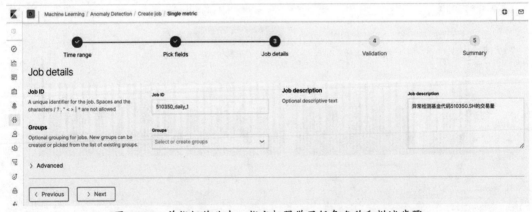

图 15-18　单指标作业中，指定机器学习任务名称和描述步骤

（3）填写信息之后，需要通过验证步骤。如果有任何错误，将以红色显示验证错误消息。在图 15-19 所示没有错误，蓝色是信息性通知。单击下一步（Next）按钮，将出现将要创建任务的内容界面。

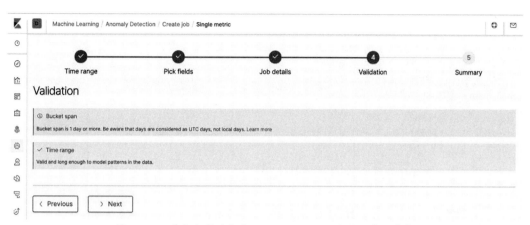

图 15-19　单指标作业任务 510350_daily_1 的验证步骤消息

（4）在创建任务之前，需要检查所有填写内容。如有任何错误，可以使用上一步（Previous）按钮，回到相应的步骤中进行编辑。图 15-20 显示给定的任务内容。单击下一步（Next）按钮，进入创建名称为 510350_daily_1 的任务。

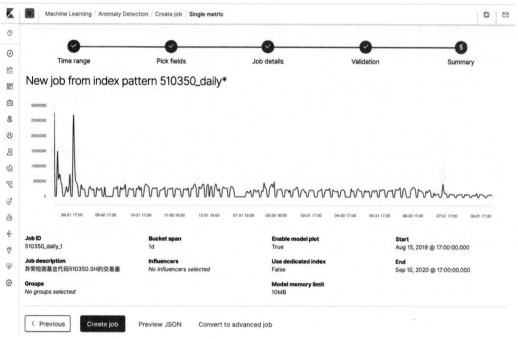

图 15-20　单指标作业中，创建任务任务 510350_daily_1 前检查任务内容的步骤

名称为 510350_daily_1 的任务创建后，如果存在异常，它将显示在折线图中。例如显示图 15-21 中的橙色条形和蓝色条形。

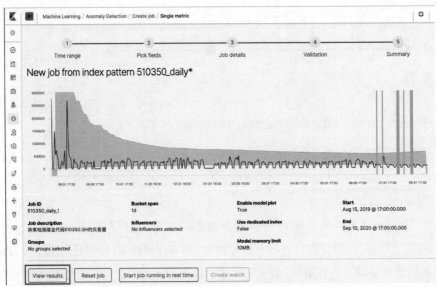

图 15-21　单指标作业创建任务任务 510350_daily_1 结果，显示异常以橙色条形和蓝色条形出现

单击查看结果（View results）按钮，可以详细检查异常检测结果。在图 15-22 中出现以橙色显示严重性为主要级别（major）发生在 July 20th,2020。原因是其 vol 值严重的低，但是奇怪的是，开头急剧的高值并未标识为异常。

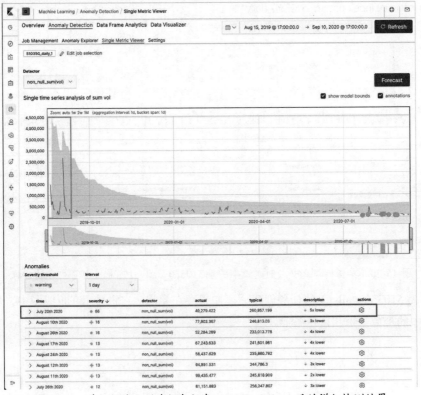

图 15-22　单指标作业创建任务任务 510350_daily_1 后的详细检测结果

除了提供异常检测外，还提供了预测功能。单击预测（Forcast）按钮，图 15-23 在弹出

窗口输入预测的天数（测试 7 天），然后按运行（Run）按钮。

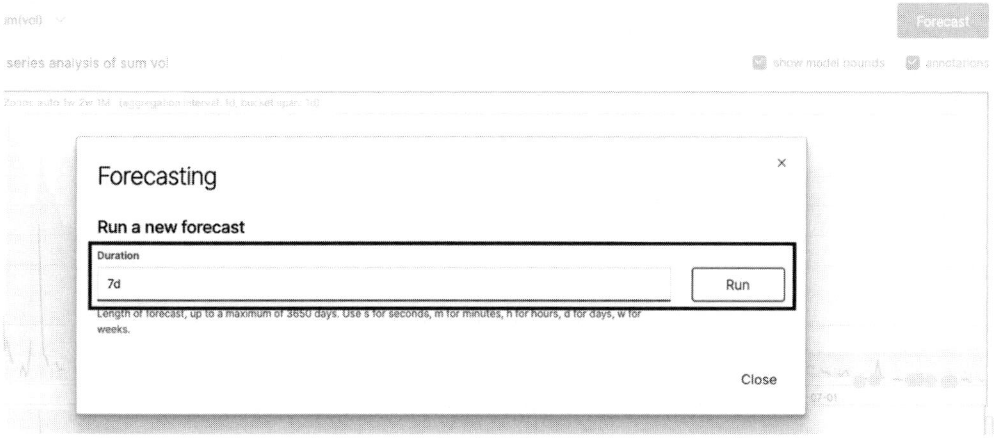

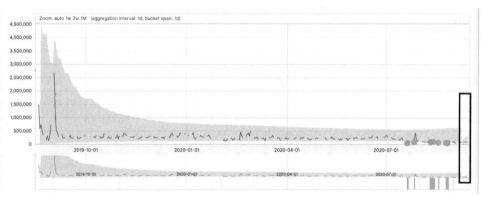

图 15-23　单指标作业提供的预测功能

如图 15-24 所示 7 天的预测结果，图中用框架标示。

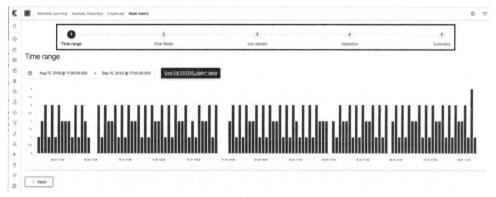

图 15-24　单指标作业预测任务 510350_daily_1 的 7 天预测结果，用框架标示

◆ 多指标作业检测

若选择多指标作业，如图 15-25 所示，也是需要四个步骤才能得出结果（summary）：

图 15-25　多指标作业的四种个步骤，并且选择使用完整数据

（1）选择使用完整数据（Use full … data）按钮，然后单击下一步（Next）按钮，将出

现选择需要分析的字段如图 15-26 所示。这个例子选择字段 vol 的总和 Sum（vol），桶间隔（Bucket span）为 1 天（1d），并且忽略空桶（Sparse data）。有一个 influencers（影响者）按钮，可猜测具备影响力的字段来分析，此测试选择多个字段如图所示。

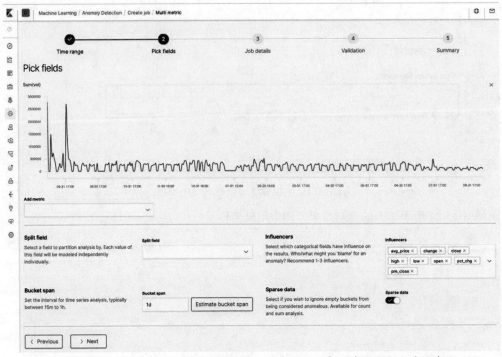

图 15-26　510350_daily_ml 多指标作业中，选择需要分析的字段和指定选项步骤

（2）选择使用完整数据和指定其他选项后，单击下一步（Next）按钮。将出现多指标作业任务使用界面，需要填写任务名称标识和描述，如图 15-27 所示。本次测试任务名称为 510350_daily_ml。

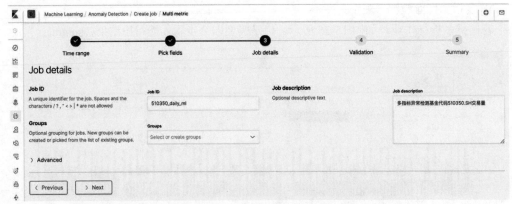

图 15-27　多指标作业中，指定机器学习任务名称和描述步骤

（3）填写信息之后，需要通过验证步骤。如果有任何错误，将以红色显示验证错误消息。在图 15-28 所示没有错误，蓝色是信息性通知。图中在黄色栏中出现一条消息以引起注意。消息是不要选太多影响者，通常不需要三个以上。单击下一步（Next）按钮，将出现将要创

建任务的内容界面。

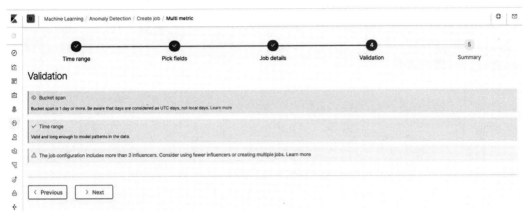

图 15-28　多指标作业任务 510350_daily_m1 的验证步骤消息

（4）在创建任务之前，需要检查所有填写内容。如有任何错误，可以使用上一步（Previous）按钮，回到相应的步骤中进行编辑。图 15-29 显示给定的任务内容。单击下一步（Next）按钮，进入创建名称为 510350_daily_m1 的任务。单击创建任务按钮（Create job）。

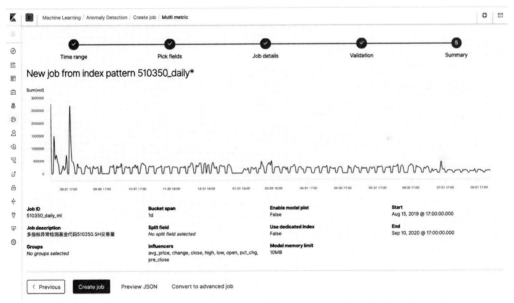

图 15-29　多指标作业中，创建任务 510350_daily_m1 前检查任务内容的步骤

名称为 510350_daily_m1 的任务创建后，如果存在异常，它将显示在折线图中。例如显示图 15-30 中的橙色条形和蓝色条形。

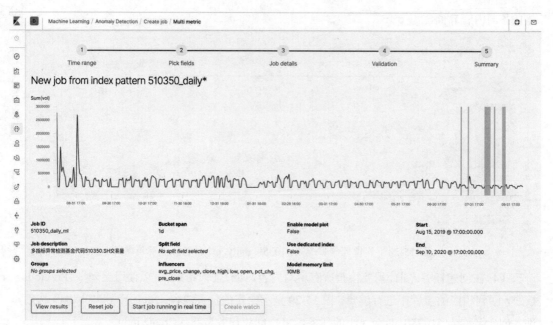

图 15-30　多指标作业创建任务任务 510350_daily_m1 结果，显示异常以橙色条形和蓝色条形出现

单击查看结果（View results）按钮，可以详细检查异常检测结果，显示见图 15-31。

图 15-31　多指标作业创建任务 510350_daily_m1 后的详细检测结果类似于单指标作业。因为找不到影响者，否则将提供更多信息

2. 监督学习数据框分析

Kibana 机器学习主界面提供进入数据框分析使用界面，如图 15-32 所示。在数据框分析面板中，单击创建任务（Create job）按钮，进入创建数据框分析任务使用界面。

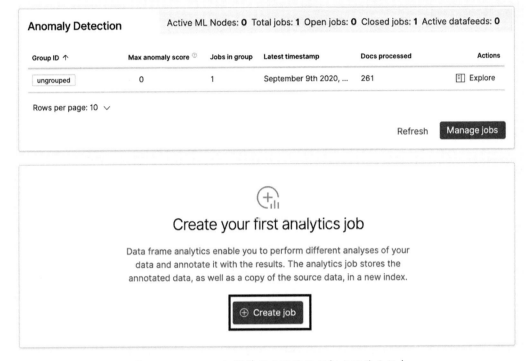

图 15-32　Kibana 机器学习主界面显示相应的作业任务

数据框分析任务使用界面如图 15-33 所示，单击创建分析任务（Create analytics job）按钮。

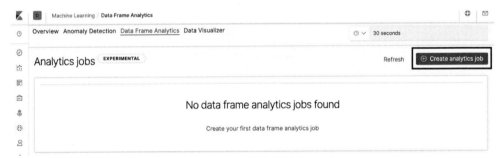

图 15-33　Kibana 数据框分析任务使用界面

在创建分析任务界面填写所需的信息，任务类型有两种，离群值检测（Outliners detection）和回归（Regression）。

◆ 离群值检测：

如图 15-34 所示，任务标识符为 analyze_510350_1、源索引为 510350_daily 及目标索引为 510350_analyze。需要再执行三个步骤才能得出结果。

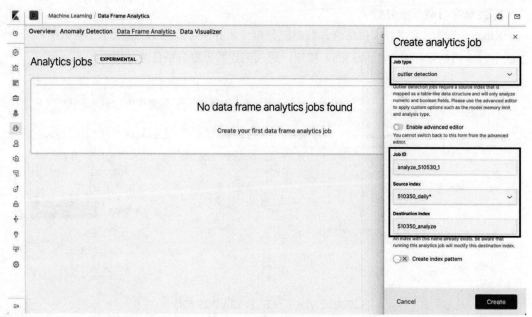

图 15-34 输入所需的信息以创建离群值检测任务

(1) 单击创建任务 (Create) 按钮,创建 analyze_510350_1 分析任务。界面变化如图 15-35 所示。analyze_510350_1 分析任务已创建并显示在列表中。创建按钮转换为开始执行按钮 (Start)。

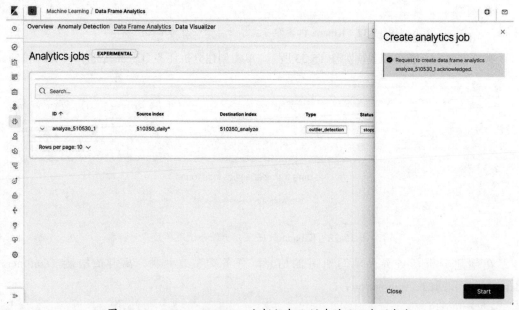

图 15-35 analyze_510350_1 分析任务已创建并显示在列表中

(2) 单击开始执行 (Start) 按钮,analyze_510350_1 分析任务开始执行,收到确认 (Acknowledged) 的消息,如图 15-36 所示。

第 15 章 使用 Elasticsearch 进行机器学习

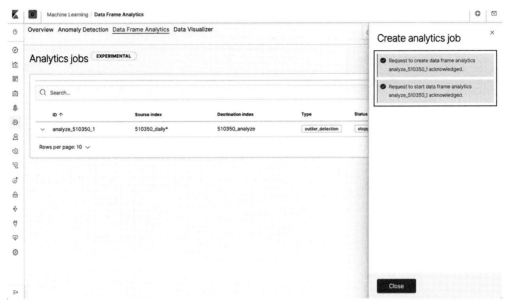

图 15-36 开始执行 analyze_510350_1 分析任务,收到确认消息

(3)单击关闭(Close)按钮,关闭当前弹出窗口后,主视窗如图 15-37 所示。

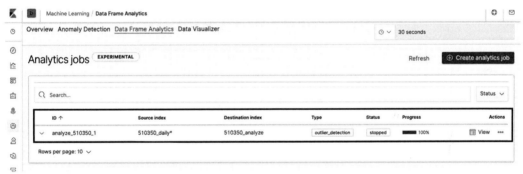

图 15-37 analyze_510350_1 分析任务的状态和支持的操作

要查看分析结果,单击图 15-37 标示的查看(View)按钮。单击该按钮后,界面如图 15-38 所示,将显示各个字段检测到的离群值。

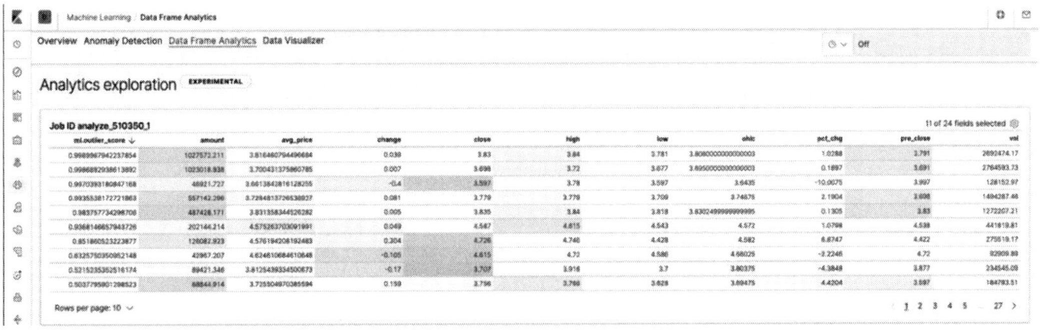

图 15-38 510350_analytics_1 分析任务检测到各个字段的离群值

- 回归检测:

如图 15-39 所示，任务标识符为 analyze_510350_2、源索引为 510350_daily、目标索引为 regression_510350 及因变量为 vol。需要再执行三个步骤才能得出结果。

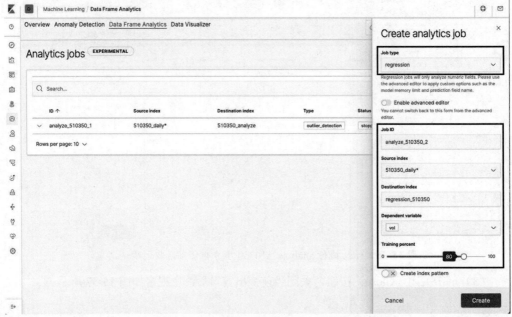

图 15-39 填写所需的信息以创建回归任务

（1）单击创建任务（Create）按钮，创建 analyze_510350_2 分析任务。界面变化如图 15-40 所示。analyze_510350_2 分析任务已创建并显示在列表中。创建按钮转换为开始执行按钮（Start）。

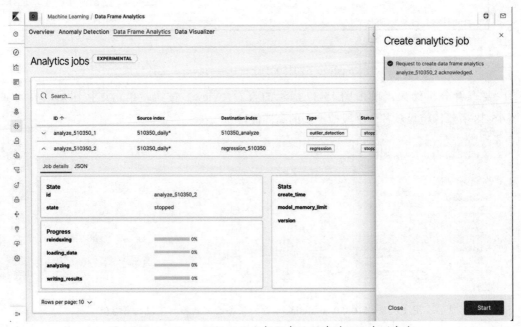

图 15-40 analyze_510350_2 分析任务已创建并显示在列表中

（2）单击开始执行（Start）按钮，analyze_510350_2 分析任务开始执行，收到确认（Acknowledged）的消息，如图 15-41 所示。

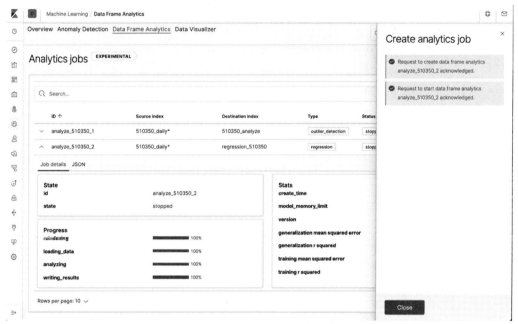

图 15-41　开始执行 analyze_510350_2 分析任务，收到确认消息

（3）单击关闭（Close）按钮，关闭当前弹出窗口。注意图 15-42 标示的查看（View）分析结果按钮已被禁用，很可能此版本尚未支持此功能。

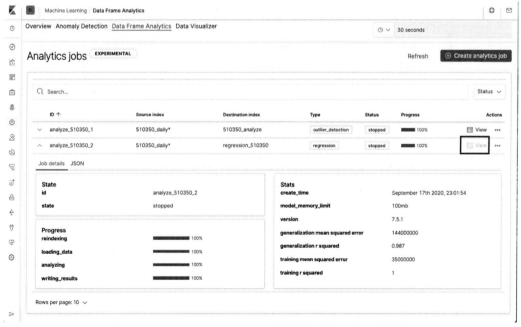

图 15-42　analyze_510350_2 分析任务的状态和支持的操作

15.3 Elasticsearch 机器学习异常检测接口

支持在 15.2.3 节项目 1 介绍的无监督学习异常检测，相关的 Elasticsearch 接口，提供在 _ml 下的四种主要资源分别为：

- _ml/anomaly_detectors(异常检测任务)
- _ml/datafeeds(数据馈送)
- _ml/calendars(特定时间段)
- _ml/filters(过滤器)

此外还提供这些主要资源下的相关资源接口：

- events(事件)
- influencers(影响)
- jobs(任务)
- record(记录)
- model_snapshots(模型快照)
- overall_buckets(存储桶结果总结)

15.3.1 异常检测任务资源

异常检测任务 (anomaly_detectors) 存储任务所需的配置信息和元数据，每个异常检测任务都有一个或多个检测器应用于特定字段。在 _ml/anomaly_detectors 资源下各个相关任务接口汇总到表 15-1 中，并简要说明。

提示：如图 15-17 所示，在 15.2.3 节项目 1 创建单指标作业任务 510350_daily_1 时指定的特定字段 Sum(vol) 及如图 15-26 所示的多指标作业任务 510350_daily_m1 时指定的多个特定字段。

表 15-1 在 _ml/anomaly_detectors 资源下各个相关接口列表

接口方法和方式	描　　述
GET _ml/anomaly_detectors/ [<job_id>|<job_id>,<job_id>|_all]	获取异常检测任务相关内容，可以给定一个或多个 (用逗号分隔) 任务标识符 <job_id>。全部则使用 _all 或不给定 <job_id>
PUT _ml/anomaly_detectors/<job_id>	创建标识符为 <job_id> 的异常检测任务
DELETE _ml/anomaly_detectors/<job_id>	删除标识符为 <job_id> 的异常检测任务
POST _ml/anomaly_detectors/<job_id> /_update	更新标识符为 <job_id> 的异常检测任务的内容
POST _ml/anomaly_detectors/ [<job_id>|<job_id>,<job_id>|_all]/_close	关闭异常检测任务，可以关闭一个、多个或全部任务。关闭之前，必须停止其数据馈送
POST _ml/anomaly_detectors/<job_id>/_open	给定一个标识符 <job_id>，打开异常检测任务
POST _ml/anomaly_detectors/<job_id>/_flush	当标识符为 <job_id> 的异常检测任务进行数据分析时，将数据刷新。刷新操作后，作业仍处于打开状态
POST _ml/anomaly_detectors/<job_id>/_data	将数据发送到标识符为 <job_id> 的异常检测任务数据。在请求主体必须为 JSON 格式，并且必须先打开任务
GET _ml/anomaly_detectors/ [<job_id>|<job_id>,<job_id>|_all]/_stats	从一个，多个或所有异常检测任务获取统计信息

续表

接口方法和方式	描　　述
POST _ml/anomaly_detectors/<job_id>/_forecast	标识符为 <job_id> 的异常检测任务的预测结果
DELETE _ml/anomaly_detectors/<job_id>/_forecast/[<forecast_id>|_all]	从标识符为 <job_id> 的任务中，删除预测标识符为 <forecast_id> 或全部 (不给定标识符或 _all) 的预测结果
GET _ml/anomaly_detectors/<job_id>/results/categories/[<category_id>]	获取任务标识符为 <job_id> 的异常检测任务的结果中全部 (不给定标识符) 或类别标识符为 <category_id> 的类别
GET _ml/anomaly_detectors/<job_id>/results/records	获取任务标识符为 <job_id> 的异常检测任务的结果中的异常记录。注意必须启动数据馈送后才产生任务的结果
GET _ml/anomaly_detectors/<job_id>/results/buckets/ [<timestamp>]	获取任务标识符为 <job_id> 的异常检测任务的结果中按存储桶分组的时间顺序视图
GET _ml/anomaly_detectors/[<job_id>|<job_id>,<job_id>|_all]/results/overall_buckets	从一个，多个或所有异常检测任务获取存储桶结果的总结
GET _ml/anomaly_detectors/<job_id>/results/influencers	获取任务标识符为 <job_id> 的异常检测任务的结果中的影响者信息
GET _ml/anomaly_detectors/<job_id>/model_snapshots[/<snapshot_id>]	获取任务标识符为 <job_id> 中，快照模型标识符为 <snapshot_id> 或全部快照模型 (不给定标识符) 的相关内容
POST _ml/anomaly_detectors/<job_id>/model_snapshots/<snapshot_id>/_update	更新任务标识符为 <job_id> 中，快照模型标识符为 <snapshot_id> 的快照模型相关的内容
DELETE _ml/anomaly_detectors/<job_id>/model_snapshots/<snapshot_id>	从标识符为 <job_id> 的任务中，删除快照模型标识符为 <snapshot_id> 的快照模型
POST _ml/anomaly_detectors/<job_id>/model_snapshots/<snapshot_id>/_revert	恢复任务标识符为 <job_id> 中，快照模型标识符为 <snapshot_id> 的快照模型

以下测试为获取所有异常检测任务的相关内容，报告当前有两个异常检测任务。其标识符 <job_id> 一个是 510350_daily_1，而另一个是 510350_daily_m1。

提示：如图 15-18 所示，在 15.2.3 节项目 1 创建单指标作业任务时指定的标识符 510350_daily_1 及如图 15-27 所示的多指标作业任务时指定的标识符 510350_daily_m1。

```
curl --request GET http://localhost:9200/_ml/anomaly_detectors?pretty=true
{
  "count" : 2,
  "jobs" : [
      ……
    {
      "job_id" : "510350_daily_m1",
      "job_type" : "anomaly_detector",
      "job_version" : "7.5.1",
      "description" : "多指标异常检测基金代码510350.SH交易量",
      "create_time" : 1600274572986,
      "finished_time" : 1600274575793,
      "analysis_config" : {
        "bucket_span" : "1d",
        "detectors" : [
          {
            "detector_description" : "non_null_sum(vol)",
            "function" : "non_null_sum",
            "field_name" : "vol",
            "detector_index" : 0
          }
        ],
        "influencers" : [
          "avg_price",
```

```
            "change",
            "close",
            "high",
            "low",
            "ohlc",
            "open",
            "pct_chg",
            "pre_close"
          ]
        },
        "analysis_limits" : {
          "model_memory_limit" : "10mb",
          "categorization_examples_limit" : 4
        },
        "data_description" : {
          "time_field" : "trade_date",
          "time_format" : "epoch_ms"
        },
        "model_snapshot_retention_days" : 1,
        "custom_settings" : {
          "created_by" : "multi-metric-wizard"
        },
        "model_snapshot_id" : "1600274575",
        "results_index_name" : "shared",
        "allow_lazy_open" : false
      ...
```

异常检测任务的控制参数比较复杂，有关任务 510350_daily_m1 的参数汇总到表 15-2 中，并简要说明。

表 15-2　异常检测任务的常用参数列表

参数	数据类型	描述
analysis_config.bucket_span	Time units	桶与桶间，时间间隔的大小
analysis_config.detectors.detector_description	string	给定的检测器说明
analysis_config.detectors.function	string	对异常检测的字段使用的分析功能
analysis_config.detectors.field_name	string	异常检测的字段名称
analysis_config.detectors.detector_index	integer	检测器为数组，detector_index 为数组的标识符，默认从 0 开始
analysis_config.influencers	array of string	以逗号分隔的影响者字段名称列表
analysis_limits.model_memory_limit	long or string	分析处理最大内存限制，在 6.1 版之后，默认值为 1024MB
analysis_limits.categorization_examples_limit	long	存储的最大示例数，默认值为 4
data_description.time_field	string	参与分析的时间戳字段名称，默认值为 time
data_description.time_format	string	将数据发送到任务时的格式，目前仅支持 JSON 格式
model_snapshot_retention_days	long	快照模型保留的时间，默认为 1 天
custom_settings.created_by	object	包含有关任务的自定义元数据，created_by 字段为创造
model_snapshot_id	string	快照模型标识符
results_index_name	string	指定机器学习目标索引的名称，默认值为 .ml-anomalies-shared
allow_lazy_open	boolean	节点容量不足下，是否可以打开该任务，默认值为 false

15.3.2　数据馈送

数据馈送（datafeeds）包含一个以时间间隔（频率）定义的查询语句。每个异常检测任务，只能关联一个数据馈送。使用 Kibana 创建异常检测任务，也同时创建并关联了一个数据馈送。异常检测任务 510350_daily_1 关联的数据馈送标识符为 datafeed-510350_daily_1，在 _ml/datafeeds 资源下各个相关数据馈送接口汇总到表 15-3 中，并简要说明。

表 15-3 在 _ml/datafeeds 资源下各个相关接口列表

接口方法和方式	描 述
GET _ml/datafeeds/[<_feed_id>|<feed_id>,>feed_id>|_all]	获取数据馈送相关内容，可以给定一个或多个 (用逗号分隔) 标识符 <feed_id>。全部则使用 _all 或不给定 <feed_id>
PUT _ml/datafeeds/<feed_id>	创建标识符为 <feed_id> 的数据馈送
DELETE _ml/datafeeds/<feed_id>	删除标识符为 <feed_id> 的数据馈送
POST _ml/datafeeds/<feed_id>/_update	更新标识符为 <feed_id> 的数据馈送的内容
GET _ml/datafeeds/[<_feed_id>|<feed_id>,>feed_id>|_all]/stats	从一个，多个或所有数据馈送获取统计信息
POST _ml/datafeeds/<feed_id>/_start	启动标识符为 <feed_id> 的数据馈送，并准备从中 Elasticsearch 检索数据
POST _ml/datafeeds/[<_feed_id>|<feed_id>,>feed_id>|_all]/_stop	停止一个、多个或全部数据馈送的操作
GET _ml/datafeeds/<datafeed_id>/_preview	预览标识符为 <feed_id> 的数据馈送

以下测试为获取所有数据馈送的相关内容，报告当前有两个数据馈送。其标识符 <feed_id>，一个是 datafeed-510350_daily_1，而另一个是 datafeed-510350_daily_m1。

提示：在 15.2.3 节项目 1 描述的所有步骤，从来没有指定数据馈送名称，可以看到是 Kibana 自动生成的。

```
curl --request GET http://localhost:9200/_ml/datafeeds?pretty=true
{
  "count" : 2,
  "datafeeds" : [
    {
      "datafeed_id" : "datafeed-510350_daily_1",
      "job_id" : "510350_daily_1",
      "query_delay" : "106306ms",
      "indices" : [
        "510350_daily*"
      ],
      "query" : {
        "bool" : {
          "must" : [
            {
              "match_all" : { }
            }
          ]
        }
      },
      "aggregations" : {
        "buckets" : {
          "date_histogram" : {
            "field" : "trade_date",
            "fixed_interval" : "8640000ms"
          },
          "aggregations" : {
            "vol" : {
              "sum" : {
                "field" : "vol"
              }
            },
```

```
          "trade_date" : {
            "max" : {
              "field" : "trade_date"
            }
          }
        }
      },
      "scroll_size" : 1000,
      "chunking_config" : {
        "mode" : "manual",
        "time_span" : "8640000000ms"
      },
      "delayed_data_check_config" : {
        "enabled" : true
      }
    },
......
```

15.3.3 特定时间段资源

特定时间段资源（calendars）用于指定预先猜测会发生异常活动的时间段，此期间所有数据将忽略，不会视为异常。机器学习模型也不会受到影响，也不会造成不良后果。在 _ml/calendars 资源下各个相关接口汇总到表 15-4 中，并简要说明。

表 15-4 在 _ml/calendars 资源下各个相关接口列表

接口方法和方式	描 述
GET _ml/calendars/[<calendar_id>/_all]	获取特定时间段资源相关内容，可以给定一个特定时间段标识符 <calendar_id>。全部则使用 _all 或不给定标识符
PUT _ml/calendars/<calendar_id>	创建标识符为 <calendar_id> 的特定时间段
DELETE _ml/calendars/<calendar_id>	删除标识符为 <calendar_id> 的特定时间段及其相关的预定事件
GET_ml/calendars/[<calendar_id>|_all]/events	获取识符为 <calendar_id> 的特定时间段内所有事件
POST_ml/calendars/<calendar_id>/events	将事件添加到标识符为 <calendar_id> 的特定时间段
DELETE _ml/calendars/<calendar_id>/events/<event_id>	将标识符为 <event_id> 的事件从标识符为 <calendar_id> 的特定时间段删除
PUT _ml/calendars/<calendar_id>/jobs/<job_id>	将标识符为 <job_id> 的异常检测任务，添加到标识符为 <calendar_id> 的特定时间段
DELETE_ml/calendars/<calendar_id>/jobs/<job_id>	将标识符为 <job_id> 的异常检测任务，从标识符为 <calendar_id> 的特定时间段删除

要测试特定时间段资源，需要创建一个试特定时间段。请按照以下步骤进行练习：

（1）在异常检测主界面中提供 settings（设置）超连结如图 15-43 所示，提供特定时间段和过滤器的操作。单击此超连结会进入设置界面。

图 15-43　异常检测主界面中提供设置超连结

（2）设置界面中提供特定时间段管理和过滤器超连结列表的超连结，如图 15-44 所示。

图 15-44　特定时间段管理和过滤器超连结列表的超连结

（3）单击特定时间段管理超连结会进入特定时间段管理界面，如图 15-45 所示。

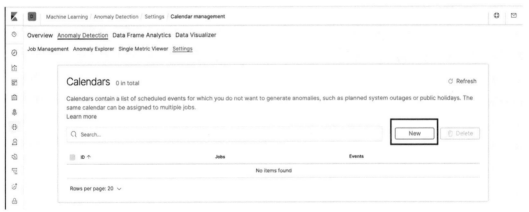

图 15-45　特定时间段管理界面

（4）单击 New（新增）按钮进入创建特定时间段管理界面，并填写新增特定时间段的信息。这个新特定时间段名称为 calendar-510350_daily_1，需要先关联一个事件，在下一步中，将创建并关联一个事件。最后按 Save（保存）按钮创建，如图 15-46 所示。

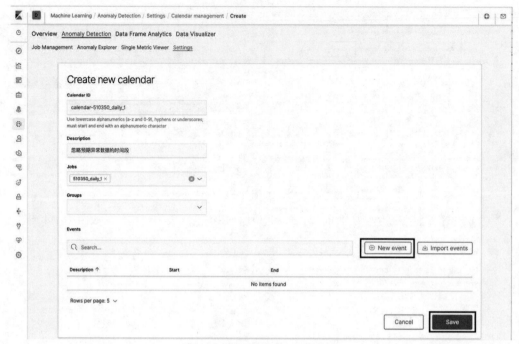

图 15-46　创建特定时间段界面

（5）单击如图 15-46 所示的 New event（新增事件）按钮，进入创建事件管理界面，如图 15-47 所示，并填写新增事件的信息。单击 Add（添加）按钮，弹出窗口将关闭。这个新增事件发生的特定时间段为 2020 年 7 月 20 日开始直到 2020 年 7 月 21 日结束。

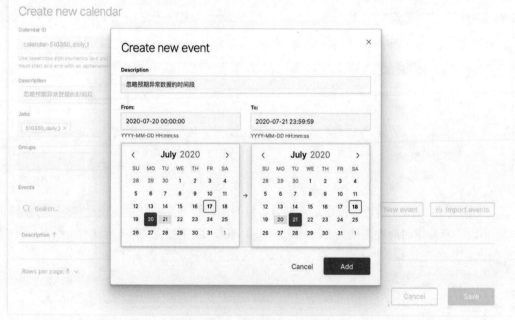

图 15-47　创建事件管理界面

（6）单击如图 15-46 所示的 Save（保存）按钮进行建，将回到特定时间段管理界面，如图 15-48 所示。列表将显示名称为 calendar-510350_daily_1 的特定时间段。

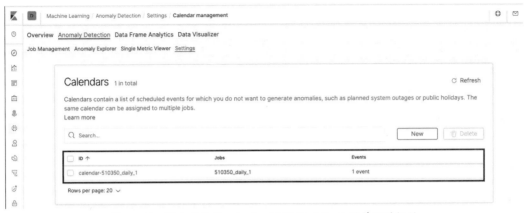

图 15-48　显示名称为 calendar-510350_daily_1 的特定时间段

（7）以下测试获取所有特定时间段的相关内容，报告当前有一个特定时间段。其标识符为 calendar-510350_daily_1。

```
curl --request GET  http://localhost:9200/_ml/calendars?pretty=true
{
  "count" : 1,
  "calendars" : [
    {
      "calendar_id" : "calendar-510350_daily_1",
      "job_ids" : [
        "510350_daily_1"
      ],
      "description" : "忽略预期异常数据的时间段"
    }
  ]
}
```

（8）以下测试获取标识符为 calendar-510350_daily_1 的特定时间段的所有事件相关内容。报告当前有一个事件，特定时间段的开始时间（start_time）为 1595228400000 和结束时间（end_time）为 1595401199000。时间单位为 Epoch 以来的毫秒数。

```
curl --request GET  http://localhost:9200/_ml/calendars/calendar-510350_daily_1/events?pretty=true
{
  "count" : 1,
  "events" : [
    {
      "description" : "忽略预期异常数据的时间段",
      "start_time" : 1595228400000,
      "end_time" : 1595401199000,
      "calendar_id" : "calendar-510350_daily_1",
      "event_id" : "HpFnnXQB5tfZj4JL3OeF"
    }
  ]
}
```

15.3.4　过滤器

如图 15-49 所示，设置界面中提供过滤器列表的超连接，单击此超连接，将进入过滤器管理界面。如图 15-49 所示，可以创建包含字符串列表的过滤器，用来过滤任务名称。读者可以自己练习。以下仅在表 15-5 简单说明与 _ml/filters（过滤器）相关的接口。

图 15-49　过滤器管理界面

表 15-5　在 _ml/filters 资源下各个相关接口列表

接口方法和方式	描　　述
GET _ml/filters/[<filter_id>]	获取过滤器资源相关内容，可以给定一个特过滤器标识符 <filter_id>。全部则不给定标识符
PUT _ml/filters/<filter_id>	创建标识符为 <filter_id> 的过滤器
POST _ml/filters/<filter_id>/_update	更新标识符为 <filter_id> 的过滤器的内容
DELETE _ml/filters/<filter_id>	删除标识符为 <filter_id> 的过滤器

15.4　Elasticsearch 机器学习数据框分析接口

支持在 15.2.3 节项目 2 介绍的监督学习数据框分析，相关的 Elasticsearch 接口主要围绕在 _ml/data_frame/analytics 资源。数据框架分析功能提供分类，离群值检测和回归算法分析数据。对源索引内数据执行分析，并将结果存储在目标索引中。如果目标索引不存在，则会自动创建新索引。各个数据框架分析接口汇总到表 15-6 中，并简要说明。

表 15-6　在 _ml/data_frame/analytics 资源下各个相关接口列表

接口方法和方式	描　　述
GET _ml/data_frame/analytics/ [<id>|<id>,<id>|_all]	获取数据框架分析任务相关内容，可以给定一个或多个（用逗号分隔）任务标识符 <id>。全部则使用 _all 或不给定 <id>
GET _ml/data_frame/analytics/ [<id>|<id>,<id>|_all|*]/_stats	获取数据框架分析任务使用情况信息，可以给定一个或多个（用逗号分隔）任务标识符 <id>。全部则使用 _all、不给定 <id> 或星号 (*)
PUT _ml/data_frame/analytics/<id>	创建标识符为 <id> 的数据框架分析任务
DELETE _ml/data_frame/analytics/<id>	删除标识符为 <id> 的数据框架分析任务
POST _ml/data_frame/analytics/<id>/_start	给定一个标识符 <id>，开始执行数据框架分析任务
POST _ml/data_frame/analytics/ [<id>|id>,<id>|_all]/_stop	停止执行数据框架分析任务，可以停止一个、多个或全部任务
POST _ml/data_frame/_evaluate	用于评估由数据框架分析创建的索引。在请求主体需要提供参数 evaluation(regression、classification 或 binary_soft_classification)、index 和 query（选项）。若是评估回归，需要提供包括参数 actual_field、predicted_field 和 metrics 的参数 regression

以下测试为获取所有数据框架分析任务的相关内容，报告当前有一个任务。其标识符 <job_id> 是 analyze_510350_1。

提示：如图 15-34 所示，在 15.2.3 节项目 2 创建离群值检测任务时指定的标识符 analyze_510350_1。

```
curl --request GET http://localhost:9200/_ml/data_frame/analytics?pretty=true
{
  "count" : 1,
  "data_frame_analytics" : [
    {
      "id" : "analyze_510350_1",
      "source" : {
        "index" : [
          "510350_daily*"
        ],
        "query" : {
          "match_all" : { }
        }
      },
      "dest" : {
        "index" : "510350_analyze",
        "results_field" : "ml"
      },
      "analysis" : {
        "outlier_detection" : {
          "compute_feature_influence" : true,
          "outlier_fraction" : 0.05,
          "standardization_enabled" : true
        }
      },
      "analyzed_fields" : {
        "includes" : [ ],
        "excludes" : [ ]
      },
      "model_memory_limit" : "50mb",
      "create_time" : 1600204286533,
      "version" : "7.5.1",
      "allow_lazy_start" : false
    }
  ]
}
```

以下测试为获取标识符为 analyze_510350_1 的数据框架分析任务的使用情况信息，只报告当前所有阶段的进度都在 100%。

提示：版本 7.5 仍未提供分析结果的信息。如果需要获得比较完整的功能，需要升级至版本 7.7。

```
curl --request GET http://localhost:9200/_ml/data_frame/analytics/analyze_510350_1/_stats?pretty=true
{
  "count" : 1,
  "data_frame_analytics" : [
    {
      "id" : "analyze_510350_1",
      "state" : "stopped",
      "progress" : [
        {
          "phase" : "reindexing",
          "progress_percent" : 100
        },
        {
          "phase" : "loading_data",
          "progress_percent" : 100
```

```
      },
      {
        "phase" : "analyzing",
        "progress_percent" : 100
      },
      {
        "phase" : "writing_results",
        "progress_percent" : 100
      }
    ]
  }
]
}
```

以下测试为评估标识符为 analyze_510350_2 的数据框架分析任务创建的目标索引,目标索引名称为 regression_510350、因变量为 vol 及字段 predicted_field 为 ml.vol_prediction。评估的指标结果 mean_squared_error 和 r_squared 的值如下:

提示:如图 15-39 所示,在 15.2.3 节项目 2 创建回归任务时指定的标识符 regression_510350。

```
    curl --request POST  http://localhost:9200/_ml/data_frame/_evaluate?pretty=true
--header "Content-Type: application/json" --data $'{"index":"regression_510350",
"evaluation":{"regression": {"actual_field":"vol","predicted_field": "ml.vol_
prediction"}}}'
    {
  "regression" : {
    "mean_squared_error" : {
      "error" : 1.390973498551504E10
    },
    "r_squared" : {
      "value" : 0.7915281145495076
    }
  }
}
```

第 16 章　构建金融数据分析服务 RESTful 接口

在本章，我们把前几章学到的知识放在一起，并把之前介绍的一些代码粘合在一起，以构建最终项目。这个端到端的实际用例将演示并整合 Elasticsearch 的许多功能。这将帮助读者了解如何将所学内容组合在一起。

16.1　基金业绩指标

在第 2 章的表 2-3 中经介绍 Tushare 提供了单位净值、累计净值和复权单位净值。究竟基金业绩指标看哪一个？有许多技术文章讨论了投资者应使用哪个指标来评估基金的绩效，一般认为衡量基金业绩更准确的净值指标是复权净值。例如来自庚白星君 pro 的雪球原创专栏的帖子"基金课堂一：用复权净值衡量基金业绩更准确"，主要原因在于对基金分红的计算方式不同导致，使用复权净值可以更真实地反映基金的投资业绩。

16.1.1　基金业绩分类

Tushare 提供的公募基金净值数据中的复权单位净值（fund_nav）与相关的基金净值样本文档索引操作，参见 9.1 节基金净值和基金持仓样本文档。由于各种类型投资风险程度差异很大，为了方便分析，所以初步目标是基于投资类型（fund_type）来分类基金。然而投资类型数据只位于公募基金数据中，因此本章采用丰富处理器，当基金净值文档索引操作的时候，将从公募基金数据索引复制投资类型。设定公募基金净值索引名称为 fund_nav_e，采用的显式映射内容如下：

```
{
        "mappings" : {"dynamic": false, "properties": {
        "ts_code":{"type":"text", "fields":{"keyword":{"type":"keyword"}}},
        "ann_date": {"type": "date", "format": "yyyyMMdd"},
        "end_date": {"type": "date", "format": "yyyyMMdd"},
        "unit_nav": {"type": "float"},
        "accum_nav": {"type": "float"},
        "accum_div": {"type": "float"},
        "net_asset": {"type": "float"},
        "total_netasset": {"type": "float"},
        "fund_type": {"type": "keyword"},
        "adj_nav": {"type": "float"}}}
}
```

另外设定公募基金数据索引命名为 fund_basic_e，采用的显式映射内容如 7.1 节的 my_hanlp_analyzer.json 文件。

16.1.2 投资类型数据丰富处理器

创建丰富数据策略 fund_nav_policy，将策略的参数 match_field 指定为字段 ts_code，若当前索引文档字段 ts_code 的内容匹配策略的参数 indices 指定的 fund_basic_e 索引内文档字段 ts_code 的内容，则按照策略中数据类型为数组的参数 enrich_fields，提取指定字段 fund_type 的数据来丰富当前文档。以下步骤是构建，测试和应用：

（1）创建丰富数据策略 fund_nav_policy

设置在索引 fund_basic_e 中匹配的字段为 ts_code，提取的字段为 fund_type。

```
curl --request PUT http://localhost:9200/_enrich/policy/fund_nav_policy?pretty=true
 --header "Content-Type: application/json" -data $'{"match":{"indices": "fund_basic_e", "match_field": "ts_code", "enrich_fields": ["fund_type"]}}'
```

（2）创建丰富数据策略 fund_nav_policy 的索引

```
curl --request POST http://localhost:9200/_enrich/policy/fund_nav_policy/_execute?pretty=true
```

（3）测试内含 fund_nav_policy 的丰富处理器摄取管道

以下的管道处理流程首先应用内含 fund_nav_policy 的丰富处理器，并产生包含字段 fund_type 的字段 target。然后使用 set 处理器将字段 target.fund_type 字段复制到 fund_type。最后删除无用的字段 target。测试结果显示字段 fund_type 显示在文档中。

```
curl --request POST http://localhost:9200/_ingest/pipeline/_simulate?pretty=true --header "Content-Type: application/json" --data $'{"pipeline":{"description":"fund_nav 丰富处理器", "processors":[{"enrich":{"policy_name":"fund_nav_policy", "field":"ts_code", "target_field":"target"}}, {"set":{"field":"fund_type", "value":"{{target.fund_type}}"}}]}, "docs":[{"_source":{"end_date" : "20200601", "ann_date" : "20200602", "accum_nav" : 1.0, "total_netasset" : 3.91080192E8, "ts_code" : "515830.SH", "accum_div" : null, "net_asset" : 3.91080192E8, "adj_nav" : 1.0, "update_flag" : 1.0, "unit_nav" : 1.0}}]}'
        {
            "docs" : [
                {
                    "doc" : {
                        "_index" : "_index",
                        "_type" : "_doc",
                        "_id" : "_id",
                        "_source" : {
                            "end_date" : "20200601",
                            "accum_nav" : 1.0,
                            "total_netasset" : 3.91080192E8,
                            "fund_type" : "股票型",
                            "adj_nav" : 1.0,
                            "update_flag" : 1.0,
                            "ann_date" : "20200602",
                            "ts_code" : "515830.SH",
                            "accum_div" : null,
                            "net_asset" : 3.91080192E8,
                            "unit_nav" : 1.0
                        },
                        "_ingest" : {
```

```
            "timestamp" : "2020-10-04T05:26:25.370Z"
          }
        }
      }
    ]
}
```

(4) 创建 fund_nav_pipeline 摄取管道

使用以上三个处理器 enrich、set 和 remove 来构建 fund_nav_pipeline 摄取管道。

```
curl --request PUT http://localhost:9200/_ingest/pipeline/fund_
nav_pipeline?pretty=true --header "Content-Type: application/json" --data
$'{"description":"fund_nav 丰富处理器 ", "processors":[{"enrich":{"policy_name":"fund_
nav_policy", "field":"ts_code", "target_field":"target"}},{"set":{"field":"fund_type",
"value":"{{target.fund_type}}"}}, {"remove":{"field":"target"}}]}'
```

(5) 文档索引时使用 fund_nav_pipeline 摄取管道的演示

将字符串"pipeline=fund_nav_pipeline"设置为文档索引 POST 请求的 URL 参数。这将在文档索引期间执行 fund_nav_pipeline 管道操作。

```
curl --request POST 'http://localhost:9200/fund_nav_e/_doc?pretty=true&
pipeline=fund_nav_pipeline' --header "Content-Type: application/json"
--data $'{"end_date" : "20200601", "ann_date" : "20200602", "accum_nav" :
1.0, "total_netasset" : 3.91080192E8, "ts_code" : "515830.SH", "accum_div":
null, "net_asset" : 3.91080192E8, "adj_nav" : 1.0, "update_flag" : 1.0,
"unit_nav" : 1.0}'
```

16.2 测试样本文件

由于当前公募基金约有一万余只，而场内基金也有一千多只，本章仅随意采用工银瑞信基金当前（2020 年 11 月 21 日）管理而仍然在场内（E）上销售的 19 只基金为测试样品。共有 15 只股票型、3 只债券型和 1 只商品型基金。其基金名称、基金代码、投资类型和文档数量如表 16-1 所示。

表 16-1 工银瑞信基金当前管理的 33 只基金清单

投资类型	基金名称	基金代码	文档数量
股票型	工银 MSCI 中国 ETF	512320.SH	52
	工银中证 800ETF	515830.SH	69
	深 100ET	159970.SZ	191
	湾创 ETF	159976.SZ	191
	工银中证 500ETF	510530.SH	206
	工银沪深 300ETF	510350.SH	299
	工银上证 50ETF	510850.SH	415
	印度基金	164824.SZ	506
	创业板 ET	159958.SZ	678
	高铁 B 端	150326.SZ	1279
	高铁 A 端	150325.SZ	1279
	传媒 B 级	150248.SZ	1324
	传媒 A 级	150247.SZ	1324
	深红利	159905.SZ	2403
	央企 ETF	510060.SH	2698

续表

投资类型	基金名称	基金代码	文档数量
债券型	工银纯债	164810.SZ	1988
	工银四季	164808.SZ	2350
	工银双债	164814.SZ	998
商品型	黄金 ETF 基金	518660.SH	102

16.2.1 准备测试环境

以下步骤使用 Elasticsearch 快照还原准备好的数据,并假设目录 /home/wai/esbackup 用作恢复本章使用的索引数据的测试环境:

(1)将给定的文件 esbackup.tar.gz 放入 /home/wai 目录,并将工作目录更改为该目录。

(2)执行以下命令提取文件并放入用于快照备份和还原的新目录 /home/wai/esbackup。

```
./tar -zxf ./esbackup.tar.gz
ls esbackup
index-0  indices  snap-mPWZGs42TPuMuZB0gyApog.dat  index.latest  meta-mPWZGs42TPuMuZB0gyApog.dat
```

(3)在 config/elasticsearch.yml 文档中添加以下一行之后,重新启动 Elasticsearch。

```
path.repo: [ "/home/wai/esbackup" ]
```

(4)执行以下 curl 命令创建快照还原存储库 esbackup。

```
curl --request PUT http://localhost:9200/_snapshot/esbackup?pretty=true --header "Content-Type: application/json" -d$'{"type":"fs", "settings":{"location":"/home/wai/esbackup", "compress":"true"}}'
HTTP/1.1 200 OK
{
  "acknowledged" : true
}
```

(5)验证快照还原储库 esbackup 创建成功。

```
curl --request GET http://localhost:9200/_snapshot/esbackup/_all?pretty=true
{
  "snapshot" : "snapshot_1",
  "uuid" : "Ns0anIH1RWORh-eveJLwaw",
  "version_id" : 7050199,
  "version" : "7.5.1",
  "indices" : [
    "registered_funds",
    "fund_basic_e",
    ".enrich-fund_nav_policy-16060010145357",
    "fund_nav_e",
  ],
  "include_global_state" : true,
  ……
  "state": "SUCCESS",
  "start_time": "2020-11-22T02:03:23.410Z",
  ……
}
```

(6)使用 snapshot_1 快照备份还原所有索引。

```
curl --request POST http://localhost:9200/_snapshot/esbackup/snapshot_1/_restore?wait_for_completion=true&pretty=true
{"accepted":true}
```

（7）检验索引是否已经还原。

```
curl --request GET http://localhost:9200/_cat/indices?pretty
green  open .enrich-fund_nav_policy-1606010145357 RIqkY8qbRAWTPhijiPB5Xg 1 0 19    0  3.9kb   3.9kb
yellow open registered_funds                     wPyZTy7jSa6QTGpGosD7Qg 1 1 38    0 12.4kb  12.4kb
yellow open fund_nav_e                           zMGq6KSYQOaNoW_CkV-m4A 1 1 18841 19016 2.2mb 2.2mb
yellow open fund_basic_e                         Uwvkw_YAQK2aFlidn8pWKQ 1 1 19   887 352.2kb 352.2kb
```

16.2.2 检验测试环境

以下使用 curl 命令向索引 fund_nav_e 发送一系列的 HTTP 请求，这些测试将使用 Java 程序语言重写成 RESTful 应用程序接口。

1. 以投资类型分类，列出各个基金的文档总数

curl 指令如下所示，运用两次词条聚合操作分类。一次用于投资类型（report_each_fund_type），另一次用于各个基金（report_each_fund）。聚合结果整理如表 16-1 的文档数量一栏。默认情况下，词条聚合只返回文档计数最多的分组。在此，其参数 size 设置为适当的大小。由于不需要获取返回的搜索结果，查询请求的参数 size 设置 0。聚合只列出商品型基金的 518660.SH（黄金 ETF 基金），文档总数为 129。

```
curl --request POST http://localhost:9200/fund_nav_e/_search?pretty=true
--header "Content-Type: application/json" -d$'{"aggs":{"report_each_fund_
type":{"terms":{"field":"fund_type","size":10}, "aggs": {"report_each_fund":
{"terms":{"field": "ts_code.keyword","size":19}}}}}, "size":0}'
{
  ……
  "hits" : {
    "total" : {
      "value" : 19,
      "relation" : "eq"
    },
    ……
  },
  "aggregations" : {
    "report_each_fund_type" : {
      ……
      "buckets" : [
        ……
        {
          "key" : "商品型",
          "doc_count" : 129,
          "report_each_fund" : {
            ……
            "buckets" : [
              {
                "key" : "518660.SH",
                "doc_count" : 129
              }
            ]
          ……
        }
  ……
}
```

2. 从创业板 ET 基金中获取最近 30 个交易日的数据

curl 指令如下所示，创业板 ET 的 ts_code 为 159958.SZ，以字段 end_date 降序排列，结

果显示30个交易日的数据，截止日期（end_date）从20201012到20201120，数据总数为705个。

```
curl  --request POST http://localhost:9200/fund_nav_e/_search?pretty=true
--header "Content-Type: application/json" -d$'{"query":{"term":{
"ts_code.keyword":"159958.SZ"}}, "sort": [{"end_date":"desc"}], "size":30}'
{
  ……
  "hits" : {
   "total" : {
     "value" : 705,
     "relation" : "eq"
   },
   "max_score" : null,
   "hits" : [
     {
       "_index" : "fund_nav_e",
       "_type" : "_doc",
       "_id" : "9LCr7XUBp1RIvERaY2o2",
       "_score" : null,
       "_source" : {
         "end_date" : "20201120",
         "accum_nav" : 1.5512,
         "total_netasset" : null,
         "fund_type" : "股票型",
         "adj_nav" : 1.5512,
         "update_flag" : 0.0,
         "ann_date" : "20201121",
         "ts_code" : "159958.SZ",
         "accum_div" : null,
         "net_asset" : null,
         "management" : "工银瑞信基金",
         "unit_nav" : 1.5512
       },
       "sort" : [
         1605830400000
       ]
     },
     ……
}
```

3. 从创业板 ET 基金中获取最近 3 个月完整的月份数据

curl 指令如下所示，创业板 ET 的 ts_code 为 159958.SZ，而最近 3 个月完整的月份数据按照 1.6 节接口用法约定说明中介绍的日期数学表示方式以锚点日期（now）开始，四舍五入到最近一个月，然后减去所需范围的月份数目。以字段 end_date 降序排列（最近），结果显示总共 60 个交易日的数据，截止日期从 20200803 到 20201030。

提示：当日为 2020 年 11 月 21 日，最近 3 个月完整的月份数据亦即从 2020 年 8 月 1 日到 2020 年 10 月 30 日。由于 8 月 1 日、8 月 2 日和 10 月 31 日是周末，所以没有数据。

```
curl  --request POST http://localhost:9200/fund_nav_e/_search?pretty=true
--header "Content-Type: application/json" -d$'{"query":{"bool": {"must":[{
"term": {"ts_code.keyword":"159958.SZ"}}, {"range":{"end_date":{"gte":
"now/M-3M", "lte":"now/M-1M"}}}]}}, "sort":[{"end_date":"desc"}], "size":100}'
{
  ……
  "hits" : {
   "total" : {
     "value" : 60,
     "relation" : "eq"
   },
```

```
      "max_score" : null,
      "hits" : [
        {
          "_index" : "fund_nav_e",
          "_type" : "_doc",
          "_id" : "A7Cr7XUBp1RIvERaY2s2",
          "_score" : null,
          "_source" : {
            "end_date" : "20201030",
            "accum_nav" : 1.5444,
            ……
          },
          ……
        {
          "_index" : "fund_nav_e",
          "_type" : "_doc",
          "_id" : "PrCr7XUBp1RIvERaY2s2",
          "_score" : null,
          "_source" : {
            "end_date" : "20200803",
            "accum_nav" : 1.6668,
      ……
}
```

4. 以投资类型分类，获取各个基金在最近3个完整的月份的月平均复权单位净值

使用项目3的搜索条件，获取最近3个月完整的月份数据。然后运用项目(1)的两次词条聚合操作分类。再在每个基金再执行名称为 monthly 的聚合操作，首先是日期直方图聚合，以一个月为固定时间间隔单位，并设置最小文档数量为1，以确保该月内有数据可用。最后执行 adj_nav_avg 的平均值聚合操作计算月平均复权单位净值。

各个聚合名称和其使用的相关聚合如下：

- 聚合名称 report_each_fund_type 使用 terms 聚合，并调用字段 fund_type 分组。
- 聚合名称 report_each_fund 使用 terms 聚合，并调用字段 ts_code 分组。
- 聚合名称 monthly 使用 date_histogram 聚合，并调用字段 end_date 分组。
- 聚合名称 adj_nav_avg 使用 avg 聚合，并调用字段 adj_nav，计算月平均复权单位净值。

以下仅显示商品型投资类型的返回结果，商品型基金三个月总共有60条数据，开始日期为20200801。

```
curl --request POST http://localhost:9200/fund_nav_e/_search?pretty=true
 --header "Content-Type: application/json" -d$'{"query":{"range":{
"end_date":{"gte":"now/M-3M", "lte":"now/M-1M"}}}, "aggs": {
"report_each_fund_type":{"terms":{"field":"fund_type"}, "aggs": {
"report_each_fund": { "terms":{"field":"ts_code.keyword", "size":50},
"aggs": {"monthly":{"date_histogram": {"field": "end_date",
 "calendar_interval": "month", "min_doc_count": 1}, "aggs": {"adj_nav_
avg":{"avg":{"field":"adj_nav"}}}}}}}}},"size":0'
{
  ……
  "aggregations" : {
    "report_each_fund_type" : {
      ……
      "buckets" : [
        ……
        {
          "key" : "商品型",
          "doc_count" : 60,
          "report_each_fund" : {
```

```
      ......
      "buckets" : [
        {
          "key" : "518660.SH",
          "doc_count" : 60,
          "monthly" : {
            "buckets" : [
              {
                "key_as_string" : "20200801",
                "key" : 1596240000000,
                "doc_count" : 21,
                "adj_nav_avg" : {
                  "value" : 1.0947048777625674
                }
              },
      ......
}
```

5. 在项目 4 上添加月度平均复权单位净值差分和其净值差分累计总和

在项目 4 的搜索条件和聚合操作的基础上，对月度平均复权单位净值使用串行差分聚合，然后再对串行差分聚合的结果进行累计总和聚合。比较这个指标，可用于评估基金业绩。

在项目 4 的各个聚合，计算月平均复权单位净值后，继续执行的相关聚合如下：

- 聚合名称 adj_nav_serial_diff 使用 serial_diff 聚合，设定滞后值 lag 为 1，并使用 buckets_path 调用 adj_nav_avg 聚合的结果。
- 聚合名称 adj_nav_serial_diff_accum 使用 cumulative_sum 聚合，并使用 buckets_path 调用 adj_nav_serial_diff 聚合的结果。

以下仅显示商品型投资类型的返回结果。

```
curl --request POST http://localhost:9200/fund_nav_e/_search?pretty=true
--header "Content-Type: application/json" -d$'{"query":{"range":{"end_date":{
"gte":"now/M-3M", "lte":"now/M-1M"}}}, "aggs": {"report_each_fund_type":{
"terms":{"field":"fund_type"}, "aggs": {"report_each_fund": { "terms":{
"field":"ts_code.keyword", "size":50}, "aggs": {"monthly":{"date_histogram":{
"field": "end_date", "calendar_interval": "month", "min_doc_count": 1},
"aggs": {"adj_nav_avg":{"avg":{"field":"adj_nav"}}, "adj_nav_serial_diff": {
"serial_diff": {"buckets_path":"adj_nav_avg","lag":1}},
"adj_nav_serial_diff_accum":{"cumulative_sum": {"buckets_path":
"adj_nav_serial_diff"}}}}}}}},"size":0}'
{
  ......
  "aggregations" : {
    "report_each_fund_type" : {
      ......
      "buckets" : [
      ......
        {
        "key" : "商品型",
        "doc_count" : 60,
        "report_each_fund" : {
          ......
          "buckets" : [
            {
              "key" : "518660.SH",
              "doc_count" : 60,
              "monthly" : {
                "buckets" : [
                  {
                    "key_as_string" : "20200801",
```

```
              "key" : 1596240000000,
              "doc_count" : 21,
              "adj_nav_avg" : {
                "value" : 1.0947048777625674
              },
              "adj_nav_serial_diff_accum" : {
                "value" : 0.0
              }
            },
            {
              "key_as_string" : "20200901",
              "key" : 1598918400000,
              "doc_count" : 23,
              "adj_nav_avg" : {
                "value" : 1.0616648870965708
              },
              "adj_nav_serial_diff" : {
                "value" : -0.03303999066599661
              },
              "adj_nav_serial_diff_accum" : {
                "value" : -0.03303999066599661
              }
            },
            ......
}
```

6. 在项目5上添加月度平均复权单位净值差分标准差

复权单位净值的偏离程度反映了基金业绩的波动情况，稳定上升才是长期投资的关键指标。在项目5的各个聚合结果后，对adj_nav_serial_diff聚合结果执行桶扩展统计(extended_stats_bucket)聚合，可以获得所有桶内的聚合结果作出扩展统计，并获得标准差。聚合名称extended_stats_serial_diff使用extended_stats_bucket聚合，并使用buckets_path调用adj_nav_serial_diff聚合的结果。

以下仅显示商品型投资类型的返回结果。

提示：由于聚合名称extended_stats_serial_diff和monthly是同级(sibling)，所以bucket_paths的格式为monthly>adj_nav_serial_diff，请参考9.6节管道聚合。

```
    curl --request POST http://localhost:9200/fund_nav_e/_search?pretty=true
--header "Content-Type: application/json" -d$'{"query":{"range":{"end_
date":{"gte":"now/M-3M", "lte":"now/M-1M"}}}, "aggs": {"report_each_
fund_type":{"terms":{"field":"fund_type"}, "aggs": {"report_each_fund": {
"terms":{"field":"ts_code.keyword", "size":50}, "aggs": {"monthly":{"date_histogram":
{"field": "end_date", "calendar_interval": "month", "min_doc_count": 1}, "aggs":
{"adj_nav_avg":{"avg":{"field":"adj_nav"}}, "adj_nav_serial_diff": {"serial_
diff": {"buckets_path":"adj_nav_avg","lag":1}}, "adj_nav_serial_diff_accum":{
"cumulative_sum": {"buckets_path":"adj_nav_serial_diff"}}}},"extended_stats_
serial_diff": {"extended_stats_bucket": {"buckets_path": "monthly>adj_nav_serial_
diff"}}}}}}},"size":0}'
    {
      ......
      "aggregations" : {
        "report_each_fund_type" : {
          ......
          {
            "key" : "商品型",
            "doc_count" : 60,
            "report_each_fund" : {
              ......
              "buckets" : [
```

```
        {
          "key" : "518660.SH",
          "doc_count" : 60,
          "monthly" : {
            "buckets" : [
              {
                "key_as_string" : "20200801",
          ……
          ]
        },
        "extended_stats_serial_diff" : {
          "count" : 2,
          "min" : -0.03303999066599661,
          "max" : -0.020581066932367165,
          "avg" : -0.026810528799181887,
          "sum" : -0.053621057598363775,
          "sum_of_squares" : 0.0015152212992837202,
          "variance" : 3.8806195150098725E-5,
          "std_deviation" : 0.006229461866814719,
          "std_deviation_bounds" : {
            "upper" : -0.014351605065552449,
            "lower" : -0.039269452532811326
          }
        }
      }
    ……
    }
```

16.3 使用 Spring Boot 构建 RESTful 接口服务

这个项目基于第 12 章 Java 客户端编程，使用 Spring Boot 构建 Java 高级别 REST 客户端的 RESTful 接口数据服务，提供类似于 16.2.2 节检验测试环境。项目名称为 poof-analytics。项目程序的结构从 Sprint Tool Suite IDE 中提取，如图 16-1 所示。

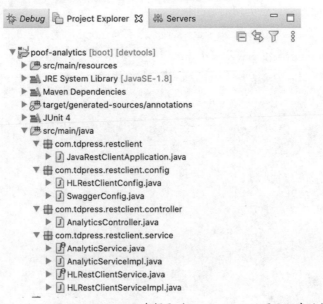

图 16-1　从 Sprint Tool Suite IDE 中提取的 poof-analytics 项目程序结构

16.3.1 AnalyticsController 类简介

图 16-2 描述了 AnalyticsController 类提供 RESTful 接口服务的方法，相对的接口名称和路径如图 16-3 所示。

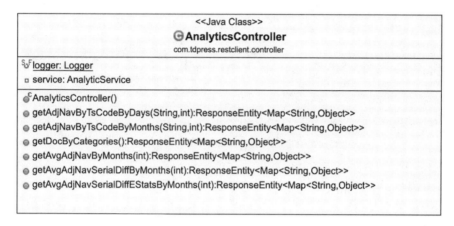

图 16-2 poof-analytics 项目的 AnalyticsController 类提供 RESTful 接口服务的方法

类似于 12.3 节使用 Swagger UI 测试低级别 REST 客户端，可按照 Maven 指令运行清理、编译、构建和执行程序项目。访问项目的 Swagger UI 页面，URL 为 http://localhost:10010/swagger-ui.html。然后按 analytics-controller 请求栏以展开面板，显示如图 16.3 所示。

图 16-3 poof-analytics 项目提供 RESTful 接口的名称、路径和简述

16.3.2 AnalyticsServiceImpl 类简介

AnalyticsServiceImpl 类实现 AnalyticsService 接口的方法以提供 AnalyticsController 类使用。它们的依赖关系可以通过图 16-4 进行初步了解。

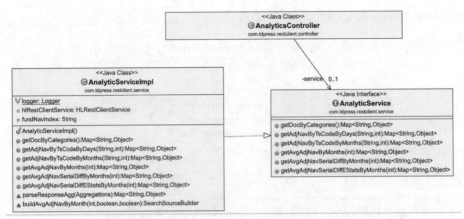

图 16-4 AnalyticsController 类和 AnalyticsServiceImpl 类的依赖关系

以下是各个接口方法的简要说明。

1. getDocByCategories 方法

这方法是对应 16.2.2 节项目 1，以投资类型分类，列出各个基金的文档总数。按照 12.4.4 节构造查询请求对复权单位净值索引（fundNavIndex）构造一个查询请求（SearchRequest），并添加两个连续的词条聚合，使结果先以投资类型分类，再以各个基金分类列出相关数据。构造好的查询请求使用类似于 12.4.6 节自定义 searchSync 与处理其响应发送至 Elasticsearch。poof-analytics 项目将高级别 REST 客户端 HLRestClient 打包成 HLRestClientService，提供 searchSync 方法同步发送 search 请求执行搜索操作。返回的数据遵循一个定义明确的结构，但也是相当复杂。Poof-analytics 项目简化了显示结果。感兴趣的读者可以参考项目中的代码，以下是相关的代码片段。

```java
public Map<String, Object> getDocByCategories() throws IOException {
    //构造 SearchRequest 查询索引 fundNavIndex 请求
    SearchRequest request = new SearchRequest(fundNavIndex);
    SearchSourceBuilder sourceBuilder = new SearchSourceBuilder();
    sourceBuilder.size(0);
    //使用 terms 聚合，并调用字段 fund_type 分组
    AggregationBuilder reportEachFundTypeBuilder =
    AggregationBuilders.terms("report_each_fund_type").
      field("fund_type").size(10);
    //使用 terms 聚合，并调用字段 ts_code 分组
    AggregationBuilder reportEachFundBuilder =
    AggregationBuilders.terms("report_each_fund").
      field("ts_code.keyword").size(50);
    reportEachFundTypeBuilder.subAggregation(reportEachFundBuilder);

    request.source(sourceBuilder);
    sourceBuilder.aggregation(reportEachFundTypeBuilder);
    RequestOptions options = RequestOptions.DEFAULT;
    //通过自定义高级别 REST 客户端 HLRestClient 同步发送 search 请求
    Map<String,Object> response =
      hlRestClientService.searchSync(request, options);
    ......
}
```

2. getAdjNavByTsCodeByDays 方法

这方法是对应 16.2.2 节项目 2，从创业板 ET 基金中获取最近 30 个交易日的数据。类似

于项目 1 对复权单位净值索引（fundNavIndex）构造一个查询请求（SearchRequest），并添加三个条件。

- 以输入参数 ts_code 的映射字段 ts_code.keyword，进行词条查询。
- 以输入参数 days 限制返回结果的笔数。
- 返回结果以字段 end_date 降序排列。

造好的查询请求使用 searchSync 方法同步发送 search 请求。返回的数据以 SearchHits 类和 SearchHit 类数组解析，并提取数据。以下是相关的代码片段。

```java
public Map<String, Object> getAdjNavByTsCodeByDays(String tsCode, int days)
  throws IOException{
    // 构造 SearchRequest 查询索引 fundNavIndex 请求
    SearchRequest request = new SearchRequest(fundNavIndex);
    SearchSourceBuilder sourceBuilder = new SearchSourceBuilder();
    // 以字段 end_date 降序排列（最近）
    sourceBuilder.sort("end_date", SortOrder.DESC);
    // 获取输入的 days(30) 个交易日的数据
    sourceBuilder.size(days);
    request.source(sourceBuilder);

    BoolQueryBuilder boolQueryBuilder = QueryBuilders.boolQuery();
    // 词条搜索输入的 tsCode
    boolQueryBuilder.must(QueryBuilders.termQuery("ts_code.keyword",
      tsCode));

    request.source(sourceBuilder.query(boolQueryBuilder));
    RequestOptions options = RequestOptions.DEFAULT;
    // 通过自定义高级别 REST 客户端 HLRestClient 同步发送 search 请求
    Map<String,Object> response = hlRestClientService.searchSync(request,
      options);
    ......
}
```

3. getAdjNavByTsCodeByMonths 方法

这方法是对应 16.2.2 节项目 3，从创业板 ET 基金中获取最近 3 个月完整的月份数据提取数据。类似于项目 2，所不同的是以输入参数 months 限制获取数据的时间范围和笔数。可参考时间范围 1.6 节接口用法约定说明的日期数学表示方式，使用锚点日期 now，限制取得数据为完整的月份。以下是相关的代码片段。

```java
public Map<String, Object> getAdjNavByTsCodeByMonths(String tsCode, int months)
  throws IOException {
    // 构造 SearchRequest 查询索引 fundNavIndex 请求
    SearchRequest request = new SearchRequest(fundNavIndex);
    SearchSourceBuilder sourceBuilder = new SearchSourceBuilder();
    // 以字段 end_date 降序排列（最近）
    sourceBuilder.sort("end_date", SortOrder.DESC);
    // 准备数据最大的总量，月份数乘以最大的月份天数
    sourceBuilder.size(31*months);
    request.source(sourceBuilder);

    BoolQueryBuilder boolQueryBuilder = QueryBuilders.boolQuery();
    // 词条搜索输入的 tsCode
    boolQueryBuilder.must(QueryBuilders.termQuery("ts_code.keyword", tsCode));
    // 开始的月份等于现在 now 减去输入的月份数,(now/M 等于完整的月份数据)
    String startPeriod = String.format("now/M-%dM", months);
    // 完整月份数据的范围是开始的月份至上个月
    boolQueryBuilder.must(
      QueryBuilders.rangeQuery("end_date").gte(startPeriod).lte("now/M-1M"));
```

```
request.source(sourceBuilder.query(boolQueryBuilder));
RequestOptions options = RequestOptions.DEFAULT;
//通过自定义高级别 REST 客户端 HLRestClient 同步发送 search 请求
Map<String,Object> response =
  hlRestClientService.searchSync(request, options);
……
}
```

4. getAvgAdjNavByMonths 方法

这方法是对应 16.2.2 节项目 4，以投资类型分类，获取各个基金在最近 3 个完整的月份的月平均复权单位净值。首先使用项目 3 的搜索条件，获取最近 3 个月完整的月份数据。然后使用类似于项目 1 的聚合操作分类，再计算各个基金的月平均复权单位净值。由于项目 5 和项目 6 是基于项目 4 的聚合操作结果上，因此创建 buildAvgAdjNavByMonth 方法，可为各个方法用来构建合适的聚合。以下是 buildAvgAdjNavByMonth 方法中的相关代码片段。

```
// 此方法是为项目 4、5 和 6 创建其聚合
SearchSourceBuilder buildAvgAdjNavByMonth(int months, boolean withSerialDiff,
    boolean withExtendedStats) {
  SearchSourceBuilder sourceBuilder = new SearchSourceBuilder();
  // 以字段 end_date 降序排列（最近）
  sourceBuilder.sort("end_date", SortOrder.DESC);
  // 不需要返回搜索结果，因此将其设置为零
  sourceBuilder.size(0);
  BoolQueryBuilder boolQueryBuilder = QueryBuilders.boolQuery();
  // 开始的月份等于现在 now 减去输入的月份数,(now/M 等于完整的月份数据)
  String startPeriod = String.format("now/M-%dM", months);
  // 完整月份数据的范围是开始的月份至上个月
  boolQueryBuilder.must(
    QueryBuilders.rangeQuery("end_date").gte(startPeriod).lte("now/M-1M"));
  // 使用 terms 聚合，并调用字段 fund_type 分组
  AggregationBuilder reportEachFundTypeBuilder =
    AggregationBuilders.terms("report_each_fund_type").field("fund_type");
  // 使用 terms 聚合，并调用字段 ts_code 分组
  AggregationBuilder reportEachFundBuilder =
    AggregationBuilders.terms("report_each_fund").field("ts_code.keyword").
    size(50);
  reportEachFundTypeBuilder.subAggregation(reportEachFundBuilder);
  // 使用 date_histogram 聚合，并调用字段 end_date 分组
  AggregationBuilder reportMonthlyBuilder = AggregationBuilders.
    dateHistogram("monthly").field("end_date").calendarInterval(
      DateHistogramInterval.MONTH).minDocCount(1);
  reportEachFundBuilder.subAggregation(reportMonthlyBuilder);
  // 使用 avg 聚合，并调用字段 adj_nav,计算月平均复权单位净值
  AggregationBuilder reportAvgBuilder =
        AggregationBuilders.avg("adj_nav_avg").field("adj_nav");
  reportMonthlyBuilder.subAggregation(reportAvgBuilder);
  // 供项目 5 使用
  if (withSerialDiff) {
    // 使用 serial_diff 聚合，设定滞后值 lag 为 1，并使用 buckets_path 调用
    //adj_nav_avg 聚合的结果
    SerialDiffPipelineAggregationBuilder reportSerialDiffBuilder =
    PipelineAggregatorBuilders.diff("adj_nav_serial_diff", "adj_nav_avg").
      lag(1);
    reportMonthlyBuilder.subAggregation(reportSerialDiffBuilder);
      // 使用 cumulative_sum 聚合，并使用 buckets_path 调用 adj_nav_serial_diff 聚合的结果
    CumulativeSumPipelineAggregationBuilder reportCumulativeSumBuilder =
      PipelineAggregatorBuilders.cumulativeSum("adj_nav_serial_diff_accum",
        "adj_nav_serial_diff");
    reportMonthlyBuilder.subAggregation(reportCumulativeSumBuilder);
```

```
    }
    //供项目 6 使用
    if (withExtendedStats) {
      //使用 extend_stats 聚合,并使用 buckets_path 调用 adj_nav_serial_diff 聚合的结果
      ExtendedStatsBucketPipelineAggregationBuilder reportExtendedStats =
      PipelineAggregatorBuilders.extendedStatsBucket("extended_stats_bucket",
        "monthly>adj_nav_serial_diff");
      reportEachFundBuilder.subAggregation(reportExtendedStats);
    }
    sourceBuilder.aggregation(reportEachFundTypeBuilder);
    //返回创建聚合
    return sourceBuilder.query(boolQueryBuilder);
}
```

为了进一步的可重用性,创建 getAvgAdjNavCommon 方法,可为项目 4、5 和 6 的操作共用许多通用代码。

```
//此方法是项目 4,5 和 6 的通用方法,并创建其适当的聚合
public Map<String, Object> getAvgAdjNavCommon(int months, boolean withSerialDiff,
    boolean withExtendedStats) throws IOException{
  //构造 SearchRequest 查询索引 fundNavIndex 请求
  SearchRequest request = new SearchRequest(fundNavIndex);
  //调用 buildAvgAdjNavByMonth 创建其聚合
  SearchSourceBuilder sourceBuilder = buildAvgAdjNavByMonth(
    months, withSerialDiff, withExtendedStats);
  request.source(sourceBuilder);
  RequestOptions options = RequestOptions.DEFAULT;
  //通过自定义高级别 REST 客户端 HLRestClient 同步发送 search 请求
  Map<String,Object> response = hlRestClientService.searchSync(request, options);
  Aggregations retAggs = (Aggregations) response.get("aggs");
  Map<String,Object> aggs = parseResponseAgg(retAggs);
  response.replace("aggs", aggs);
  return response;
}
```

以下是 getAvgAdjNavByMonths 方法的代码。使用 getAvgAdjNavCommon 方法,输入参数月份 months,并设置参数 withSerialDiff 和 withExtendedStats,两者都为 false。

```
//项目 4,以投资类型分类,获取各个基金在最近 3 个完整的月份的月平均复权单位净值
public Map<String, Object> getAvgAdjNavByMonths(int months)
  throws IOException{
  return getAvgAdjNavCommon(months, false, false);
}
```

5. getAvgAdjNavSerialDiffByMonths 方法和 getAvgAdjNavSerialDiffEStatsByMonths 方法

这两个方法是对应 16.2.2 节项目 5,在项目 4 上添加月度平均复权单位净值差分和其累计总和。使用月平均复权单位净值 (adj_nav_avg) 来构建月度平均复权单位净值差分 (adj_nav_serial_diff) 和其累计总和 (adj_nav_serial_diff_accum)。以下是 getAvgAdjNavSerialDiffByMonths 方法的代码。使用 getAvgAdjNavCommon 方法,输入参数月份 months,并设置参数 withSerialDiff 为 true 及参数 withExtendedStats 为 false,只打开用来构建差分聚合的开关。。

```
//在项目 4 上添加月度平均复权单位净值差分和其净值差分累计总和
public Map<String, Object> getAvgAdjNavSerialDiffByMonths(int months)
  throws IOException{
  return getAvgAdjNavCommon(months, true, false);
}
```

6. getAvgAdjNavSerialDiffEStatsByMonths 方法

这方法是对应在项目 5 上添加月度平均复权单位净值差分标准差。使用月度平均复权单位净值差分结果来构建桶扩展统计聚合 (extended_stats_bucket)。扩展统计聚合结果包括标准差 (std_deviation)。以下是 getAvgAdjNavSerialDiffEStatsByMonths 方法中的相关代码片段。使用 getAvgAdjNavCommon 方法，输入参数月份 months，并设置参数 withSerialDiff 和 withExtendedStats，两者都为 true。同时打开用来构建差分聚合的开关和差分标准差聚合的开关。

```
// 在项目 5 上添加月度平均复权单位净值差分标准差
public Map<String, Object> getAvgAdjNavSerialDiffEStatsByMonths(int months)
    throws IOException{
    return getAvgAdjNavCommon(months, true, true);
}
```

16.3.3 使用 Swagger UI 测试 poof-analytics 项目

以下使用网页浏览器进行各个 RESTful 接口服务互动测试。

1. 测试 /api/analytics/get_docs_by_categories 接口

测试以投资类型分类，列出各个基金的文档总数。该测试没有输入参数，点击"Execute"按钮，向下滑动屏幕后可以看到响应结果。屏幕截图如图 16-5 所示，正如 16.2.2 节项目 1 列出商品型基金的 518660.SH(黄金 ETF 基金)，文档总数为 129。

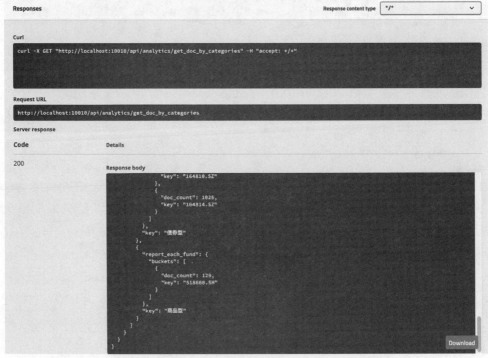

图 16-5　以投资类型分类，列出各个基金的文档总数的测试响应结果

2. 测试 /api/analytics/get_adj_nav_by_ts_code_by_days 接口

点击"获取指定的基金，最近 30 个交易日数据"一栏以展开面板，默认基金代码为

159958.SZ(创业板 ET) 和最近交易日的数据为 30 日,填写数据后屏幕截图如图 16-6 所示。

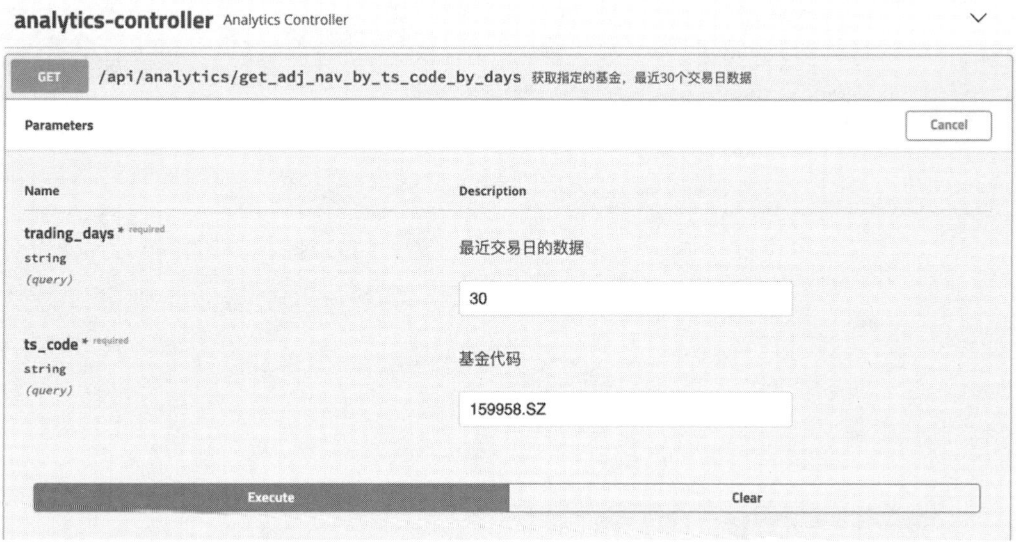

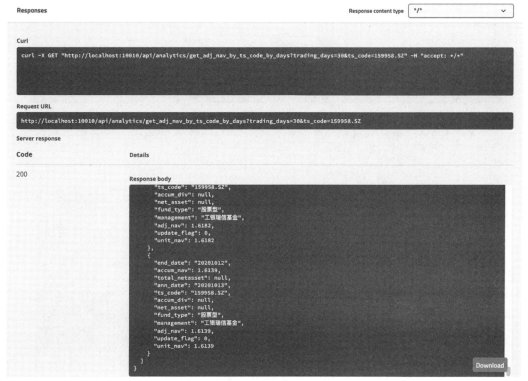

图 16-6　创业板 ET 最近 30 个交易日数据接口测试

击"Execute"按钮,向下滑动屏幕后可以看到响应结果列出创业板 ET 最近 30 个交易日的数据。屏幕截图如图 16-7 所示。30 个交易日的最早的数据,正如 16.2.2 节项目 2 所述,截止日期(end_date)从 20201012 开始。

图 16-7　创业板 ET 最近 30 个交易日数据接口测试响应结果

3. 测试 /api/analytics/get_adj_nav_by_ts_code_by_months 接口

测试获取指定的基金最近 3 个月（完整月份的数据）交易日数据。默认基金代码为 159958.SZ(创业板 ET) 和完整月份的数据为 3。填写数据后屏幕截图如图 16-8 所示。

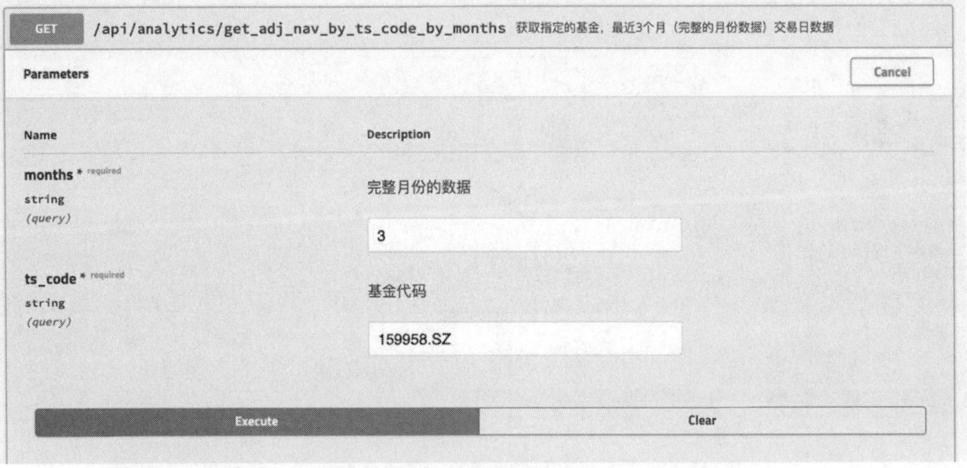

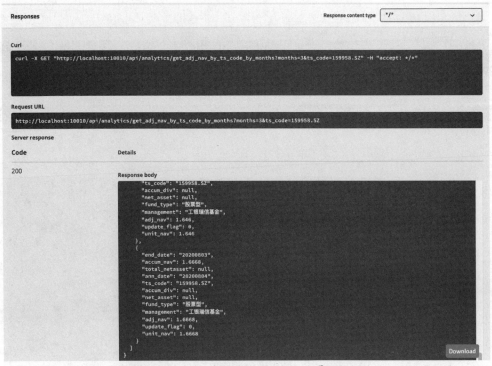

图 16-8 创业板 ET 最近 3 个月（完整月份的数据）交易日数据接口测试

点击"Execute"按钮，向下滑动屏幕后可以看到响应结果列出创业板 ET 最近 3 个月（完整月份的数据）交易日数据。屏幕截图如图 16-9 所示。最近 3 个月的最早的数据，正如 16.2.2 节项目 3 所述，截止日期（end_date）从 20200803 开始。

图 16-9 创业板 ET 最近 3 个月（完整月份的数据）交易日数据接口测试响应结果

4. 测试 /api/analytics/get_monthly_avg_adj_nav 接口

测试以投资类型分类，获取各个基金在最近 3 个月（完整的月份数据），每个月的平均复权单位净值。默认完整月份的数据为 3，点击"Execute"按钮，向下滑动屏幕后可以看到响应结果。屏幕截图如图 16-10 所示，列出商品型基金的 518660.SH（黄金 ETF 基金）的 8 月、9 月和 10 月的月平均复权单位净值。正如 16.2.2 节项目 4 所述，商品型基金三个月总共有 60 条数据，开始日期为 20200801。

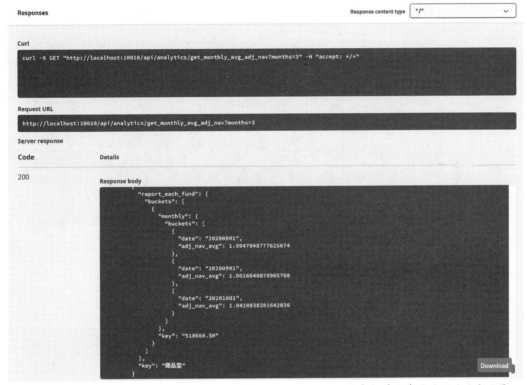

图 16-10　以投资类型分类，获取各个基金在最近 3 个月的月平均复权单位净值的测试响应结果

5. 测试 /api/analytics/get_monthly_avg_adj_nav_serial_diff 接口

测试以投资类型分类，获取各个基金在最近 3 个月（完整的月份数据），每个月的平均复权单位净值、其序列差分和其序列差分的累计总和。默认完整月份的数据为 3，点击"Execute"按钮，向下滑动屏幕后可以看到响应结果。屏幕截图如图 16-11 所示，列出商品型基金的 518660.SH(黄金 ETF 基金）的 8 月、9 月和 10 月的月平均复权单位净值、其序列差分和其序列差分的累计总和。结果显示正如 16.2.2 节项目 5 所述。

6. 测试 /api/analytics/get_monthly_avg_adj_nav_serial_diff_estats 接口

测试以投资类型分类，获取各个基金在最近 3 个月，每个月的平均复权单位净值、其序列差分、其序列差分的累计总和及其序列差分的扩展统计。默认完整月份的数据为 3，点击"Execute"按钮，向下滑动屏幕后可以看到响应结果。屏幕截图如图 16-12 所示，列出商品型基金的 518660.SH（黄金 ETF 基金）的 8 月、9 月和 10 月的月平均复权单位净值、其序列差分和其序列差分的累计总和。结果显示正如 16.2.2 节项目 6 所述。

图 16-11 以投资类型分类，获取各个基金在最近 3 个月的月平均复权单位净值、其序列差分及其序列差分的累计总和的测试响应结果

图 16-12 以投资类型分类，获取各个基金在最近 3 个月的月平均复权单位净值、其序列差分、其序列差分的累计总和及其序列差分的扩展统计的测试响应结果。

16.3.4 聚合结果解析简介

每种特定的聚合结果有其单独的解析方法，详细信息必须参考官方网站 www.javadoc.io/doc/org.elasticsearch/elasticsearch/7.5.1，然后可以了解其一对一的对应。原则上根据聚合请求的构造，相应地分析层次结构中的聚合结果。在 16.3.2 节 AnalyticsServiceImpl 类简介的项目 4 getAvgAdjNavByMonths 方法中，曾经介绍过 getAvgAdjNavCommon 方法，该方法调用通用解析方法 parseResponseAgg，解析本章中所使用的聚合结果。表 16-2 描述在本章所使用的聚合，其名称、类型、层次及其解析的方式

表 16-2　在本章所使用的聚合，其名称、类型、层次及其解析的数据类型。

聚合名称	层次	聚合类型	解析的数据类型
report_each_fund_type	顶层	词条聚合	ParsedStringTerms
report_each_fund	第一层	词条聚合	ParsedStringTerms
monthly	第二层	日期直方图聚合	ParsedDateHistogram
adj_nav_avg	第三层	平均值聚合	ParsedAvg
adj_nav_serial_diff	第三层	串行差分聚合	ParsedSimpleValue
adj_nav_serial_diff_accum	第三层	累计总和聚合	ParsedSimpleValue
extended_stats_bucket	第一层	桶扩展统计聚合	ParsedExtendedStatsBucket

以下是 parseResponseAgg 方法的代码概要。

```java
// 解析本章中所使用的聚合结果的通用解析方法
Map<String, Object> parseResponseAgg(Aggregations retAggs) {
    Map<String, Object> aggs = new HashMap<String, Object>();
    Map<String, Object> reportEachFundTypeMap =
        new HashMap<String, Object>();
    aggs.put("report_each_fund_type", reportEachFundTypeMap);
    // 解析各个顶层聚合结果
    for (Aggregation level0Agg : retAggs) {
        // 对于顶层 report_each_fund_type 聚合进行解析
        if (level0Agg.getName().equals("report_each_fund_type")) {
            // 对字符串的词条聚合 /report_each_fund_type,
            // 使用 ParsedStringTerms 数据类型解析
            ParsedStringTerms fundTypeAgg = (ParsedStringTerms) level0Agg;
            List<Map<String, Object>> reportEachFundTypeBuckets =
                new ArrayList<Map<String, Object>>();
            // 对顶层词条聚合 report_each_fund_type 所产生的存储桶进行解析
            for (Terms.Bucket fundTypeAggBucket:fundTypeAgg.getBuckets()){
                // 对 report_each_fund_type 产生的存储桶取得下一层聚合结果
                for (Aggregation level1Agg:fundTypeAggBucket.getAggregations()){
                    List<Map<String, Object>> reportEachFundBuckets =
                        new ArrayList<Map<String, Object>>();
                    Map<String, Object> reportEachFundBucketMap =
                        new HashMap<String, Object>();
                    // 对于第一层词条聚合 report_each_fund 聚合进行解析
                    if (level1Agg.getName().equals("report_each_fund")) {
                        // 对字符串的词条聚合 report_each_fund,
                        // 使用 ParsedStringTerms 数据类型解析
                        ParsedStringTerms fundAgg = (ParsedStringTerms) level1Agg;
                        // 对第一层词条聚合 report_each_fund 所产生的存储桶进行解析
                        for (Terms.Bucket eachFundAggBucket:fundAgg.getBuckets()){
                            Map<String, Object> eachFundMap =
                                new HashMap<String, Object>();
                            eachFundMap.put("key", eachFundAggBucket.getKeyAsString());
                            // 对 report_each_fund_type 产生的存储桶取得下一层聚合结果
                            for (Aggregation level2Agg:
```

```java
        eachFundAggBucket.getAggregations()){
    Map<String, Object> reportMonthlyMap =
      new HashMap<String, Object>();
    // 对第二层 date_histogram 聚合 monthly 进行解析
    if (level2Agg.getName().equals("monthly")) {
      List<Map<String, Object>> reportMonthlyBuckets =
        new ArrayList<Map<String, Object>>();
      reportMonthlyMap.put("buckets", reportMonthlyBuckets);
      // 对 date_histogram 聚合 monthly 使用 ParsedDateHistogram
      // 数据类型解析结果
      ParsedDateHistogram dateHistoAgg =
        (ParsedDateHistogram) level2Agg;
      // 对 date_histogram 聚合 monthly 所产生的存储桶进行解析
      for (Histogram.Bucket monthlyBucket :
        dateHistoAgg.getBuckets()) {
      Map<String, Object> monthlyAggMap =
        new HashMap<String, Object>();
      monthlyAggMap.put("date", monthlyBucket.getKeyAsString());
      // 对各个第三层聚合进行解析
      for (Aggregation level3Agg:monthlyBucket.getAggregations()){
        if (level3Agg.getName().equals("adj_nav_avg")) {
          // 平均值聚合使用 ParsedAvg 数据类型解析结果
          ParsedAvg avgAgg= (ParsedAvg) level3Agg;
          monthlyAggMap.put("adj_nav_avg", avgAgg.getValue());
        }
        else if (level3Agg.getName().equals("adj_nav_serial_diff")) {
          // 串行差分聚合使用 ParsedSimpleValue 数据类型解析结果
          ParsedSimpleValue serialDiffAgg=
            (ParsedSimpleValue) level3Agg;
          monthlyAggMap.put("adj_nav_serial_diff",
            serialDiffAgg.getValueAsString());
        }
        else if (level3Agg.getName().
          equals("adj_nav_serial_diff_accum")) {
          // 累计总和聚合使用 ParsedSimpleValue 数据类型解析结果
          ParsedSimpleValue accumAgg= (ParsedSimpleValue) level3Agg;
          monthlyAggMap.put("adj_nav_serial_diff_accum",
            accumAgg.getValueAsString());
        }
      }
      reportMonthlyBuckets.add(monthlyAggMap);
      }
      eachFundMap.put("monthly", reportMonthlyMap);
    } else if (level2Agg.getName().equals("extended_stats_bucket")) {
      // 桶扩展统计聚合使用 ParsedExtendedStatsBucket 数据类型解析结果
      ParsedExtendedStatsBucket extendedStatsAgg =
        (ParsedExtendedStatsBucket) level2Agg;
      Map<String, Object> reportEStatsMap =
        new HashMap<String, Object>();
      reportEStatsMap.put("min", extendedStatsAgg.getMin());
      reportEStatsMap.put("max", extendedStatsAgg.getMax());
      reportEStatsMap.put("avg", extendedStatsAgg.getAvg());
      reportEStatsMap.put("std_deviation",
        extendedStatsAgg.getStdDeviation());
      reportEStatsMap.put("variance", extendedStatsAgg.getVariance());
      eachFundMap.put("extended_stats", reportEStatsMap);
    }
  }
  reportEachFundBuckets.add(eachFundMap);
  }
}
Map<String, Object> reportEachFundMap =
```

```
                    new HashMap<String, Object>();
                reportEachFundBucketMap.put("buckets", reportEachFundBuckets);
                reportEachFundMap.put("key", fundTypeAggBucket.getKey());
                reportEachFundMap.put("report_each_fund", reportEachFundBucketMap);
                reportEachFundTypeBuckets.add(reportEachFundMap);
            }
        }
        reportEachFundTypeMap.put("buckets", reportEachFundTypeBuckets);
    }
  }
  return aggs;
}
```